MICROCOMPUTER INTERFACING

MICROCOMPUTER INTERFACING

A Practical Guide for Technicians, Engineers, and Scientists

Joseph J. Carr

Prentice-Hall International, Inc.

A Simon & Schuster Company
Englewood Cliffs, New Jersey 07632

Printed in the United States of America
10 9 8 7 6 5 4 3 2 1

ISBN 0-13-587619-2

Prentice-Hall International (UK) Limited, *London*
Prentice-Hall of Australia Pty. Limited, *Sydney*
Prentice-Hall Canada Inc., *Toronto*
Prentice-Hall Hispanoamericana, S.A., *Mexico*
Prentice-Hall of India Private Limited, *New Delhi*
Prentice-Hall of Japan, Inc., *Tokyo*
Simon & Schuster Asia Pte. Ltd., *Singapore*
Editora Prentice-Hall do Brasil, Ltda., *Rio de Janeiro*
Prentice-Hall, Inc., *Englewood Cliffs, New Jersey*

CONTENTS

4

IBM PC/XT/AT PLUG-IN CARD AND CHIP INTERFACES *19*

5

APPLE II PLUG-IN CARD INTERFACING *47*

6

GENERATING DEVICE- AND BOARD-SELECT PULSES *51*

17

DATA CONVERSION TECHNIQUES: A/D AND D/A *330*

18

ANALOG OUTPUT DISPLAY DEVICES *368*

19

CONTROLLING EXTERNAL CIRCUITS *421*

PREFACE

The microcomputer literally revolutionized the electronic instrument and control system design fields, and it did so in record time. Where instrument designers were once exclusively analog engineers, the instrument designer today has to be a synergist who can integrate the principals of sensor selection, analog circuit design, computer hardware selection and/or design, and software design and operation. Today, even small instruments are based on microcomputer chips, and for that reason we are going to consider these devices in some detail. Later on, you will learn how to interface the various computers to external circuits, peripherals, and sensors that are used in instrumentation and control applications.

There are three basic forms of machine considered in this text. First, of course, are the IBM PC, IBM XT, and IBM AT machines (plus the clones) that are now the standard of the industry. But also covered are the Apple II computer and the Z80 chip. The reason for these seeming anachronisms is that my market survey showed that many small laboratories, plus a lot of plant process control computers, were still Apple II machines purchased earlier. The reason for the Z80 chip being included is that a large number of Z80 single-board computers are used embedded into products that don't look like computers on the outside, but are nonetheless "dedicated" personal computers. Others could have also been selected as well, e.g. the Intel 8048/50/51 series of "single-chip" computers.

The purpose of this book is to provide a practical, workbench basis for both electronics technology professionals and those technical people whose expertise and sophistication is in fields other than electronics or computers. It was the author's experience that not a few laboratory scientists in research institutions are more than basically competent in the field of electronic instrumentation.

Joseph J. Carr

MICROCOMPUTER INTERFACING

1

THE MICROCOMPUTER

Its Role in Modern Instrument Design

At one time, definitions of computers were simpler. As a first-year engineering student at Old Dominion University in Norfolk, Virginia, I was allowed to use an IBM 1601/1602 machine. *That* was a computer! No doubt about it. It took up an entire room on the second floor of the engineering school's building. Today, an engineering student can sit at a small desk with an IBM PC/XT (complete with video CRT display, printer, 40-megabyte or larger hard disk drive, and two floppy disk drives) that has more computing power than that old 1601. Many engineering school students find that the cost of the typical small system is affordable, and they purchase their own computer. Indeed, while engineering students a generation ago treated the computer room with fear and awe approaching reverence, today many engineering schools require students to bring at least an XT-class machine to school with them. The cost of the modern microcomputer is less than one-tenth what one of the lesser machines of only a decade ago cost, not counting the fact that "then-year" dollars were bigger than today's dollars.

TYPES OF COMPUTER

Before attempting to define the role of the microcomputer, let's first try to define the microcomputer. Terminology tends to become sloppy both from our own laziness and from the fact that once-genuine distinctions have become blurred as the state-of-the-art advances; terminology in the computer field is often overcome by events. For example, consider the terms *micro-*

computer and *minicomputer*. Laziness tends to make some people use these terms interchangeably, and modern chips (for example the 80286, 80386, or 68xxx) make such usage reasonable. But, for our purposes, we require sharply focused meanings for these two terms and others: microprocessor, microcomputer, single-chip computer, single-board computer, minicomputer, and mainframe computer.

Microprocessor

The microprocessor is a *large-scale-integration* (LSI) *integrated circuit* (IC) that contains the *central processing unit* (CPU) of a programmable digital computer. The CPU section of a computer contains the *arithmetic/logic unit* (ALU), which performs the basic computational and logical operations of the computer. The CPU also houses the *control logic* section (which performs housekeeping functions) and may or may not have several *registers* for the temporary storage of data. All CPUs have at least one temporary storage register called the *accumulator*, or *A-register*. The principal attribute of a microprocessor is that it will execute instructions sequentially. These instructions are stored in coded binary form in an external *memory*.

Microcomputer

A microcomputer is a full-fledged programmable digital computer that is built around a microprocessor chip (i.e., an integrated circuit); the microprocessor acts as the CPU for the computer. In addition to the microprocessor chip, the microcomputer typically will have some additional chips, the number varying from two to hundreds depending upon the design and application. These external chips may provide such functions as memory (both temporary and permanent) and input/output (I/O). The microcomputer may be as simple as a KIM-1 or as complex as a 30-board professional machine with all the electronic data-processing (EDP) options. The trend today in IBM PC/XT/AT and compatible machines is to stuff more and more capability into less and less space.

Single-Chip Computer

For several years we had no excuse for interchanging the terms *microprocessor* and *microcomputer*; a microprocessor was an LSI chip and a microcomputer was a computing machine. But the 8048 and similar devices dissolved previously well-defined boundaries because they were both LSI ICs and computers. A typical single-chip computer may have a CPU section, two types of internal memory (temporary and long-term permanent storage), and at least two I/O ports. Some machines are even more complex. The single-chip computer does, however, require some external components before it can do work. By definition, the microcomputer already has at least a minimum of components needed to perform a job.

Single-Board Computer

The single-board computer (SBC) is a programmable digital computer, complete with input and output peripherals, on a single printed circuit board. Popular examples are the Rockwell AIM-65 and Z80 family of machines. The single-board computer might have either a microprocessor or a single-chip computer at its heart. The peripherals on a single-board computer are usually of the most primitive kind (AIM-65 is a notable exception). Most single-board computers have at least one interface connector that allows either expansion of the computer or interfacing into a system or instrument design. The manufacturers of early SBCs, such as the KIM-1 and others, probably did not envision their wide application. These computers were primarily touted as trainers, that is, for use in teaching microcomputer technology. But for simple projects such computers also work well as a mini development system. More than a few SBC trainers have been used to develop a microcomputer-based product, only to wind up being specified as a "component" in the production version. In still other cases, the commercially available SBC was used as a component in prototype systems and then, in the production version, a special SBC (of lower cost) was either bought or built.

Minicomputer

The minicomputer predates the microcomputer and was originally little more than a scaled-down version of larger data-processing machines. The Digital Equipment Corporation (DEC) PDP-8 and PDP-11 machines are examples. The minicomputer uses a variety of small-scale (SSI), medium-scale (MSI), and large-scale-integration (LSI) chips.

Minicomputers traditionally were more powerful than microcomputers, although that distinction has also blurred. For example, they had longer-length binary data words (12 to 32 bits instead of 8 bits found in microcomputers) and operated at faster speeds of 6 to 12 megahertz (MHz) instead of 1 to 3 MHz. But this is an area of past distinctions now that 33-MHz 80386 machines are available and 10-MHz 8088 machines are considered the "low end" of high technology. Digital Equipment Corporation, for example, offers the LSI-11 microcomputer that acts like a minicomputer. Similarly, 32-bit microcomputers are available, as are 20- to 33-MHz devices. It is sometimes difficult to draw the line of demarcation when an 80386-based microcomputer is in the same-sized cabinet as a minicomputer, and minicomputers can be bought in desk-top configurations.

Mainframe Computer

The larger computer that comes to mind when most people think of computers is the *mainframe computer.* These computers are used in large-scale data-processing departments. Microcomputerists who have an elitist mentality sometimes call mainframe computers "dinosaurs." But, unlike their reptilian namesakes, these dinosaurs show no signs of extinction and are, in fact, an evolving species.

Advantages of Microcomputers

That microcomputers have certain advantages is attested by the fact that so many are sold. But what are these advantages, and how are they conferred?

The most obvious advantage of the microcomputer is *reduced size*; compared with dinosaurs, microcomputers are mere lizards! A 16-bit microcomputer with 640 Kbytes of random-access memory can easily fit inside a tabletop cabinet.

Another advantage is cost. Prices have dropped precipitously in the last few years. While the early microcomputers (circa 1977) were both very expensive and of limited capability, the machine today is a tremendous advance in technology at a low cost. It is now reasonable for nearly everyone to own personal or desktop computers (and indeed they do), and their spread throughout industry and science has set records.

The LSI microcomputer chip is more complex than a discrete-components circuit that does the same job. The interconnections between circuit elements, however, are much shorter (micrometers instead of millimeters). Input capacitances are thereby made lower. The metal-oxide semiconductor (MOS) technology used in most of these ICs produces very low current drain; hence the overall power consumption is reduced. A benefit of reduced power requirements is reduced heating. While a minicomputer may require a pair of 100 cubic feet per minute (cfm) air blowers to keep the operating temperature within specifications, a microcomputer may be able to use a single 40-cfm muffin fan or no fan at all.

Another advantage of the LSI circuit is reduced component count. Although this advantage relates directly to reduced size, it also affects reliability. If the LSI IC is just as reliable as any other IC (and so it seems), then the overall reliability of the circuit is increased dramatically. Even if the chip reliability is lower than for lesser ICs, we still achieve superior reliability due to fewer interconnections on the printed circuit board, especially if IC sockets are used on the ICs. Some of the most maddening troubleshooting problems result from defective or dirty IC sockets.

MICROCOMPUTER INTERFACING

The design of any device or system in which a microcomputer or microprocessor is used is the art of *defining the operation* of the device or system, *selecting the components* for the device or system, *matching and integrating* these components (if necessary), and *constructing* the device or system. These activities are known collectively as *interfacing*.

But let's get down to a more basic level. Most readers of this book are technically minded people with some knowledge of electronics and computer technology. For most readers, therefore, interfacing consists of selecting and matching components and then connecting them into a circuit that does a specific job. These are the matters that are addressed in later chapters.

MICROCOMPUTERS IN INSTRUMENT AND SYSTEM DESIGN

Designers in the past used analog electronic circuits, electromechanical relays (which sometimes precipitate a maintenance nightmare), and other devices in order to design instruments, process controllers, and the like. These circuit techniques had their limitations and produced some irritating results; factors like thermal drift loomed large in some of these circuits.

In addition, the design was cast in cement once the final circuit was worked out. Frequently, relatively subtle changes in a specification or requirement produced astonishing changes in the configuration of the instrument; analog circuits are not easily adaptable to new situations in many cases. But with the advent of the microcomputer, we gained the advantage of flexibility and solved some of the more vexing problems encountered in analog circuit design. The software stored in the memory of the computer tells it what to do and can be changed relatively easily. We can, for example, store program code in a *read-only memory* (ROM), which is an integrated-circuit memory. If a change is needed, the software can be modified and a new ROM installed. If the microcomputer was configured in an intelligent manner, then it is possible to redesign only certain interface cards (or none at all) to make a new system configuration.

An engineer of my acquaintance built an anode heat computer for medical x-ray machines. A microprocessor computed the heating of the anode as the x-ray tube operated and sounded a warning if the limit of safety was exceeded, thus saving the hospital the cost of a $10,000 x-ray tube. But different x-ray machines require different interfacing techniques, a problem that previously had meant a new circuit design for each machine. However, by intelligent engineering planning, the anode heat computer was built with a single interface card that married the "universal" portion of the instrument with each brand of x-ray machine. Thus the company could configure the instrument uniquely for all customers at a minimum cost.

Another instrument that indicates the universality of the microcomputer is a certain cardiac output computer. This medical device is used by intensive-care physicians to determine the blood-pumping capability of the heart in liters per minute. A bolus of iced or room-temperature saline solution is injected into the patient at the "input" end of the right side of the heart (the heart contains two pumps, a right and left side, with the right side output feeding the left side input via the lungs). The temperature at the output end of the right side is monitored and the time integral of temperature determined. This integral, together with some constants, is used by the computer to calculate the cardiac output.

These machines come in two versions, research and clinical. The researcher will take time to enter certain constants (that depend upon the catheter used to inject saline, temperature, and other factors) and will be more vigorous in following the correct procedure. But in the clinical setting, technique suffers owing to the need of caring for the patient, and the result is a perception of "machine error," which is actually operator error. To combat this problem, the manufacturer offers two machines. The research instrument is equipped with front panel controls that allow the operator to select a wide range of options. The clinical model allows no options to the operator and is a "plug and chug" model. The interesting thing about these instruments is that they are *identical* on the inside! All that is different is the front panel and the position of an on-board switch. The manufacturer's program initially interrogates the switch to see if it is open or closed. If it is open, then it "reads" the keyboard to obtain the constants. If, on the other hand, it is closed, the program branches to a subprogram that assumes certain predetermined constants. The cost savings to the company of using a single design for both instruments are substantial.

2

MICROPROCESSOR/ MICROCOMPUTER FUNDAMENTALS

Most readers of this book are familiar with the basic operation of the programmable digital computer. Even so, there are many for whom the computer is still little more than a complicated "black box." All too often, microcomputer books assume a level of familiarity with computers that is simply not there in all readers, even though most engineers, scientists, and technicians are "computer literate" these days. Thus we have a need for a brief chapter that discusses the programmable digital computer as a generic device.

The "machine" of this chapter is a so-called universal or generic computer, representative of the entire set of programmable digital computers and not any one manufacturer's offerings. It is a hypothetical machine, not available anywhere. It, like an androgynous being, has elements of all kinds within its body. We call this hypothetical "mind experiment" computer the Generic Automatic Computer, or "GAC" for short.

THE GENERIC AUTOMATIC COMPUTER (GAC)

Figure 2-1 shows the block diagram of GAC. We will first describe each of the blocks in this diagram and then progress to a discussion of how GAC processes a typical computer program.

Like any programmable digital computer, GAC has three main parts: the central processing unit (CPU), memory, and input/output (I/O). There are other functions found in certain specific machines, but many of these are

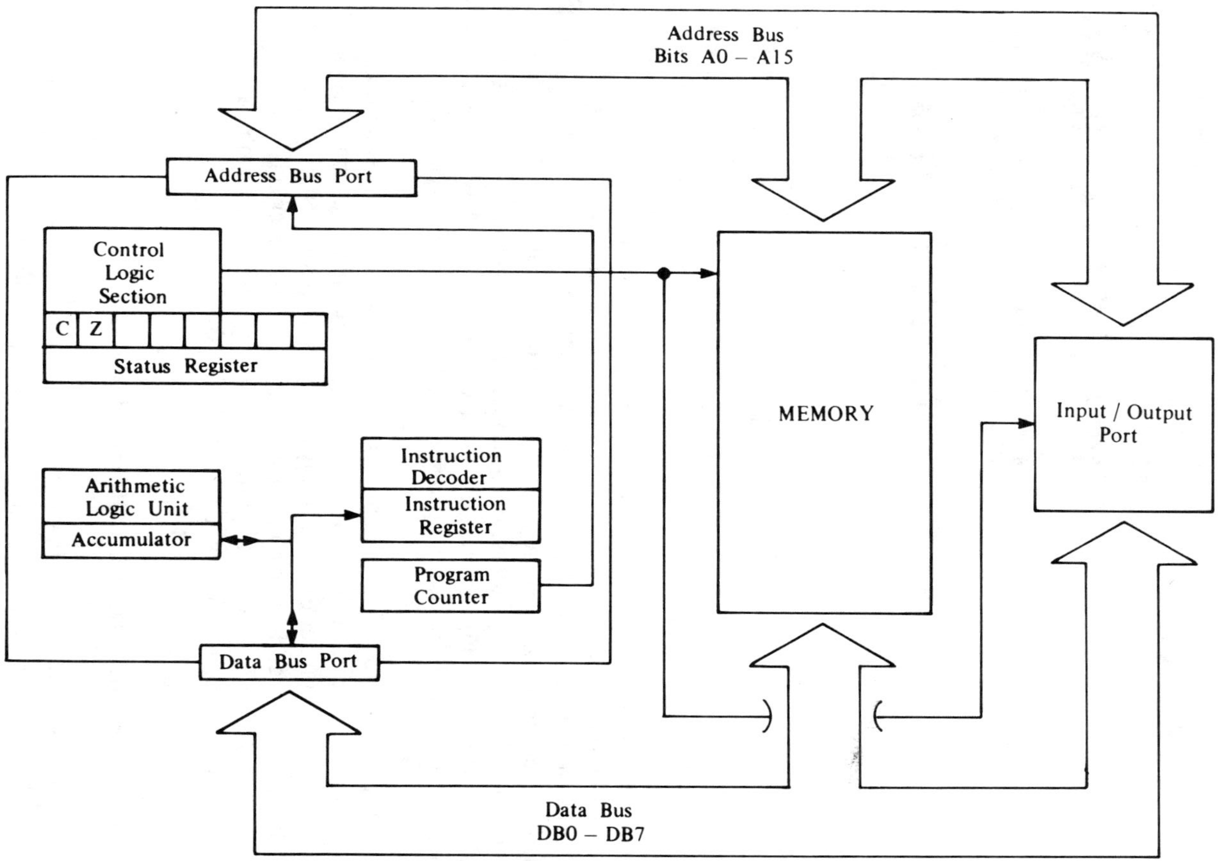

Figure 2-1 Block diagram of a typical computer.

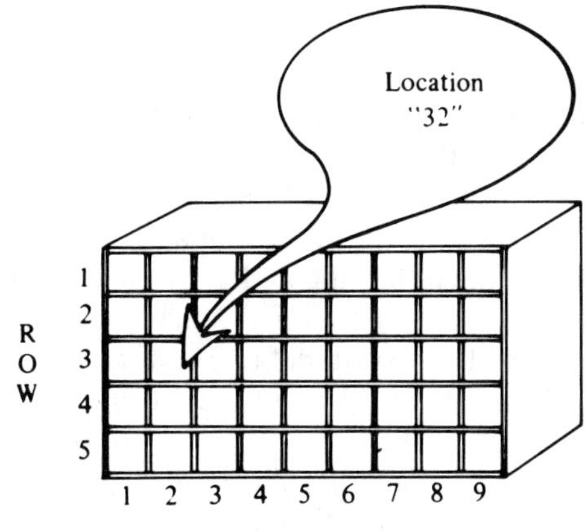

Figure 2-2 Analogy of computer memory.

either special applications of these main sections or are too unique to be considered in a discussion of a general "universal" machine.

The CPU controls the operation of the entire machine. It consists of several necessary subsections which are described in greater detail below.

Memory can be viewed as an array of "pigeonholes" or "cubbyholes" (Figure 2-2) such as those used by postal workers to sort mail. Each pigeonhole represents a specific address on the letter carrier's route. An address in the array can be uniquely specified (identifying only one location by designating its row and column). If we want to specify the memory location at row 3 and column 2, then we would create a row-3, column-2 address number, which in this case is "32."

Each pigeonhole represents a unique location in which to store mail. In the computer, the memory location stores not pieces of paper, but a single binary "word" of information. In an 8-bit microcomputer, for example, each memory location will store a single 8-bit binary word (e.g., 11010101). The different types of memory device are not too terribly important to us here, except in the most general terms. To the CPU, and our present description of how it operates, it doesn't matter much whether memory is random-access read/write memory (RAM), read-only memory (ROM), dynamic, or static.

There are three main lines of communication between the memory and the CPU: the address bus, the data bus, and control logic signals. These avenues of communication control the interaction between memory and I/O, on the one hand, and the CPU on the other, regardless of whether the operation is a read or write function. The address bus consists of parallel data lines, one for each bit of the binary word that is used to specify the address location. In a simple 8-bit microcomputer, for example, the address bus contains 16 bits. A 16-bit address bus can uniquely address up to 2^{16}, or 65,536, different 8-bit memory locations. This size is specified as "64K" in computerese, not "65K" as one might expect. The lowercase letter "k" is used almost universally in science and engineering to represent the metric prefix kilo, which represents the multiplier 1,000. Thus when someone tells you that their computer has 64K of memory, you might expect it to contain 64 × 1,000, or 64,000 electronic pigeonholes. But you would be wrong in that assumption. For long ago computerists noted that 2^{10} is 1,024, so they determined that their "k" would be 1,024, not 1,000! This means that a 64K computer will contain a total of 64 × 1,024, or 65,535, electronic pigeonholes in which to stuff data. To differentiate "big K" (1,024) from "little k" (1,000), standard computerist shorthand uses uppercase "K" rather than lowercase "k."

Be aware that the size of memory that can be addressed doubles for every bit added to the address bus. Hence adding one bit to our 16-bit address bus creates a 17-bit address bus that is capable of addressing up to 128K of memory. Some so-called 8-bit machines that have 16-bit address buses are made to look larger by tactics which create a pseudo-address bus. In those machines, several 64K memory banks are used to simulate continously addressable 128K, 256K, or more machines.

The IBM PC/XT-class machines are designed around the 8088 microprocessor chip. This chip is capable of addressing more than one million 8-bit locations in memory, yet most of these machines actually address only 640K of memory. According to industry legend, when the originators of MS-DOS were setting basic program parameters they decided to use 10 times the normal maximum memory that was then current (64K), because it was inconceivable that anyone would ever need more in a personal computer.

The data bus is the communications channel over which data travels between the main register (called the accumulator or A-register) in the CPU and the memory. The data bus also carries data to and from the various input and/or output ports. If the CPU wants to read the data stored in a particular memory location, then that data is passed from the memory location to the accumulator (in the CPU) by way of the data bus. Memory write operations are exactly the opposite: Data from the accumulator is passed over the data bus to a particular memory location.

The last memory signal is the control logic or timing signal. These signals tell memory if it is being addressed and whether the CPU is requesting a read or a write operation. The details of the control logic signals vary considerably from one microprocessor chip to another, so there is little that we can say at this point. Elsewhere in this book you will be introduced to both the 6502, Z80, and 8088/8086 standard signals (which are representative of different architectures), but for others I recommend the chip manufacturer's literature.

The input/output (I/O) section of GAC is the means by which the CPU communicates with the outside world. An input port will bring data in from the outside world and then pass it over the internal data bus to the CPU, where it is stored in the accumulator. An output port is exactly the opposite: it passes accumulator data from the data bus to the outside world. In most cases, the "outside" world consists of either peripherals (e.g., printers, video monitors) or communications devices (e.g., modems).

In some machines, there are separate I/O instructions that are distinct from the memory-oriented instructions. For example, in the Z80 machine there are separate instructions for "write to memory" and "write to output port" operations. In Z80-based computers, the lower 8 bits (A0 to A7) of the address bus are used during I/O operations to carry the unique address of the I/O port being called (the accumulator data will pass over both the data bus and the upper 8 bits—A8 to A15—of the address bus). Since there are 8 bits in the unique I/O address, we can use up to 2^8, or 256, different I/O ports that are numbered from 000 to 255. In the 6502-based computer, there are no distinct I/O instructions. In those machines (e.g., Apple II), the I/O components are treated as memory locations. This technique is called memory-mapped I/O. Input and output operations then become memory-read or memory-write operations, respectively.

Central Processing Unit (CPU)

The CPU is the heart and soul of GAC. Although there are some differences among specific machines, all will have at least the features shown in our GAC diagram of Figure 2-1. The principal subsections of GAC are: the accumulator or A-register, the arithmetic/logic unit (ALU), the program counter (PC), the instruction register (IR), the status register (SR) or processor status register (PSR), and the control logic section.

The accumulator is the main register in the CPU. With a few exceptions, all of the instructions use the accumulator as either the source/destination of data or the object of some action (for example, unless otherwise indicated, an ADD instruction always performs a binary addition operation between some specified data and the data in the accumulator).

Although there are other registers in the CPU (the Z80 is loaded with them!), the accumulator is the main register. The main purpose of the accumulator is the temporary storage of data being operated on or transferred within the machine. Note that data transfers to and from the accumulator are nondestructive. In other words, it is a misnomer to tag such operations as "transfers," because in actuality they are "copying" operations. Suppose, for example, the program calls for us to "transfer" the hexadecimal number 8FH from the accumulator to memory location A008H. After the proper instruction ("STA A008H" in 6592 assembly language) is executed, the hex number 8FH will be found in both the accumulator and memory location A008H.

It is important to remember that accumulator data changes with every new instruction! If we have some critical datum stored in the accumulator, it is important that we write it to some location in memory for permanent storage. This function is sometimes performed on the "external stack" or in some portion of memory set aside by the programmer as a "pseudo-stack."

The ALU contains the circuitry needed to perform the arithmetic and logical operations. In most computers, the arithmetic operations consist of addition and possibly subtraction, while the logical operations consist of AND, OR, and exclusive-OR (XOR). Note that even subtraction is not always found! In some computers, there is no hardware arithmetic function other than addition. The subtraction function is performed in software using two's complement arithmetic (a method of making the computer think it is actually adding). Multiplication and division are treated as multiple additions or subtractions unless the designer has thought to provide a hardware multiply/divide capability.

The program counter (PC) contains the address of the next instruction to be executed. The secret to the success of any programmable digital computer such as GAC is its ability to fetch and execute instructions sequentially. Normally, the PC will increment appropriately (1, 2, 3, or 4) while executing each instruction (i.e., 1 for a one-byte instruction, 2 for a two-byte instruction, and so on). For example, the "LDA, N" instruction on the 6502 microprocessor loads the accumulator with the number "N." In a program listing, we will find the number "N" stored in the next sequential memory location from the code for the LDA portion (called the operations code, or "op-code" for short), as follows:

```
0205 LDA
0206 N
```

At the beginning of this operation, the PC contains "0205," but after execution the PC will contain the number "0206" because "LDA, N" is a 2-byte instruction.

There are several ways to modify the contents of the PC. One way is to let the program execute sequentially: the PC contents will increment for each instruction. We can also activate the reset line, which forces the PC to either location 0000H, or to some other specific location (often at the other end of the memory, for example FFFAH in the 6502). Another method is to execute either a JUMP or a conditional-JUMP instruction. In this latter case, the PC will contain the address of the "jumped-to" location after the instruction is executed. Finally, some computers have a special instruction that will load the PC with a programmer-selected number. This "direct entry" method is not available on all microprocessors, however.

The instruction register is the temporary storage location for the instruction codes that were stored in memory. When the instruction is fetched from memory by the CPU, it will reside in the instruction register until the next instruction is fetched.

The instruction decoder is a logic circuit that reads the instruction register contents and then carries out the intended operation.

The control logic section is responsible for the "housekeeping" chores in the CPU. It issues and/or responds to control signals between the CPU and the rest of the universe. Examples of typical control signals are: memory requests, I/O requests (in non-memory-mapped machines), read/write signalling, and interrupts.

The status register, also sometimes referred to individually as the status flags, is used to indicate to the program and sometimes the outside world the exact status of the CPU at any given instant. Each bit of the status register represents a different function. Different microprocessor chips use different sets of status flags, but all will have at least a carry flag (C-flag) and a zero flag (Z-flag). The C-flag indicates when an arithmetic or logical operation results in a "carry" from the most significant bit of the accumulator (B7), while the Z-flag indicates when the result of the present operation was zero (00000000_2); typically, Z = 1 when the result is zero, and C = 1 when a carry occurred.

We have now finished our tour of the CPU of GAC. This discussion in general terms also describes most typical microprocessors. Although various manufacturers use different names for the different sections, and some will add sections, almost all microprocessors are essentially the same inside the CPU.

Operation of GAC

A programmable digital computer such as GAC operates by sequentially fetching, decoding, and then executing instructions stored in memory. These instructions are stored in the form of binary numbers. In some early machines there were two memory banks, one each for program instructions and data. The modern computer, however, uses a single memory bank for both instructions and data.

How, one might legitimately ask, does dumb ol' GAC know whether any particular binary word fetched from memory is data, an instruction, or a binary representation of an alphanumeric character (e.g., ASCII)? The answer to this instruction is key to the operation of GAC: cycles!

GAC operates in cycles. A computer will have at least two discrete cycles: instruction fetch and execution (in some machines the process is more sophisticated, and cycles are added). While the details differ from one machine to another, the general operation is similar for all of them.

Instructions are stored in memory as binary numbers (op-codes). During the instruction fetch cycle, an op-code is retrieved from the memory location specified by the program counter and stuffed into the instruction register inside the CPU. The CPU assumes that the programmer was smart enough to arrange things according to the rules, so that the datum fetched from location ABCD during some instruction cycle in the future is, in fact, an instruction op-code. It is the responsibility of the programmer to arrange things in a manner so as to not confuse the poor, dumb GAC.

During the first cycle, an instruction is fetched and stored in the instruction register (IR). During the second (i.e., next) cycle, the instruction decoder circuit inside the CPU will read the IR and then carry out the indicated operation. The second cycle is called the execution cycle, while the first cycle was the instruction fetch/decode cycle. The CPU then enters the next instruction fetch cycle, and the process is repeated. This process is repeated over and over as long as GAC is operating. Each step is synchronized by an internal clock that is designed to make things remain rational.

From the above description, you might be able to glimpse a truth concerning what a computer can or cannot do. The CPU can shift data around, perform logical operations (AND, OR, XOR), add two N-bit numbers (sometimes—but not always—subtract as well), all in accordance with a limited repertoire of instructions encoded in the form of binary words. Operations in GAC (and all other microcomputers) are performed sequentially through a series of discrete steps. The secret to whether or not any particular problem is suitable for computer solution depends entirely on whether or not a plan of action (called an algorithm) can be written that will lead to a solution through sequentially executed steps. Most practical instrumentation, control, measurement, or data-processing problems can be so structured—a fact that accounts for the meteoric rise of the microcomputer in those fields.

3

SELECTING THE RIGHT MICROCOMPUTER

The terms *microcomputer* and *microprocessor* cover a variety of different devices that have different capabilities. There is no such thing as a "universal" computer, or computer configuration, that is all things to all applications. Matters to consider when buying a microcomputer for instrumentation purposes are the architecture of the machine (i.e., register-, memory-, or I/O-oriented), the instruction set (computationally strong or I/O operations strong), and the associated hardware requirements. If the IBM PC/XT series of machines is selected, then a wide variety of software products and add-on circuit cards is available. Indeed, the IBM PC class of machine is now the standard for the industry.

Which machine do you select for *your* application? All are microcomputers, yet they have vastly different properties. Obviously, one would not use the same machine for such vastly different chores as data processing and, say, burglar alarm or environmental systems monitoring.

It is, therefore, essential that you evaluate the task to be performed and also discern any *reasonable* future accretions to the system. Keep in mind that all projects tend to grow in scope as time passes. Some of this growth is legitimate; some growth occurs because people tend to enlarge a project into additional functions that were neither intended nor advisable; some growth is due to *your* poor planning in the early stages of the project. Try to anticipate future needs and plan for them. A more or less valid rule of thumb is to follow the 50 percent rule regarding initial capacity: the current requirements should occupy only *one-half* of the machine resources (memory size, processing time, and number of I/O ports).

It is claimed that a smart designer will provide twice or three times the memory actually required for the presently specified chore but will not under any circumstances tell the programmer. Programmers tend to use up all the memory available. Perhaps if they think there is somewhat less available to them, they will find more efficient means to solve their problems.

The decisions made during planning phases of a project will affect future capabilities in large measure. If adequate means for expansion are not provided, extraordinary problems will surface later. One sure sign of poor planning in a microprocessor-based instrument is the use of extra "kluge boards" hanging onto the main printed circuit board.

The key to good planning is evaluation of system requirements. How many I/O ports are needed? How much memory? How fast will the processor have to operate? What kind of displays and/or input devices are needed? How many? What size of power supply is needed? In a small system, a bank of seven-segment LED numerical readouts can draw as much current as the rest of the computer.

Perhaps one of the earliest hardware decisions regards the microprocessor chip that will be selected. You will find such decisions are often made more or less on emotional grounds rather than technical ones; users get attached to a type—often the first type they learned to program. Just as it is with photographic and high-fidelity equipment, microprocessors and microcomputers attract "true believers." Sometimes, however, an emotionally satisfying choice turns out later to have been utterly stupid for the need at hand.

Typical of the factors that must be considered, especially if the microcomputer is being used as a part of another instrument, are the following: power consumption, speed-power product, size, cost, reliability, and maintainability. These factors are not here arranged in any hierarchy, but they should be ranked by importance in your design planning. For example, if you are designing a computerized bedside patient monitor for a hospital or other medical user, power consumption and speed-power product assume less important roles than in, say, a space shuttle computer where available power is limited, heat dissipation is tightly controlled, and data transfer rates are extremely high. Similarly, in the bedside monitor we can tolerate lower-reliability (hence lower-cost) equipment because repair service is readily available, and replacement units can be procured and stocked against the possibility of a failed unit. For a NASA satellite, on the other hand, once launched there is slim or no possibility of repair. For that computer, it might be worthwhile to build according to high-reliability specifications. Maintainability is less important to the satellite, but of critical importance to the hospital's biomedical equipment technician or clinical engineer. Cost, of course, is also very important to the hospital user.

Speed-power product can become important in many applications. Processing speed, often as measured by a system clock, is usually related to power consumption. In most semiconductor devices, the operating speed relates to internal resistances and capacitances that form frequency-limiting resistor-capacitor (RC) time constants. Reducing internal resistance in order to increase operating speed (i.e., from a short RC time constant) also causes increased power consumption.

Processing speed as measured from program execution time, however, is another matter. This time limitation depends upon the efficiency of instruction execution and the nature of the instructions available to the programmer. There are cases, for example, where the 1-MHz 6502 device (used in the Apple II series of machines) will execute a program slightly faster than the 2-MHz Z80 machine used in some controllers or single-board computers today (in addition to obsolete CP/M machines).

A measure of programming speed is the *benchmark program* such as the often-touted Norton rating. Such a program attempts to standardize evaluation comparisons by having the different microcomputers under test perform some standard task and noting the amount of time required. There are numerous pitfalls in this approach, however, because the selection of the task and the programming approach used to solve the problem or task can significantly affect results through bias in favor of one machine over another. The benchmark program should, therefore, be representative of the tasks to be performed by the end product.

Factors that can seriously affect processing speed are the nature of the instruction set and the architecture of the microcomputer. If the task is heavy on I/O operations, for example, it may be wise to use a microprocessor with a good repertoire of input/output instructions. A number-crunching data-processing task, on the other hand, requires strong shift-left and shift-right instructions and a coprocessor capability. Some microcomputers have hardware multiply and divide coprocessing capability that is much faster than software implementations of these functions. Since these arithmetic functions tend to be time-consuming in software, they are a major consideration if your computer will have to make many such computations.

If the application requires a 16-bit word (or anything greater than the 8 bits normally found on "traditional" microprocessors), the computer will have to be either a 16-bit machine or be programmed to process 16 bits by sequentially grabbing 2 bytes at a time. The IBM PC and IBM XT machines are based on the 8088 chip, so they are 16-bit machines, while the IBM AT class machines are 32-bit machines.

Support can also be a driving factor in the selection of the microcomputer. First, there is the matter of software and/or hardware available on the open market for that microcomputer. The IBM PC/XT/AT, the Macintoshes, plus the Z80 and 6502 machines, for example, have immense amounts of software available. The old CP/M operating system worked on Z80 machines, but it is now obsolete. It was, however, seminal for modern operating systems such as MS-DOS (or PC-DOS).

If you are going to include either a microcomputer or a microprocessor in the design of a product, be sure to consider second sourcing. Most major microprocessor chips are now multisourced, as are products such as motherboards and plug-in boards. There are also supposedly improved alternate chips. For example the NEC V-20 series of devices is available to replace the slower 8088 devices used in the IBM PC. The reason for requiring a second source is that all companies from time to time have problems that prevent timely deliveries. If you are locked into a source that is a sole source, and they should have such problems, you will be in a bind that is difficult to resolve. Your own production will be brought to a halt by someone else's problems.

It is often more difficult to obtain single-board computers that are second sourced. These products are often unique in their design, so only one company will make them. There are options, however, and these should be considered. Some standard-bus single-board computers are made by several companies, so even if the products are not exactly interchangeable, they are close enough to make conversion less damaging to your schedules. Also, some single-board original equipment manufacturers (OEMs) advertise that they will give you the drawings to allow you to become your own second source once you purchase a minimum number of machines (usually 100 to 200).

4

IBM PC/XT/AT PLUG-IN CARD AND CHIP INTERFACES

The microcomputer revolution spurred interest in developing digitally oriented instrumentation and control systems. Some computers, like the IBM PC, XT, and AT series (and their clones), have a series of six or eight plug-in slots (Figure 4-1A) to accommodate add-on hardware and firmware features. A "plug-in slot" is nothing more than a printed circuit board card-edge connector that has pins designated according to some standardized protocol. One of the chief benefits of selecting a computer with plug-in slots is the ability to configure the machine as a special-purpose instrument almost at will. Although some care is needed to prevent incompatibilities within a single machine (between plug-ins of various makers), it is generally true that each slot is memory-mapped into special space, so there is little chance of a problem.

The modules that plug into the slots are printed circuit boards. One or more connectors, switches, or controls that must be available to the outside world are located on a metal tab on the rear of the card (Figure 4-1B). It is through these external connectors that the computer interfaces with the rest of the world. The external connectors are fitted to one or more of a series of slots cut into the rear panel of the computer (Figure 4-1C).

The utility of plug-in modules is that the computer can be custom configured according to the needs of each customer. For example, the same basic IBM PC/XT/AT or compatible might serve both the department secretary as his word processor or the scientist in the laboratory down the hall as her data collection and logging system. And once the data is collected, the same class of machine, differently configured, can massage the data statistically as needed and print out the results in both alphanumeric and graphical format.

Figure 4-2 shows two plug-in modules for the IBM PC/XT machines (and those clones that have IBM-like slots). The printed circuit module in

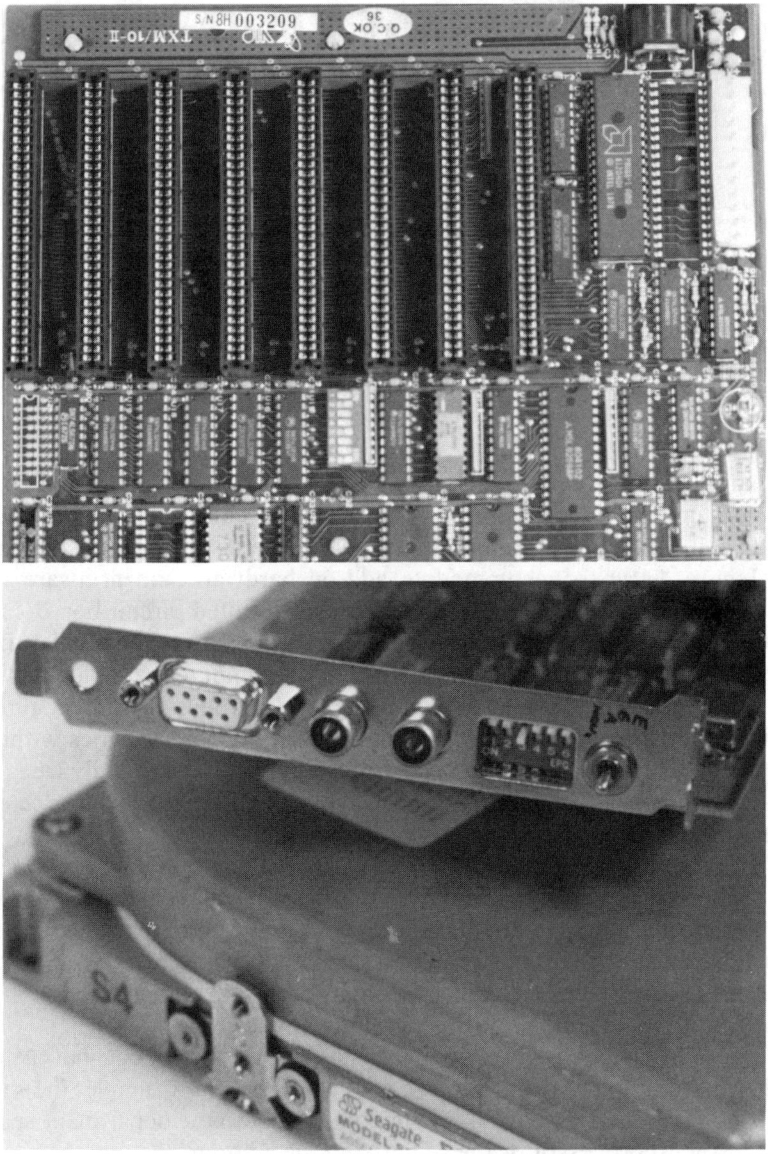

(a)

(b)

Figure 4-1 A) Expansion slots on IBM PC/XT machines, B) Typical rear panel connector for IBM PC/XT machines, C) Expansion slot cutouts on rear panel.

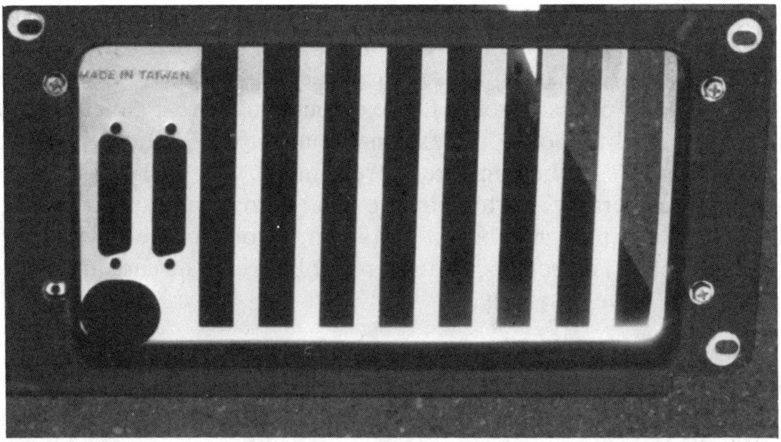

(c)

Figure 4-1c

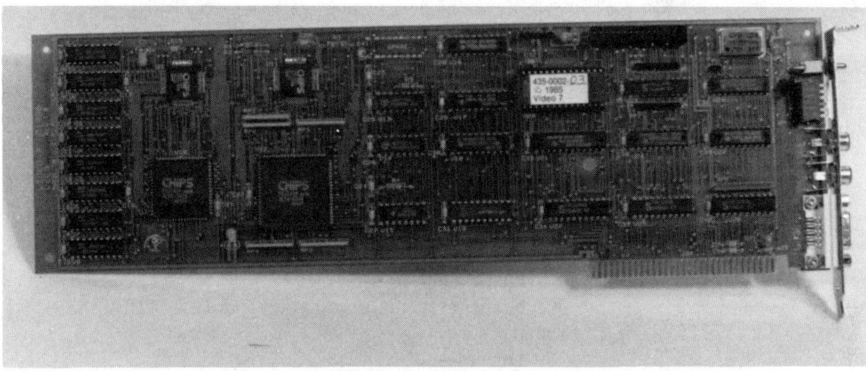

(a)

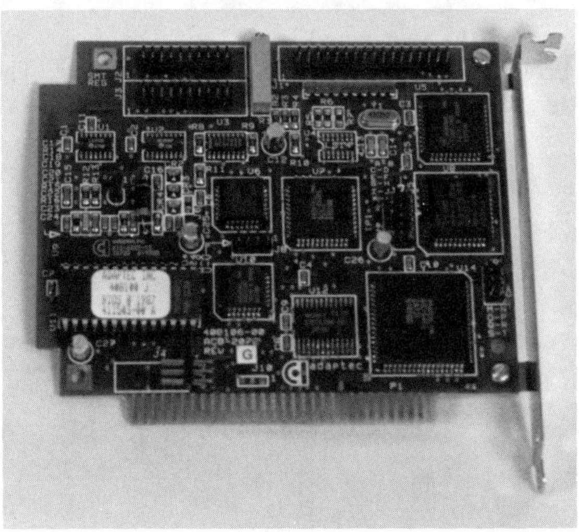

(b)

Figure 4-2 A) Full-size expansion card, B) Half-card.

Figure 4-2A is the full-card size used originally in the IBM PC. The card's external connector end is attached to the cutouts on the rear panel, while the "rear" of the card (opposite the external connectors) is anchored to card guides on the inside of the computer's front panel. The half-size card (Figure 4-2B) is considerably shorter than the other card, and in some computers one or two of the total number of slots can accommodate only this size card. The problem is mechanical interference between the card and the rear end of the disk drive assembly.

COMPUTER PLUG-IN INTERFACE CONNECTIONS

The IBM PC/XT-class machines have several different interface connections, both internal and external. These are described in the sections below.

Keyboard Connector

The keyboard connector uses a five-pin DIN connector with the following pin-outs:

Pin No.	Function
1	+Keyboard clock
2	+Keyboard data
3	−Keyboard reset (not used by keyboard)
4	Ground
5	+5 V DC power

The *expansion port* on the rear panel is a 62-pin DB-series connector that looks like a big brother of the familiar RS-232C connector. The signals on this connector are used to drive the expansion unit, hard disk drives, and similar devices.

Pin No.	Function
("E" means "Extended")	
1	+E IRQ6
2	+E DRQ2
3	+E DIR
4	+E ENABLE
5	+E CLK
6	−E MEM IN EXP
7	+E A17
8	+E A16
9	+E A5
10	−E DACK0
11	+E A15
12	+E A11

Pin No.	Function
13	+E A10
14	+E A9
15	+E A1
16	+E A3
17	+E DACK1
18	+E A4
19	−E DACK2
20	−E IOW
21	+E A13
22	+E D5
23	+E DRQ1
24	+E DRQ3
25	(reserved)
26	+E ALE
27	+E T/C
28	+E RESET
29	+E AEN
30	+E A19
31	+E A14
32	+E A12
33	+E A18
34	+E MEMR
35	−E MEMW
36	+E A0
37	−E DACK3
38	+E A6
39	−E IOR
40	+E A8
41	+E A2
42	+E A7
43	+E IRQ7
44	+E D6
45	+E I/O CH RDY
46	+E IRQ3
47	+E D7
48	+E D1
49	−E I/O CH RDY
50	+E IRQ2
51	+E D0
52	+E D2
53	+E D4
54	+E IRQ5
55	+E IRQ4
56	+E D3
57	GROUND
58	''
59	''
60	''
61	''
62	''

I/O Channel Connector (On System Board)

The versatility of the IBM PC is due in large part to the ability of the user to configure the computer into a large number of different machines through the use of plug-in printed wiring boards. The slots for the cards (Figure 4-1A) are female printed circuit card-edge connectors mounted on the system board inside the IBM PC. The slots are installed close to the rear panel of the IBM PC such that interfacing connectors to the outside world can be accommodated on each plug-in card. The cards are much larger than Apple II plug-in cards, so they can contain a lot more (which means more complex) circuitry. The pinouts of the I/O channel connector (Figure 4-3) are shown in the following table:

Pin No.	Function
(Wiring side of inserted PWB)	
B1	GROUND
B2	+RESET DRV
B3	+5 V
B4	+IRQ2
B5	−5 V
B6	+DRQ2
B7	−12 V
B8	(reserved)
B9	+12 V
B10	GROUND
B11	−MEMW
B12	−MEMR
B13	−IOW
B14	−IOR
B15	DACK3
B16	+DRQ3
B17	−DACK1
B18	+DRQ1
B19	−DACK0
B20	CLOCK
B21	+IRQ7
B22	+IRQ6
B23	+IRQ5
B24	+IRQ4
B25	+IRQ3
B26	−DACK2
B27	+T/C
B28	+ALE
B29	+5 V
B30	+OSC
B31	GROUND
(Component side)	
A1	−I/O CH CK

Pin No.	Function
A2	+D7
A3	+D6
A4	+D5
A5	+D4
A6	+D3
A7	+D2
A8	+D1
A9	+D0
A10	+I/O CH RDY
A11	+AEN
A12	+A19
A13	+A18
A14	+A17
A15	+A16
A16	+A15
A17	+A14
A18	+A13
A19	+A12
A20	+A11
A21	+A10
A22	+A9
A23	+A8
A24	+A7
A25	+A6
A26	+A5
A27	+A4
A28	+A3
A29	+A2
A30	+A1
A31	+A0

These signals and many other details of the IBM PC are given in the *IBM PC Technical Reference Manual* (IBM part number 6025005).

IBM PC Cassette Connector (Used on Older Machines Only)

The cassette connector is designed for storing and retrieving data on an audiocassette. This method of data storage was popular when personal computers first appeared, when it was a low-cost alternative to the then-expensive 5.25-inch disk drives. Because audiocassette storage is both time-consuming and of low reliability, and because disk drives are a lot cheaper (and more effective) than they once were, everyone uses disks today.

A five-pin DIN connector is used for the cassette port. This same style connector is also used for the keyboard connector. The pinouts for the cassette port are shown in the table on page 27.

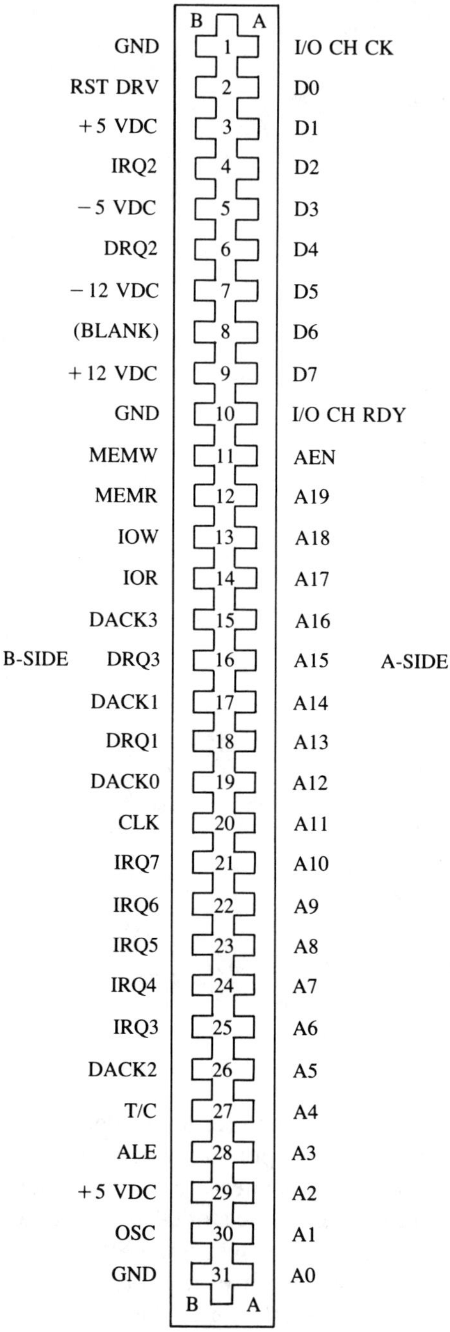

IBM-PC Compatible
Plug-In Slot

Figure 4-3 Pin-outs for IBM PC/XT expansion slot.

Pin No.	Function
1	Motor control (common from relay)
2	Ground
3	Motor control (n.o. relay)
4	Data in (1–2 kbaud) from earphone
5	Data output to microphone or auxiliary

THE 8088/8086 CHIPS

The 8088 and 8086 microprocessors come in a 40-pin dual in-line package (DIP) integrated circuit package (Figure 4-4), the same package that was used on other microprocessors by Intel and others over the past decade. The 8086/8088 devices have more pin functions than can be accommodated in just 40 pins, however, so some pins serve several purposes through a system of multiplexing. The 8088 pins are grouped in the following categories:

1. Address bus
2. Data bus
3. Control, timing, and status
4. Power/ground

Figure 4-4 shows the 8088 package with pin labelling. Note that some of the address bus shares pins with data bus lines. The 8088 uses a 20-bit address bus to directly address up to 1 megabyte (1,024,000 bytes) of RAM or ROM. The lower 8 bits of the address bus serve double duty as an 8-bit data bus, with a multiplexing scheme used to determine which use occurs at any given time; these pins are designated AD0 through AD7. The address bus pins are divided into three groups, described in the following sections.

AD0 through AD7. These pins produce the lower 8 bits (bits A0–A7) of the address bus plus the entire 8 bits of the data bus. These pins are used as address bus lines during the beginning of each machine cycle and switch to data bus lines later in each machine cycle (they become inputs on read operations and outputs on write operations). In order for an address to be nontransient to the external world, these pins must be latched.

A8 through A15. These pins are address bus bits A8 through A15, and they serve no other purpose. During the "hold acknowledge" and "interrupt acknowledge" operations A8 through A15 float "tristate" (i.e., exhibit a high impedance).

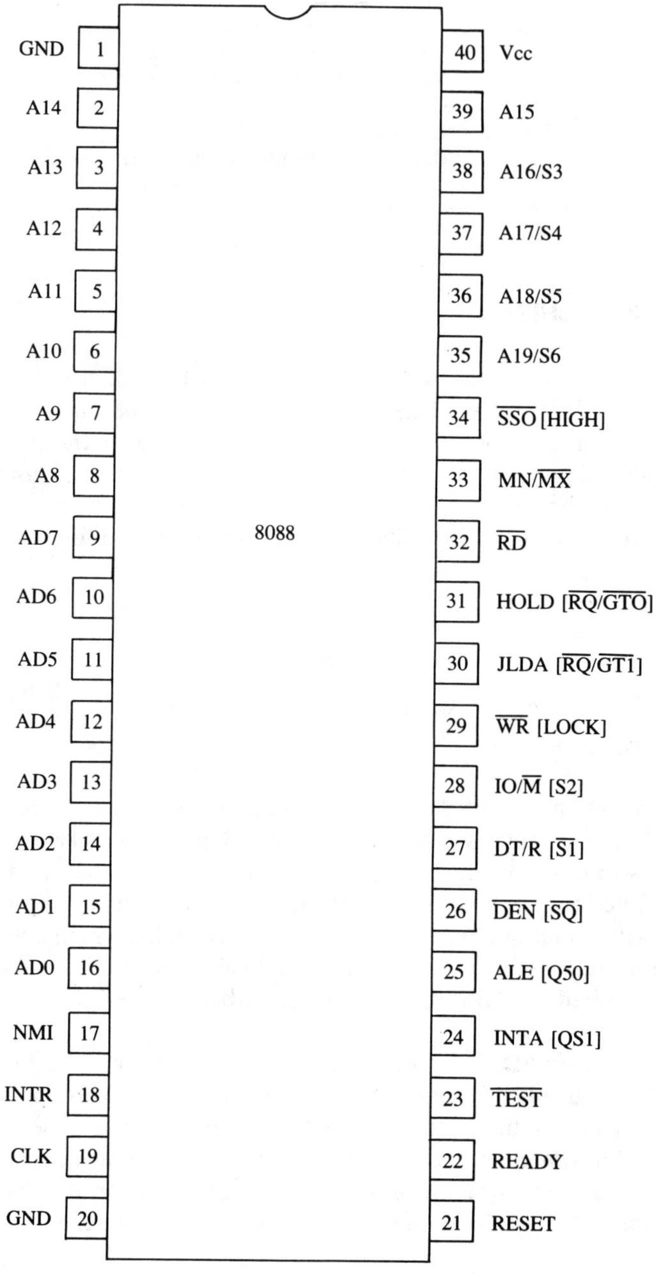

Figure 4-4 8088 chip pin-outs.

A16/S3 through A19/S6. These pins are address bus bits A16 through A19, and they also serve as the status signals. Multiplexing determines which function is active at any given time. In order for an address to be nontransient to the external world, these pins must be latched.

During the first part of the machine cycle these pins are used for the upper 4 bits of the address bus. Later on in the machine cycle, however, these pins become status signals, as shown in the following table. Signals S3 and S4 are encoded to indicate which memory segment is being accessed by the 8088 CPU during the current machine cycle and can be used in certain memory selection techniques. Signal S5 tells the outside world the status of the interrupt enable flag, while S6 is always LOW and tells the world that an 8088 is present on the system.

A16/S3	A17/S4	Function
0	0	Alternate (relative to ES segment)
1	0	Stack (relative to SS segment)
0	1	Code/None (relative to the CS segment, or default = 0)
1	1	Data (relative to DS segment)

Power/Ground. These pins accept the DC power supply connections. Pin number 40 accepts +5 V DC (regulated), the same level as for TTL devices. Pin numbers 1 and 20 are ground. Intel recommends that, like all digital circuits, the 8088 be equipped with decoupling capacitors at the power terminal to reduce the effect of noise on the power system.

CONTROL, TIMING, AND CONTROL STATUS SIGNALS

The signals in this group are of primary interest to people performing interfacing chores, so this section should be read carefully. Almost all interfacing, whether involving input/output, memory, or other devices, requires manipulation of certain of the control signals.

The timing subgroup contains only one member: the clock line (pin number 19). The clock frequency must be 5 MHz or less for most 8088 devices and 8 MHz or less for the 8088-2. Although you can design any suitable TTL-compatible oscillator, Intel recommends their own 8284A clock generator.

The control status lines tell the outside world what is going on and control external logic. These lines are defined below.

RD (pin number 32). This active-LOW output tells the outside world when the CPU is reading data from either memory, an I/O port, or a peripheral. RD will be in the high-impedance tristate during hold acknowledge operations.

WR (pin number 29). This active-LOW output tells the outside world when the CPU is writing data from either memory, an I/O port, or a peripheral. WR will be in the high-impedance tristate during hold acknowledge operations.

ALE (pin number 25). This active-HIGH output is used to latch the multiplexed address bus lines. This signal will be used to drive the enable line of some external latch device.

IO/M (pin number 28). This signal indicates whether the CPU is addressing memory or I/O space. IO/M will be HIGH for I/O operations and LOW for memory operations.

RESET (pin number 21). This active-HIGH input to the CPU is essentially a hardware "JUMP" instruction that causes the CPU to cease executing the present instruction and jump immediately to location FFFF0H. The CPU will also reset the status flag register to 0000H, disable both the interrupt flags and single-step mode, reset to 0000H registers DS, ES, SS, and IP, and set register CS to FFFFH.

MN/MX (pin number 33). This input to the CPU allows the 8088 to configure either in the minimum mode or the maximum mode. When MN/MX is HIGH, the 8088 is in the minimum mode, and when LOW the 8088 is in the maximum mode. The control signals previously described are used for minimum mode, while those following are used for maximum mode.

DT/R (pin number 27). This data transmit/receive output signal is used to control data bus drivers such as the Intel 8286 or 8287 devices. When a HIGH is output from DT/R, the 8088 is demanding a transmit-to-the-system condition from the bus transceiver/drivers; when a LOW is output, the 8088 is demanding that the transceiver/drivers receive data from the system bus. The DT/R output will float at high impedance (tristate) during hold acknowledge operations.

DEN (pin number 26). This active-LOW data enable output signal controls the output enable inputs of the 8286 and 8287 bus driver devices during active periods. This output will disable the bus drivers when address data is on the bus. The DEN signal will float tristate at high impedance during hold acknowledge operations.

INTR (pin number 18). This active-HIGH interrupt request line is sampled by the 8088 CPU during the last clock cycle of each instruction. If it is found HIGH during this time, then the 8088 will respond in one of two ways. If the interrupt enable bit in the internal 8088 flag register is reset, then the 8088 will ignore the request. If, on the other hand, the interrupt enable flag is set, then the 8088 will cease operation of the current program and branch to an interrupt subroutine. An interrupt vector table in system memory tells the 8088 where to find the program that services the interrupt. In most cases, the interrupt program is designed to return execution to the next instruction that would have been executed if the interrupt had not occurred. INTR is an example of a maskable interrupt.

INTA (pin number 24). This active-LOW interrupt acknowledge output line will go LOW during timing periods T2, T3, and T4 during the interrupt acknowledge cycle. External devices know by this signal being active that the 8088 CPU is responding to an interrupt request. This signal never goes tristate high impedance.

SSO (pin number 34). This active-LOW status output signal is used with IO/M and DT/R to specify the type of bus activity in a program (see the following table).

IO/M	DT/R	SSO	Function
1	0	0	Interrupt acknowledge
1	0	1	Read input/output (I/O) port
1	1	0	Write input/output (I/O) port
1	1	1	Halt
0	0	0	Code access
0	0	1	Read memory
0	1	0	Write memory
0	1	1	Passive

Note: HIGH = logical 1, LOW = logical 0

HOLD (pin number 31). This active-HIGH hold input signal is used to indicate to the 8088 that another device is requesting control over the local bus. The HOLD signal is used in conjunction with the HLDA output signal.

HLDA (pin number 30). This line is used for active-HIGH hold acknowledge output. When the 8088 CPU receives a HOLD request (see the previous paragraph), it will issue an active-HIGH output on HLDA to notify the other device that its request is granted. During the time when the 8088 has relinquished control it will float its own address and data bus lines tristate (high impedance).

NMI (pin number 17). This active-HIGH edge-triggered interrupt request input results in a type-2 interrupt. This request is not maskable by the programmer, and it will cause the 8088 to begin executing a subroutine pointed to by a vector table in memory. An active-HIGH edge-triggered input responds on the LOW-to-HIGH, i.e., positive-going, transition of the clock period at the end of the last instruction execution.

READY (pin number 22). This active-HIGH ready signal helps accommodate slow memory or peripheral devices.

TEST (pin number 23). This active-LOW signal is used to synchronize the CPU with some external event. There is a wait-for-test instruction (WAIT) that will cause the 8088 CPU to enter the "wait state" until an active-LOW signal is received on the TEST input.

8088 TIMING AND CONTROL SIGNALS

If your interests or needs in a computer are limited to programming in a high-level language, then you do not need to know anything at all regarding control and timing signals on the 8088 microprocessor chip. But if you are

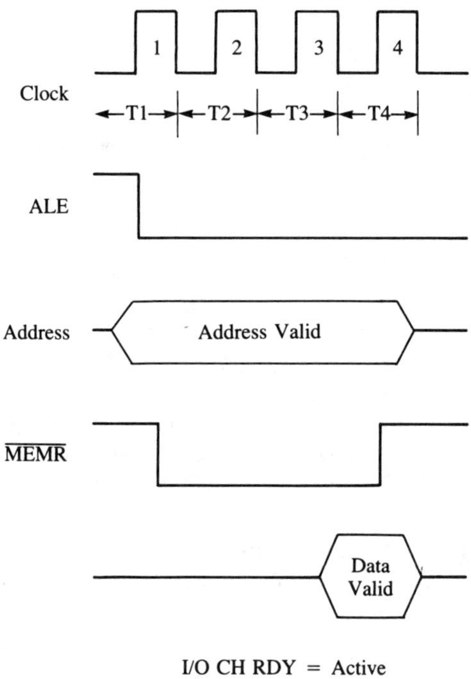

Figure 4-5 8088 read cycle timing.

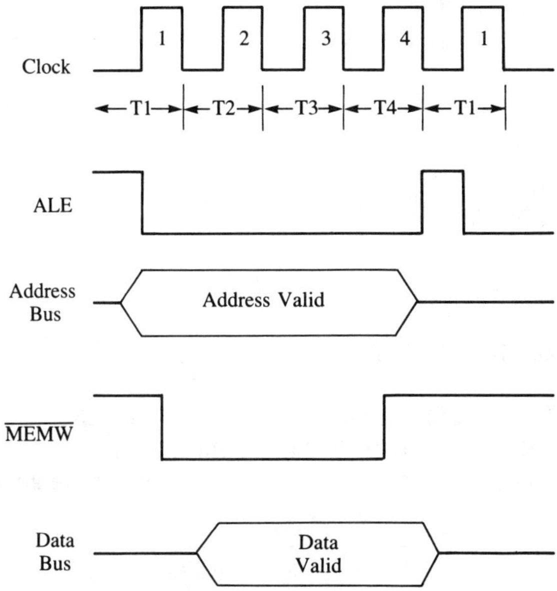

I/O CHNL RDY = Active

Figure 4-6 8088 write cycle timing.

interested in interfacing the 8088 or 8088-based computers, or in designing 8088-based computers, or in designing 8088-based instruments and control systems, then those timing signals (see Figures 4-5 and 4-6) assume critical importance. On the 8088, the following signals are used:

Data bus
Address bus
Read (RD)
Write (WR)
Address Logic Enable (ALE)
Input-Output/Memory (IO/M)
Reset
Minimum/Maximum Mode (MN/MX)
Data Transmit/Receive (DT/R)
Data Enable (DEN)
Interrupt Request (INTR)
Interrupt Acknowledge (INTA)
Status Output (SSO)

Hold (HOLD/HLDA)
Nonmaskable Interrupt (NMI)
Ready
Test (TEST)

These signals form inputs and outputs that are either control signals or indicate the status/operation of the 8088 CPU. The control signals are defined as follows.

DATA AND ADDRESS BUSES

There are two main buses on the 8088: data and address. The data bus is an 8-bit parallel pathway to transfer data in and out of the 8088. It shares the following package pins with the lower-order 8 bits of the address bus:

Designation	Pin No.
AD0	16
AD1	15
AD2	14
AD3	13
AD4	12
AD5	11
AD6	10
AD7	9

Pins AD0 through AD7 are time-multiplexed between address and data bus functions. During the T1 clock period these pins are used by the address bus. Since the address data on AD0–AD7 disappears at the beginning of the T2 clock period, the external circuitry must use the coincident ALE signal to latch and hold the address data.

During the period from the end of clock pulse T2 until the beginning of T4, the AD0 through AD7 pins operate as the data bus. These pins will be inputs during read operations and outputs during write operations.

When active, the AD0–AD7 pins are TTL-compatible, but they go to the high-impedance tristate condition during local bus hold acknowledge and interrupt acknowledge sequences.

A8–A15. These pins form the second byte of the 20-bit address. Unlike the 8086, these address lines remain for the entire machine cycle T1–T4. In other words, they are not multiplexed with any other functions.

During interrupt acknowledge and local bus hold acknowledge periods these pins go to the high-impedance tristate condition. These pins are as follows:

Designation	Pin No.
A8	8
A9	7
A10	6
A11	5
A12	4
A13	3
A14	2
A15	39

A16/S3–A19/S6. These pins supply the upper half-byte (i.e., 4 bits) of the 20-bit address bus. Like the lower byte, these pins are multiplexed with another function. In the case of these pins, the alternate functions are status indications.

During the Tl clock cycle, the high-order 4 bits of the address are output through A16/S3, A17/S4, A18/S5, and A19/S6. Like their companion multiplexed pins, these must be latched by external circuitry during the time that ALE is HIGH. From T2 through T4 these pins are used as status indicators. Pin A19/S6 is always LOW during the status segment, while A18/S5 indicates the status of the internal interrupt flag. If the interrupt enable flip-flop is set, then A18/S5 is HIGH, and if the flip-flop is reset, then the A18/S5 pin is LOW.

Pins A16/S3 and A17/S4 form a 2-bit coded status indicator that obeys the following rules:

A16/S3	A17/S4	Use
0	0	Alternate (relative to ES segment)
0	1	Code/None (relative to CS segment or default of zero)
1	0	Stack (relative to SS segment)
1	1	Data (relative to DS segment)

The pins for A16/S3 through A19/S6 are as follows:

Designation	Pin No.
A16/S3	38
A17/S4	37
A18/S5	36
A19/S6	35

As is true with the other bus pins, this group floats in the high-impedance tristate condition during interrupt acknowledge and local bus hold acknowledge operations.

READ/WRITE SIGNALS

The read and write lines are active-LOW output signals. During input and memory read operations, the RD line is LOW, while during output and memory write operations WR is LOW. In both cases, the active periods are during the T2–T3 clock cycles.

The RD and WR signals are used in conjunction with IO/M and the decoded address selection to turn on a specific memory location and to tell whether a read or write is taking place.

During interrupt acknowledge and hold acknowledge, RD and WR are in the high-impedance tristate condition.

ADDRESS LOGIC ENABLE

Part of the 8088 address bus is time-multiplexed with other functions. A0–A7, for example, are multiplexed with the data bus. The ALE is an active-HIGH output that tells external circuitry when the address is valid. If the 8282 latch is used, then ALE is connected to the active-LOW STB input. In that case, the address data is latched when ALE drops LOW at the end of clock period T1 (see either Figure 4-1 or 4-2).

The ALE output from the 8088 goes HIGH during T1, which is the period when the multiplexed portions of the address are output. We could use a device such as the 74100 or 74LS373 to latch the address. In that case, data is transferred from inputs to outputs when the strobe line(s) goes HIGH, and it latches on the negative-going transition of the strobe.

Unlike certain other 8088 pins, the ALE does not float at high impedance at any time during the machine cycle.

INPUT-OUTPUT MEMORY

The IO/M signal is an output that tells the world whether any given operation is an I/O or memory operation. This signal is HIGH for input/output operations and LOW for memory read/write operations. The IO/M signal remains valid over the entire T1–T4 cycle, but it goes into the high-impedance tristate condition during hold acknowledge operations.

RESET

The RESET input is a "hardware JUMP to FFFF0" instruction. When RESET goes HIGH, the CPU terminates the present action, loads the program counter with FFFF0, resets the flag register (including interrupt flip-flop) to 0000, resets registers DS, ES, SS, and IP to 0000, and sets register CS to FFFF.

Normally, RESET is activated on two occasions: 1) when the operator of an 8088 computer presses a reset button, and 2) at power-on. The effect of the reset operation is to initialize the computer so that it always starts at the beginning. If there was no power-on reset function, the computer would most likely start at whatever "noise" address happened to wind up in the program counter during power application. A HIGH on the input produces a LOW on the output, and a LOW on the input produces a HIGH on the output.

MINIMUM/MAXIMUM MODE SELECT

The MN/MX signal is an input that tells the 8088 whether to operate in the minimum mode or the maximum mode. MN/MX is HIGH for minimum mode and LOW for maximum mode. The minimum mode is the simpler configuration, and it is the mode that is normally selected. In this mode, the operation of the chip is more nearly like other microprocessor devices. The maximum mode is used in multiprocessing applications. The 8088 is paired with the Intel 8288 Bus Controller. The 8288 arbitrates the operation and controls the system buses. Several 8088 pins assume multiple uses in the maximum mode.

DATA TRANSMIT/RECEIVE

DT/R is a directional signal that tells external devices (such as bidirectional bus drivers/transceivers) which direction data is to flow from the 8088. When DT/R is HIGH, the external world knows that data is flowing from the transceiver to the system bus (a "transmit" is a write operation). A LOW on DT/R indicates that the 8088 wants to receive (i.e., read) data.

DATA ENABLE

The DEN pin is an active-LOW output that is used to turn on and off the tristate output enable on the bus transceivers. The purpose of the signal is to prevent inappropriate contention for the bus during periods when the 8088 is transmitting addresses.

INTERRUPT REQUEST

The INTR line is an active-HIGH input that permits maskable interrupt capability. This line is interrogated by the 8088 during the last clock period (T4) of the machine cycle. If INTR is HIGH, and the internal 8088 interrupt flip-flop is set (i.e., HIGH), then the 8088 will cease executing the present program and jump to a program called an "interrupt service subroutine."

The INTR input is program maskable by resetting the internal interrupt flip-flop. In this case, the 8088 will ignore the INTR signal.

INTERRUPT ACKNOWLEDGE

This signal is an active-LOW output that goes active during T2–T4 of each interrupt response machine cycle. It can be used as a read signal for interrupt servicing.

STATUS OUTPUTS

The SSO signal is used in conjunction with DT/R and IO/M to indicate to the outside world the type of bus activity that is taking place. The encoding for these outputs is as follows:

IO/M	DT/R	SSO	Use
0	0	0	Code access
0	0	1	Read memory
0	1	0	Write memory
0	1	1	Passive
1	0	0	Interrupt acknowledge
1	0	1	Read I/O port
1	1	0	Write I/O port
1	1	1	Halt

HOLD/HOLD ACKNOWLEDGE

The HOLD line is an active-HIGH input that tells the 8088 that another controller wants control of the local bus. This situation is sometimes called direct memory access, or DMA. The external device that wants control will place a HIGH on the HOLD line. At the end of the present machine cycle the 8088 will relinquish control and issue a hold acknowledge (HLDA) active-HIGH output signal. During the period when the 8088 has given up control it will float in the high-impedance tristate condition the following lines:

AD0–AD7
A8–A15
A16–A19
RD
WR
IO/M
INTA
DT/R
DEN

In addition, ALE is held LOW and HLDA is held HIGH. When the external device no longer needs control of the buses it will force HOLD to LOW, and the 8088 will regain control at the beginning of the next cycle.

NONMASKABLE INTERRUPT

This active-LOW input is a positive edge-triggered (i.e., LOW-to-HIGH) interrupt request line that will initiate a type-2 interrupt in the 8088. It differs from INTR in that NMI cannot be masked by the programmer. If NMI goes LOW, then the 8088 CPU will enter the interrupt sequence at the beginning of the next machine cycle (note: NMI must be held LOW for not less than two clock periods). This form of interrupt is vectored, and the address of the interrupt service subroutine is found at location 008H in memory.

READY

This active-HIGH input serves to add wait states to the machine cycle. This facility permits the fast-working 8088 to accommodate slower memory, I/O, or peripheral devices.

TEST

The TEST signal is an active-LOW input that works with the "wait for test" instruction. TEST allows the 8088 to be synchronized with external events.

BUS TIMING IN MINIMUM MODE

Let's return to Figures 4-1 and 4-3 for a description of timing signal dynamics over the course of a machine cycle. Like all programmable digital computers, the 8088 operates in a synchronous manner. That is, all action takes place in step with a chain of square waves called a clock signal ("CLK" in Figures 4-1 and 4-2). Each pulse produced by the clock is called a T-state (T1, T2, T3, and T4).

The basic unit of time for the activity of the 8088 is the machine cycle, or bus cycle. Each machine cycle consists of the four T-states T1 through T4 (see Figure 4-1 or 4-2).

During the first half of T-state T1 the 8088 will output the 20-bit address on AD0–AD7, A8–A15, and A16/S3–A19/S6. At this same time, the ALE line goes HIGH. Before the end of T1, however, ALE will drop LOW again (while all address lines are still valid). The address bits on AD0–AD7 and A16/S3–A19/S6 are connected to an external data latch IC. These bits are transferred to the outputs of their respective latches when ALE is HIGH—and latched when ALE drops LOW again. Thus the system bus will see a permanent address over the entire machine cycle even though the multiplexed bits go invalid at the end of T-state T1. Address bits A8–A15 remain valid over the entire machine cycle T1–T4.

Certain other control signals activate at the beginning of T1. The IO/M signal will go LOW if the operation is a memory write and HIGH if it is an input/output operation; the WR goes LOW if the operation is a write to either memory or an output port (Figure 4-1); the RD goes LOW if it is a memory read or input operation; the DT/R will be HIGH for writes and outputs and LOW for reads and inputs (T1–T4); DEN will be LOW during T2–T4.

The difference between DT/R and the WR and RD signals is that DT/R remains active over the entire T1–T4 machine cycle, while RD and WR are only active during T2–T3 when the data is stable on AD0–AD7. The DT/R signal is used to control external bidirectional bus drivers and should not be used as a *de facto* R/W line. If that were done, then external devices might accept address signals as if they were valid data signals.

Comparing Figures 4-1 and 4-2 shows subtle differences in bus timing for read and write operations. The duration of the data transmission periods and the states of RD, WR, and DT/R should be examined.

8088 ADDRESSING MODES

An *addressing mode* is a scheme for finding the location of an operand somewhere in memory. Normally, users of high-level languages and assemblers will not need to concern themselves with 8088 addressing modes because the language takes care of that chore for them. Knowledge of addressing modes becomes desirable when trying to understand the chip itself or when programming in machine language or with an unsophisticated assembler.

8088 ADDRESSING

Before considering addressing modes, let's first review the way the 8088 forms a general address. In order to address over one megabyte of memory the 8088 relies on a 20-bit address (note: $2^{20} = 1,048,576$). This address is formed from two components: a 16-bit logical address and a 4-bit segment address. The segment is automatically selected.

The logical address is always 16 bits in length, which allows addressing of 64K (i.e., 65,536 bytes). As in less sophisticated microprocessors, the 16-bit address permits easy movement within any 64K segment of memory.

There are four segment registers: code, stack, data, and extra. The code segment register is used to compute the address during instruction fetch operations. The effective location of the instruction is the sum of the code segment register contents and the 16-bit logical address supplied by the instruction pointer. Similarly, operations on the external memory stack use addresses formed by summing the stack segment register and the 16-bit logical address supplied by the stack pointer. When fetching data, the address is formed by the contents of either the data or extra segment register and a 16-bit logical address formed in any of several ways which will be described shortly.

The segment register value is not used directly, incidentally, but is automatically multiplied by 16 before use. Multiplication by 16 in a binary arithmetic system means that the data is shifted 4 bits to the left.

8088 ADDRESSING MODES

The 8088 and 8086 microprocessors use two dozen different addressing modes. Intel breaks these modes into the following seven categories:

Immediate
Register
Direct
Register Indirect
Indexed
Based
Based and Indexed with Displacement

The Immediate Mode contains the data operand within the instruction itself (usually the second or third byte). The operand will occupy the second byte in 8-bit cases and both the second and third bytes in 16-bit cases. The 8088 will assume an 8-bit case is used if the W-field (i.e., the least significant bit) of the op-code is zero (W = 0) and a 16-bit case if it is one (W = 1).

In the Register Mode, the data operand is found in a register specified by the instruction.

The Direct Mode places the data operand at a memory location specified by an address specified by the instruction.

In the Register Indirect Mode the data operand is found at a memory location specified by the contents of an 8088 internal register.

The Indexed Mode forms the address of the data operand by summing the contents of an index register with the immediate data supplied by the instruction.

The Based Mode is the same as the Indexed Mode, except that the base register is used instead of the index register.

The Based and Indexed with Displacement Mode is the most complex of the 8088 addressing modes. The address of the operand data is formed by summing the contents of the base register, index register, and immediate (displacement) data provided by the instruction.

8088 STATUS FLAGS

Status flags are bits in an internal register of the CPU that indicate the status of the CPU and/or last relevant operation that was performed. For example, the zero flag in a computer CPU will be set (i.e., placed in a logical 1 or HIGH condition) if the result of an operation is zero and reset (to logical 0 or LOW) if the result was nonzero. Some programs use these flags to test for the existence of a condition. In a counter routine, for example, a binary number proportional to a desired duration is loaded into a register. The counter program will loop through a time-wasting routine that decrements the register once for each pass through and then test for zero. Eventually, the binary number in the register will decrement to zero and the program "breaks out" to continue the main program.

Figure 4-7 shows the organization of the 8088 flag register. There are nine 1-bit status flags in the 16-bit flag register. These are as follows:

Bit	Designation	Use
0	CF	Carry flag
2	PF	Parity flag
4	AF	Auxilliary carry flag
6	ZF	Zero flag
7	SF	Sign flag
8	TF	Trap flag

Bit	Designation	Use
9	IF	Interrupt enable flag
10	DF	Direction flag
11	OF	Overflow flag

The unlisted bits (i.e., B1, B3, and B12 through B15) are not used at present but serve to leave room for future expansion. Each of the used bits can be either logical 1 (HIGH) or logical 0 (LOW), depending upon the situation.

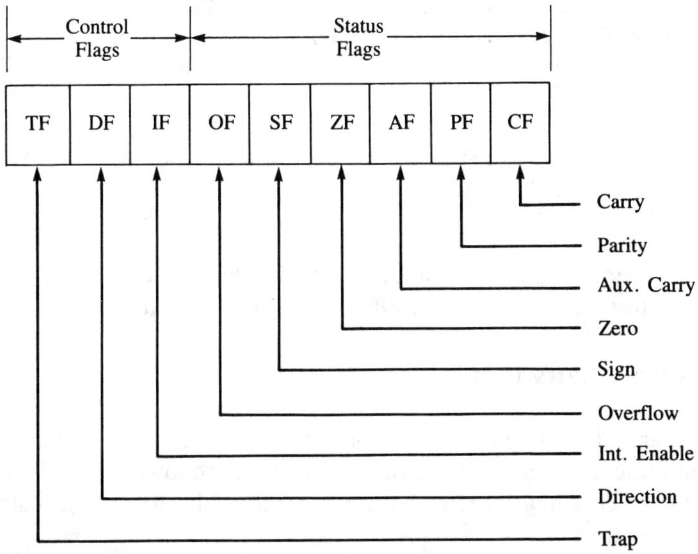

Figure 4-7 8088 flag register allocations.

8088 FLAG OPERATION

Carry (CF) Flag

The carry flag is set HIGH if an operation causes either a carry-out (addition) from or a borrow (subtraction) into the register. It is primarily the arithmetic and shift/rotate forms of instruction that affect carry flag status. Regardless of whether an operation is 8-bit or 16-bit, it is always the highest-order bit that will affect the carry flag. The following instructions will produce a change in the carry flag:

AAA	AAD	AAM	AAS	ADC	ADD	AND	CLC	CMC	CMP
CMPS	DAA	DAS	DIV	IDIV	IMUL	IRET	MUL	NEG	
OR	POPF	RCL	RCR	ROL	ROR	SAHF	SAL	SAR	
SBB	SCAS	SHR	STC	SUB	TEST	XOR			

Several instructions (CLC, CMP, and STC) have as their sole function alteration of the carry flag. The CLC instruction clears the carry flag (makes it LOW); CMP complements the carry flag (makes it HIGH if it is LOW, and LOW if it is HIGH); STC sets the carry flag (makes it HIGH).

Parity (PF) Flag

Parity refers to the number of "1" data bits and is defined as either "even" or "odd." An example of even parity is 01010011_2 (there is an even number of 1s), while an example of odd parity is 00010011_2 (there is an odd number of 1s). This concept is useful in checking data transmissions for errors. The 8088 parity flag is set HIGH for a result with even parity and reset LOW for results with odd parity. The following instructions affect the parity flag:

AAD	AAM	ADC	ADD	AND	CMP	CMPS	DAA	DAS
DEC	INC	IRET	NEG	OR	POPF	SAHF	SAL	SHL
SAR	SBB	SCAS	SHR	SUB	TEST	XOR		

Few other microprocessors have a parity flag, a factor that could cause one to lean towards the 8088 (or 8086) in some applications.

Auxilliary Carry (AF) Flag

The auxilliary carry flag is used in 4-bit binary coded decimal (BCD) arithmetic and indicates a carry from, or borrow into, the lower-order half-byte of a data word. This flag affects the lower 8 bits. The following instructions affect the AF flag:

AAA	AAS	ADC	ADD	CMP	CMPS	DAA	DAS	DEC
INC	IRET	NEG	POPF	SAHF	SBB	SCAS	SUB	

Zero (ZF) Flag

The zero flag is set HIGH if a result is zero and reset LOW if the result of an operation is nonzero. The following instructions affect the zero flag:

AAD	AAM	ADC	ADD	AND	CMP	CMPS	DAA	DAS
DEC	INC	IRET	NEG	OR POPF	SAHF	SAL	SHL	SAR
SBB	SCAS	SHR	SUB	TEST	XOR			

There are several uses for the zero flag in programming. The most obvious is in arithmetic operations. Less obvious is in timer applications. We can load a number into a register proportional to the time duration required. That register is then sequentially decremented and tested for zero. When zero is detected (ZF = HIGH) then the process stops. The programmer cannot directly affect the ZF flag, but she or he can adopt program strategies that indirectly affect the flag.

Sign (SF) Flag

The sign flag duplicates the high-order bit of the result. In other words, if the high-order bit is HIGH (logical 1), then the sign flag is also HIGH; if the high-order bit is LOW (logical 0), then the sign flag is also LOW. In arithmetic operations the sign flag indicates positive (SF = LOW) or negative (SF = HIGH) quantity. The following instructions affect the sign flag:

AAD	AAM	ADC	ADD	AND	CMP	CMPS	DAA	DAS
DEC	DIV	INC	IRET	NEG	OR	POPF	SAHF	SAL
SHL	SAR	SBB	SCAS	SHR	SUB	TEST	XOR	

The sign flag cannot be directly affected, although the programmer can use certain strategies to set or reset the flag.

Trap (TF) Flag

The purpose of the trap flag is to allow single-step operation of the 8088. If the trap flag is set HIGH, then the 8088 single-steps one instruction at a time and stops between instructions. If the trap flag is reset LOW, then the CPU operates normally. The following instructions affect the trap flag:

INT IRET POPF

The trap flag is useful in debugging programs.

Interrupt Enable (IF) Flag

The interrupt enable flag determines whether externally generated maskable interrupt requests are honored. If the interrupt enable flag is reset LOW, then the 8088 will not respond to external interrupts, but if the interrupt flag is set HIGH, then the external interrupt function is enabled. The interrupt enable flag can be set or reset under program control. This flag does not affect requests on the nonmaskable interrupt input (NMI, pin number 17) but only those on the interrupt request input (INTR, pin number 18). The following instructions affect the interrupt enable (IF) flag:

CLI INT INTO IRET POPF STI

The CLI instruction clears the flag (sets it LOW), while the STI sets the flag (sets it HIGH).

Direction (DF) Flag

The DF flag causes string instructions to process instructions in a directional sequence according to its condition. If DF is set HIGH, then instructions process from high address to low address. If the DF flag is reset LOW, then processing occurs from low address to high address.

Overflow (OF) Flag

The OF flag is normally reset LOW, but it will become set HIGH if an arithmetic overflow occurs.

5

APPLE II PLUG-IN CARD INTERFACING

The Apple IIe is the last in a line of 6502-based personal computers that started in the late 1970s. The unique feature of the Apple II that made it popular with small laboratories and process control people was that it was first of all a real computer that could be programmed in a high-order language (at least to the extent that BASIC is a high-order language). In addition, it has a series of plug-in slots for printed circuit board add-ons that effectively allows users to customize the computer without a lot of external boxes hanging onto the machine.

The Apple IIe is still found almost everywhere, and although the market is now declining, there seem to be about as many retail outlets for this machine as for any other personal microcomputer except IBM PC clones. The Apple IIe is still sold new, and there are several clones on the market. One of the advantages of the Apple IIe is that it is cheap, and at the same time it uses an architecture that permits plug-in cards to be customized and installed in the computer.

The Apple II family of machines is based on the 6502 microprocessor chip. The Apple II is so popular that it has spawned not only imitators (some of which use seemingly exact copies of the Apple II printed wiring board layout), but also counterfeits. Some unscrupulous overseas manufacturers offered for sale exact duplicates of the Apple IIe without bothering to obtain a license from the U.S. manufacturer.

The Apple II is a single-board computer housed in a small case about the size of an inexpensive typewriter. There are eight slots on the mother-

Author's Note: The use of @ in the following table and throughout this text indicates the logical NOT.

TABLE 5-1 Plug-In Slot Pin-Outs

Pin No.	Designation	Function
1	@I/O SELECT	This active-LOW signal is LOW if and only if one of the 16 addresses assigned to that particular connector is called for in the program. The 6502 used in the Apple II uses memory-mapped I/O, so each I/O port number is represented by a memory location in the range from C800H to C8FFH. Reference the Apple II memory map in the manual for specific locations.
2	AO	Address bus bit O.
3	A1	Address bus bit 1.
4	A2	Address bus bit 2.
5	A3	Address bus bit 3.
6	A4	Address bus bit 4.
7	A5	Address bus bit 5.
8	A6	Address bus bit 6.
9	A7	Address bus bit 7.
10	A8	Address bus bit 8.
11	A9	Address bus bit 9.
12	A10	Address bus bit 10.
13	A11	Address bus bit 11.
14	A12	Address bus bit 12.
15	A13	Address bus bit 13.
16	A14	Address bus bit 14.
17	A15	Address bus bit 15.
18	R/@W	Control signal from 6502 microprocessor is HIGH during read operations and LOW during write operations.
19	(NC)	No connection.
20	@I/O STR	Active-LOW signal that lets the outside world know that an input or output operation is taking place. This line will go LOW whenever an address in the range C800H to C8FFH is on the address bus.
21	@RDY	Active-LOW input. If this line is LOW during the phase-1 clock period, the CPU will halt (i.e., enter a wait state) during the following phase-1 clock period. If @RDY remains HIGH, normal instruction execution will occur on the following phase-2 clock signal.
22	@DMA	Active-LOW direct memory access line allows external devices to gain access to the data bus and apply an 8-bit data word to the address it places on the address bus.
23	INTOUT	Interrupt output. Signal that allows prioritizing of interrupts from one plug-in card to another. The INTOUT line of each lower-order card runs to the INTIN pin of the next card in sequence. See pin 28.

TABLE 5-1 (Continued)

Pin No.	Designation	Function
24	DMAOUT	Direct memory access version of INTOUT.
25	+5 V	+5-V DC power supply available from main board to plug-in card.
26	GND	Ground.
27	DMAIN	Direct memory input; signal that allows prioritizing of DMA functions.
28	INTIN	Interrupt input. See DMAOUT (pin 24).
29	@NMI	Active-LOW nonmaskable interrupt line. When brought LOW, this line will cause the CPU to be interrupted at the completion of the present instruction cycle. This interrupt is not dependent upon the state of the CPU's interrupt flip-flop flag.
30	@IRQ	Interrupt request. This active-LOW input will cause the CPU to interrupt at the end of the present instruction cycle, provided that interrupt flip-flop is reset.
31	@RES	Reset line. This active-LOW input will cause the program to return to the Apple II monitor program.
32	@INH	Active-LOW input that disconnects the ROMs of the monitor in order to permit custom software stored in ROMs on the plug-in board to be executed.
33	−12 V	−12-V DC power from main board to plug-in board.
34	−5 V	−5-V DC power from main board to plug-in board.
35	(NC)	No connection.
36	7M	7-MHz clock signal.
37	Q3	2-MHz clock signal.
38	01	Phase-l clock signal.
39	USER1	Similar to @INH except that it disables all ROMs, including C800H to C8FFH, used for I/O functions.
40	02	Phase-2 clock signal.
41	@DEVICESEL	Active-LOW signal that indicates one of the 16 addresses assigned to that connector is being selected.
42	D7	Data bus bit 7.
43	D6	Data bus bit 6.
44	D5	Data bus bit 5.
45	D4	Data bus bit 4.
46	D3	Data bus bit 3.
47	D2	Data bus bit 2.
48	D1	Data bus bit 1.
48	D0	Data bus bit 0.
50	+12 V	+12-V DC power from main board to plug-in boards.

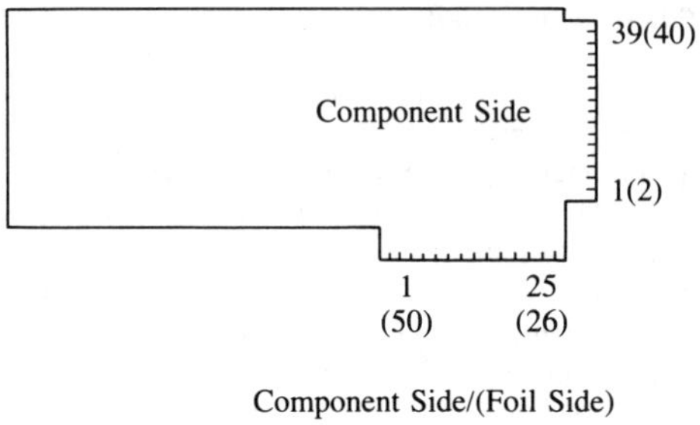

Figure 5-1 Apple II expansion card.

board that will accommodate plug-in printed wiring boards for accessories, interface devices, and special "firmware" software (i.e., on ROMs), and special function printed circuit boards. The basic computer comes with 64K of memory, but we can configure it with up to 128K of 8-bit memory by using up one of the plug-in sockets. Some Apple II series machines come with 128K built-in.

A feature of the Apple II is the use of software to replace hardware complexity. The original Apple II machines were limited to 48K because memory allocations above the 48K boundary were used for the monitor program and for housekeeping functions like driving the floppy disk system.

The connectors for each of the plug-in cards have 50 pins, with pins 1 through 25 on the component side of the inserted printed wiring boards and 26 through 50 on the foil side of the card. Several companies offer either plug-in accessory cards (e.g., I/O cards or analog-to-digital [A/D] converter cards) or blank interfacing cards on which you may build your own circuitry. The Apple II plug-in card pinouts are described in Table 5-1. Figure 5-1 shows a typical plug-in interfacing card that can be used either for prototyping circuits for Apple II plug-in cards or for building one-of-a-kind devices to be plugged into an Apple II. Additional information can be obtained by reference to *Apple Interfacing* by Jonathan A. Titus *et al.* (Howard W. Sams & Co., Inc., Cat. No. 21862).

6

GENERATING DEVICE- AND BOARD-SELECT PULSES

One aspect of any programmable digital computer is the use of a main data bus operated in a synchronous manner. In this type of arrangement, the data bus is common to a large variety of devices. The data bus, for example, services the computer memory, all input/output (I/O) ports, and many peripherals that may be connected directly to the computer as if they were either memory or I/O ports in their own right.

The secret to the operation of any bus system is *synchronization*. The CPU will designate the I/O port or the memory location as well as the type of operation that is to take place. During the period the operation is being executed, only the affected device is actively connected to the data bus. For example, let's assume that an I/O port is being designated. On the output side, we want it to accept data to be sent out to the peripheral connected to it only when the computer is executing a pertinent instruction. We would not want the port to be active at all other times, because not everything that passes along the data bus is intended for that output port. Indeed, only a few pieces of data will be destined for any one port in most cases. We want that output port to accept data only when commanded to do so.

On the other hand, we would not want the input side of the I/O port to be active except when commanded. Not only do we wish to avoid sending inappropriate data into the computer, we also want to avoid having a constantly active port capturing the data bus and thereby distorting the data transmitted over the bus.

The answer to the problem of synchronization is the generation of device-select pulses to designate and turn on the memory location, I/O port, or peripheral designated by the computer CPU chip.

All microcomputers generate several control signals that are used to synchronize operations. In the Z80 microprocessor[1] for instance, we have the @WR, @RD, @IORQ, and @MREQ signals, which are detailed as follows:

@WR	Active-LOW output signal that indicates a write operation is taking place.
@RD	Active-LOW output that indicates that a read operation is taking place. Neither @RD or @WR tell whether the operation is an I/O operation or a memory operation. At least one additional signal is needed.
@IORQ	Input/output request. This active-LOW output indicates when an I/O operation is taking place.
@MREQ	Memory request. This active-LOW output indicates that a memory operation is taking place.

In the Z80 microprocessor, two signals are needed to define fully the type of operation that is taking place, while one additional signal (valid address) is needed to designate the specific port or memory location. For memory operations, the coincidence of @MREQ and @WR indicates a write operation to memory (all directional designations in microcomputers are from the CPU point of view, not that of the outside observer). Similarly, an @MREQ and @RD indicate that a read from memory is taking place. The two different types of I/O operation are, of course, input and output. These operations are denoted by the coincidence of @IORQ/RD and @IORQ/WR, respectively. We may, therefore, use these signals and the address bus to specify uniquely a particular memory location or I/O port.

Each microprocessor chip has its own particular set of control signals, and it is necessary for the designer to learn and understand their use in order to generate correctly the device-select signals. In some systems, there will be no discrete I/O ports; instead, memory locations are designated as I/O ports. Such systems are called memory-mapped I/O computers. Other devices, such as the Z80 microprocessor, use a separate I/O structure. Since this is the more complex of the two schemes, we will construct our examples principally around the Z80 device.

In the Z80 microprocessor there is a 16-bit address bus, allowing a total of 65,536 (i.e., 64K) different memory locations to be addressed. In addition, during I/O operations the unique address of the I/O port is passed along the lower 8 bits of the address bus (A0–A7), while the data in the accumulator at that instant are passed along the higher 8 bits (A8–A15). Since 8 bits are used to designate the I/O ports, we may specify a total of 256 different discrete ports numbered from 000 to 255. Figure 6-1 shows the structure of the Z80 controller chip as needed to generate the device-select signals.

[1] For additional details, see Joseph J. Carr, *Z80 User's Manual*, Reston Publishing Co., Reston, Va., 1981.

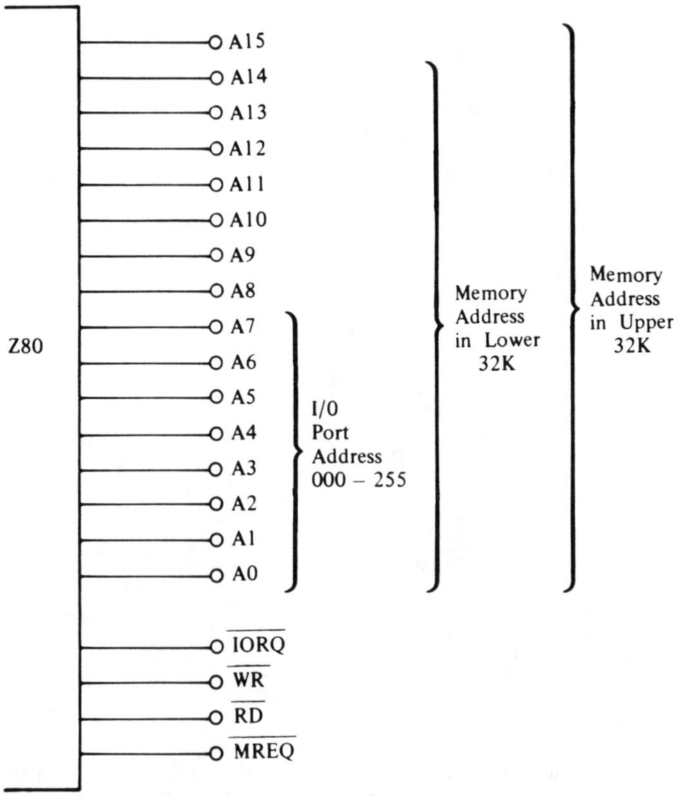

Figure 6-1 Address allocation for a popular controller chip (Z80).

ADDRESS DECODING

The microcomputer uses a 16-bit binary word passed along the 16-bit parallel address bus to indicate memory locations. In the Z80, eight of these bits are also used to designate I/O port addresses. The problem for the designer is to create a circuit that will uniquely decode the required address, that is, generate a signal that exists if and only if the correct address is passed along the bus.

Several different techniques may be used for the decoding of the address bus. One method that is based on the properties of the NAND gate is shown in Figure 6-1. Recall the properties of the NAND gate: (1) if any *one* input is LOW, then the output is HIGH; and (2) all inputs must be HIGH for the output to be LOW.

We must, therefore, create a situation in which *all* inputs of the NAND gate are HIGH when the correct address is passed along the bus. At all

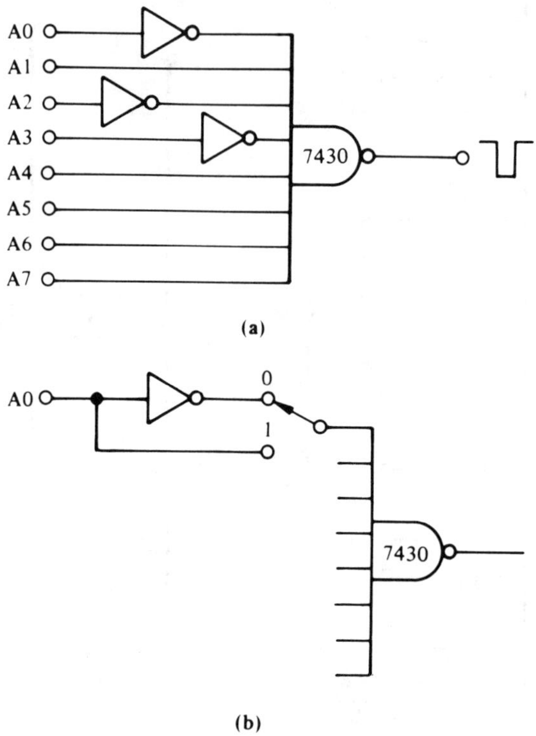

Figure 6-2 Simple address decoding: A) Using inverters only on LOW lines, B) Using inverter and switch on every line to make it HIGH/LOW-selectable.

other times at least one NAND gate input will be LOW, thereby forcing the output HIGH. The correct indication of the proper address will be a LOW on the output of the NAND gate.

In most 8-bit microcomputers we will have to decode either 8 bits or 16 bits of the address bus. The TTL 7430 device is an 8-bit NAND gate and so is almost ideally suited to this type of service.

In the circuit of Figure 6-2A, we are attempting to decode the address 11110010. Five bits of the address bus will naturally be HIGH when the correct address is present and so require no additional treatment. There are, however, 3 bits that will be LOW when the correct address is present: A0, A2, and A3. The lines for these bits are not connected directly to the address bus, but are first *inverted.* By connecting an inverter in each of these three lines, the input to the NAND gate will be 11111111 when the data word on the address bus is 11110010.

Hardwiring the inverters into the circuit sometimes unnecessarily limits our selection to the addresses selected in advance. We can, however, take one of several tacts that overcome this problem. We could, for example, place an inverter in each input line and then use either a switch or movable jumpers to determine whether a 1 or a 0 on the particular address bus will generate the required HIGH on the NAND gate input. This method is shown in Figure 6-2B.

We can also select I/O port numbers that require fewer inverters. If we select port 0, then we must decode 00000000, which requires eight inverters. If, on the other hand, we select port 255, we need no inverters, because the correct code will be 11111111 (FF in hex). Any address in the higher end of the permissible range will require substantially fewer inverters if the scheme of Figure 6-2 is used.

The key to making our decoder work on all locations or addresses other than FF_{16} is the use of inverters. We must, however, sometimes use a certain economy of design in order to achieve a lower cost or, perhaps, a lower components count. The most obvious option is to use one or more hex inverter IC devices to get the inverters that we need. Each hex inverter contains six independent inverter stages. To get eight inverters, then, we must use all six stages from one hex inverter IC and two stages from a second hex inverter IC. This means a potential waste of four inverter stages. The key to our design economy may well be the use of wasted sections of various ICs to gain the inverters that we need. Unless some printed writing board layout problem prevents it, we can use unused inverter stages from other ICs or make inverters from NAND, NOR, and XOR gates that may be left over when the IC was only partially used elsewhere in the circuit. Figure 6-3 shows the use of NAND, NOR, and XOR gates. In the TTL line, we find the 7400 NAND gate contains four independent two-input NAND gates; the 7402 contains four independent NOR gates; the 7486 device contains four independent XOR gates.

There are two tactics that will result in making either a NAND gate or a NOR gate into an inverter. For both types of gate, we can connect both inputs together to make an inverter. We can also make the NAND gate into an inverter by connecting one input permanently HIGH (i.e., tie it to +5 V DC through a 1- to 5-kiloohm (kΩ) resister). The resister effectively disables that particular input. The other input follows the normal rules for NAND gates: When the other input is LOW, then the output is HIGH; when the other input is HIGH, the output is LOW.

The NOR gate function is exactly the opposite of the NAND gate function. As might be suspected in such a case, therefore, we can make the NOR gate into an inverter by permanently grounding one input. Recall the rules for the operation of the NOR gate: (1) if either input is HIGH, the output is LOW; (2) if both inputs are LOW, then the output is HIGH. When one input is permanently LOW, therefore, the NOR gate will operate as an inverter.

The XOR gate is a little different from the other types. The output of

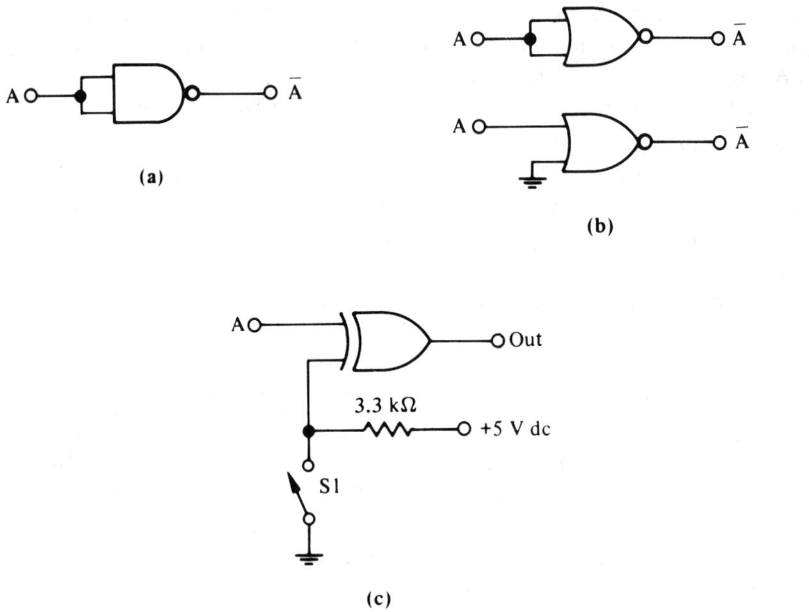

Figure 6-3 Inverter schemes: A) NAND gate, B) NOR gate, C) Selectable INVERT/NONINVERT using XOR gate.

the XOR gate will be HIGH only when the data applied to the two inputs are *different*. In other words, the output is HIGH if either input is HIGH or LOW, but not when both are at the same level at the same time. The truth table for the XOR logical operator is as follows:

Input A	Input B	Output
0	0	0
0	1	1
1	0	1
1	1	0

Figure 6-3C shows the use of an XOR gate as either an inverter or a noninverting buffer, depending on the setting of switch S1. When the switch is open, one input is held HIGH. In this case, a LOW on the other input means that the two inputs are different, so the output is HIGH: it is inverted. When switch S1 is closed, however, that input is held LOW. A LOW on the other input will produce a LOW output, and a HIGH on that input will produce a HIGH output: no inversion takes place. We can use the

XOR gate version to replace the switches used in Figure 6-2B. Connecting the control input to ground will pass the 1 on the data bus on to the NAND gate decoder input. Similarly, holding the control input HIGH will cause an inversion of the address bit. This method will be used in another illustration shortly.

We do not always need all 8 bits of an address to specify an I/O port. The Z80 microprocessor can uniquely address 256 different ports. While that may sound impressive, we will very rarely actually use that many ports. In most projects, the designer will find no more than one or two different ports (e.g., a keyboard and a teletypewriter or printer). In those cases, we can often achieve economy of design by ignoring most of the higher-order ports and only worry about a few. We can replace the eight-input NAND gate with a simpler four- or two-input NAND gate that may be a wasted section of an IC that is already in the design.

I/O Port	Code			
No.	A3	A2	A1	A0
0	0	0	0	0
1	0	0	0	1
2	0	0	1	0
3	0	0	1	1

(a)

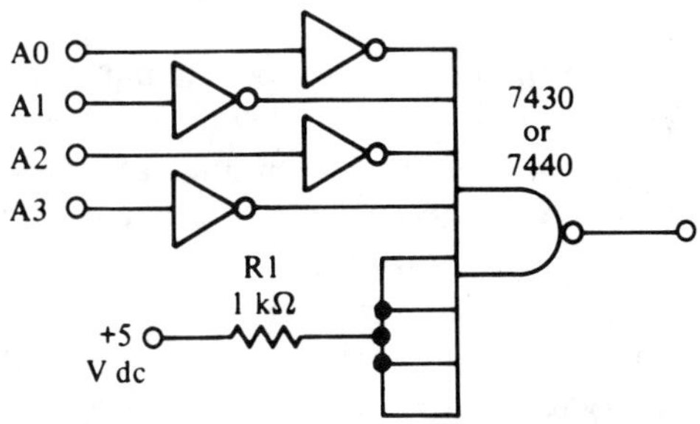

(b)

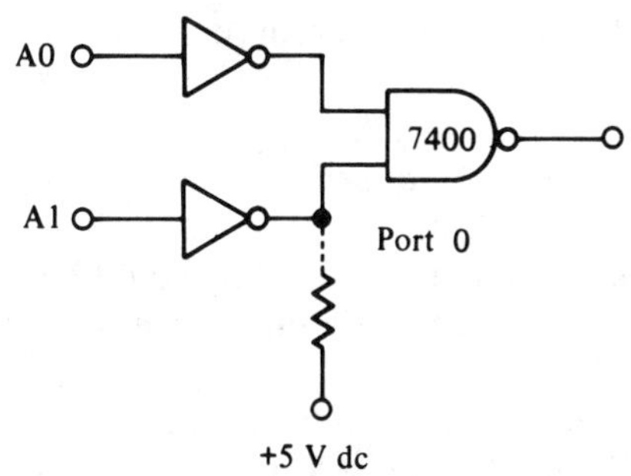

(c)

Figure 6-4 NAND gate address decoding: A) Four-bit, B) Two-bit.

Figure 6-4A shows the binary codes for the four lowest-order addresses. Note that only 2 bits are needed to uniquely decode these addresses: A0 and A1. Bits A2 and A3 are always LOW. As long as we are not going to use any higher-numbered ports, only these 2 bits are needed.

Figure 6-4B shows the use of a multi-input NAND gate to decode the first few addresses within the permissible range. The 7430 is an 8-bit device, so we can tie four to six inputs permanently HIGH and use only the required inputs. For the circuit shown, the code 0000 will cause the output of the NAND gate to drop LOW. We can delete or keep each inverter as needed. With four address lines, up to 16 devices can be selected with this circuit. The simplest case is shown in Figure 6-4C, in which a simple two-input NAND gate is used. This circuit will decode all four lower-order I/O ports. For the circuit as shown, with both inverters wired into place, the output of the 7400 will drop LOW only for port 0 (for which the code is 00). With two inputs, we can decode up to four ports (ports 0 through 3), depending upon whether or not the inverter is wired into the circuit.

In a further simplification, port 0 and port 1 can be decoded with only one input, the other input being wired permanently HIGH. In that case, we would use the inverter for port 0 and delete the inverter (or wire around it) for port 1.

Figure 6-5 shows a circuit that can be switch selected to decode any port from port 0 to port 15. Since 16 different ports are possible, a total of 4 bits are needed: A0–A3.

The circuit uses XOR gates (see Figure 6-3C) as either inverters or noninverting followers depending upon the setting of the respective bit-select switches (S1 through S4). The rules for Figure 6-3C apply in this circuit as well. The 7440 device is a four-input TTL NAND gate.

Thus far, all our address decoders have involved the use of NAND gates and inverters to generate a signal that is unique to the selected address. This method is not always the most viable, especially where lower-order addresses are called out. The circuit in Figure 6-6 is based on a different type of TTL integrated circuit, the 7442 *BCD-to-1-of-10 decoder.*

The 7442 device is not specifically intended as an address decoder, despite the use of the word "decoder" in its type designation. Originally, it was intended to drive decimal numerical readout devices. In older types of digital decimal counters, the readout display would be either a column of 10 incandescent lamps or a 10-digit Nixie tube. In either case, the binary word from the counter would be in the 8-4-2-1 binary coded decimal (BCD) format. BCD is a limited version of 4-bit binary in which only the first 10 states are allowed. The 7442 decodes the BCD word at the 4-bit input and issues an output that uniquely specifies the decimal value of the BCD word. The active output will drop LOW when its BCD word is applied. We can, therefore, use the 7442 device to select any of 10 discrete addresses or I/O port numbers, as shown in Figure 6-6.

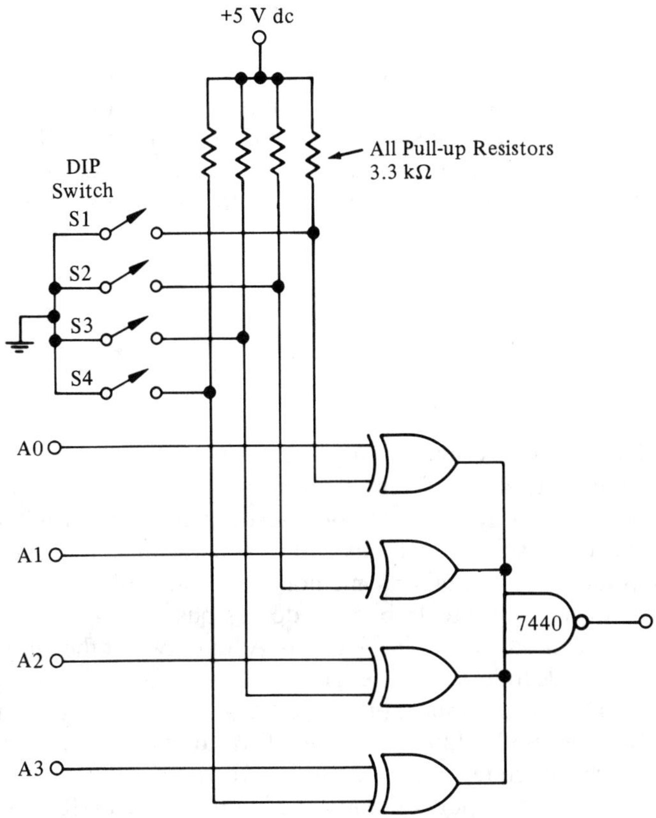

Figure 6-5 Programmable 4-bit address decoder.

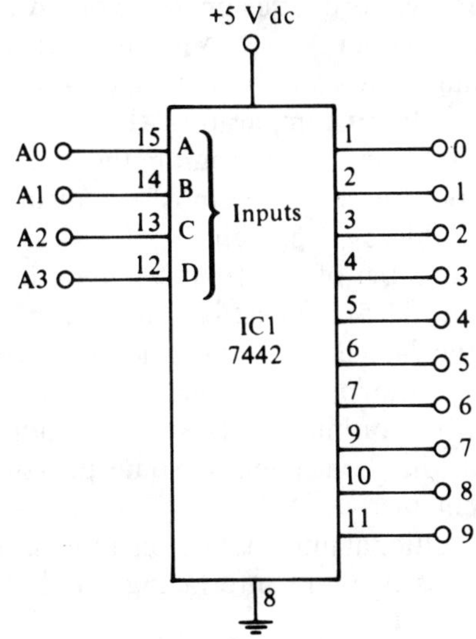

Figure 6-6 7442 chip used as a selectable 1-of-10 address decoder.

In Figure 6-6, the first 4 bits of the address bus are connected to the BCD inputs of the 7442 device. Any correct address from 0 to 9 can be decoded by this circuit. The selection is made by connecting the enable line of the selected address to the 7442 output terminal that corresponds to its address. A 16-line decoder can be made with the 74154 device, which will be discussed in the next section.

16-BIT DECODERS

The decoder circuits presented thus far have been 8-bit designs. They are, therefore, limited to 256 different combinations. Only a few microcomputers will use only 256 bytes of memory, so the 8-bit decoder will be insufficient for that purpose. We will have to be able to decode up to 16 bits in order to uniquely address all 64K memory locations. Figure 6-7 shows two methods for decoding up to 16 bits of address bus.

The first method, shown in Figure 6-7A, uses two of the circuits shown earlier; two 8-bit decoders will select from 16 bits. The active-LOW outputs are connected to the two inputs of a 7402 NOR gate. According to the rules of the NOR gate, both inputs must be LOW for the output to be HIGH. Since each 8-bit decoder output is an active-LOW select signal, we will achieve the NOR-gate input condition needed to create a HIGH output only when the correct address is present on bits A0–A15 of the address bus.

A second method is shown in Figure 6-7B. This circuit uses the 7485 *4-bit binary word comparator* integrated circuit. This device compares two 4-bit binary words, designated word A and word B, and issues outputs that indicate whether A = B, A is greater than B, or A is less than B. In addition, cascading inputs and outputs allows use of additional 7485 devices to make 8-, 12-, or 16-bit comparators. If we apply the lines of the address bus to one set of 7485 inputs and program the alternate inputs for the required address, the output (pin 6) of the most significant 7485 will go HIGH only when the correct address is present.

A number of different switch options are available to make the address selection. The cheapest is to use jumper wires. When the jumper wire is in place, the 7485 input is permanently LOW. If, on the other hand, the jumper is left out, the input is permanently LOW. Alternatively, we may use either thumbwheel or binary DIP switches mounted on the printed circuit board, or even on the front panel if some pressing design reason indicates such an arrangement.

Additional information on address decoding is given in Chapter 5 when we discuss memory interfacing. Given in that discussion are methods for

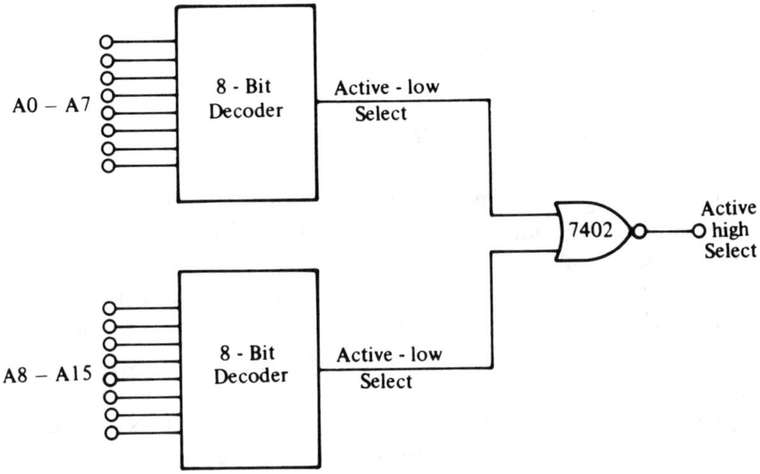

(a)

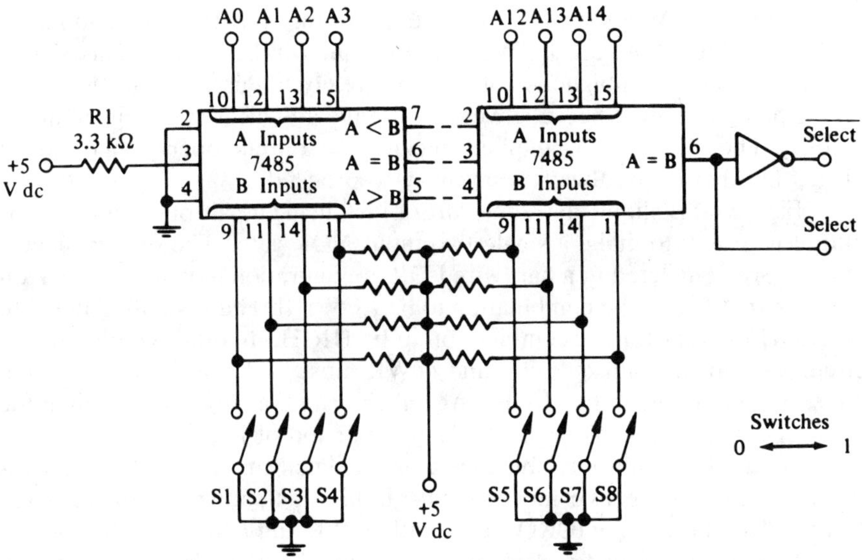

(b)

Figure 6-7 16-bit address decoders: A) Using two of the above 8-bit circuits, B) Using magnitude comparator chips.

minimizing the component count by selecting memory in banks rather than having a large (impossibly large) array of decoder circuits.

GENERATING *IN* AND *OUT* SIGNALS

The IN and OUT signals are generated by the proper convergence of control signals and the correct address. For example, let's assume that we want to generate an IN1signal (i.e., a signal to indicate that input port 1 is to be activated). In the Z80 chip, this signal would require the @IORQ and @RD signals to be LOW, as well as an address decoder SELECT output to indicate that address 1 (i.e., 00000001 on the lower 8 bits of the address bus) is present.

In some rare cases, we might want to generate an IN or OUT signal that is not dependent upon the address bus. In those cases, the signal would be active whenever the correct operation, an input or output, is taking place. Figure 6-8 shows several circuits for generating such signals. Again, the Z80 control signal system is used. In Figure 6-8A, we see the method used to generate a *data direction* signal that is HIGH for the OUT condition and LOW for the IN condition. Recall that the @IORQ and @WR signals must both be LOW for the OUT condition. In this situation, the two inputs of the 7402 NOR gate are LOW, so the output will be HIGH. This signal, however, is not unambiguous—it does not really tell us when an IN operation is being executed because we do not have an @RD (read) signal in the picture. The IN status is implied because no output operation is taking place, and that is not always a safe way to proceed.

Figure 6-8B shows the same sort of idea using a pair of open-collector output inverters to make a single two-input NOR gate. The open-collector outputs are connected in a hardwired-OR configuration in which a LOW on either output forces the combination to be LOW. It requires both outputs to be HIGH in order for the combination to be HIGH. In other words, for the circuit as shown, both @IORQ and @WR must be LOW in order for the combination output to be HIGH. Again, the same ambiguity regarding the input operation exists as in the NOR-gate version of Figure 6-8A.

To create a truly unambiguous control signal for IN and OUT operations, we must have two independent circuits, as shown in Figure 6-8C. Here we combine the @IORQ signal with @WR and @RD signals in a pair of NOR gates in order to create a pair of unambiguous data direction signals. NOR gate G1 is connected exactly as in Figure 6-8A and creates an active-HIGH signal that denotes an output operation (OUT). If both @IORQ and @WR are LOW, indicating that a write operation is taking place, the output of G1 will be HIGH. If we want an active-LOW OUT signal (i.e., @OUT), it will be necessary to place an inverter at the output of G1.

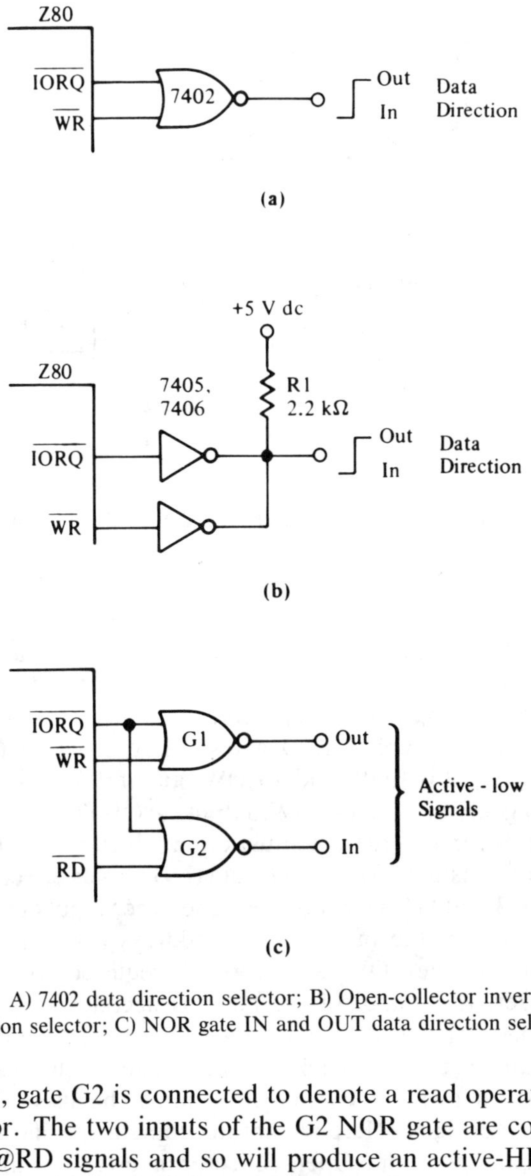

Figure 6-8 A) 7402 data direction selector; B) Open-collector inverters used as data direction selector; C) NOR gate IN and OUT data direction selectors.

Similarly, gate G2 is connected to denote a read operation to the Z80 microprocessor. The two inputs of the G2 NOR gate are connected to the @IORQ and @RD signals and so will produce an active-HIGH IN signal. Again, an inverter is required for an @IN signal.

The only truly unambiguous signal that will command one and only one device or port to turn on is that which takes into account the address of the device. Figure 6-9 shows the previous circuits combined with an address decoder to perform as a unique OUT/IN signal generator. Gates G1 and G2 are exactly as shown in Figure 6-8C and require no further discussion except to relabel their respective output signals. To avoid confusion, we will label

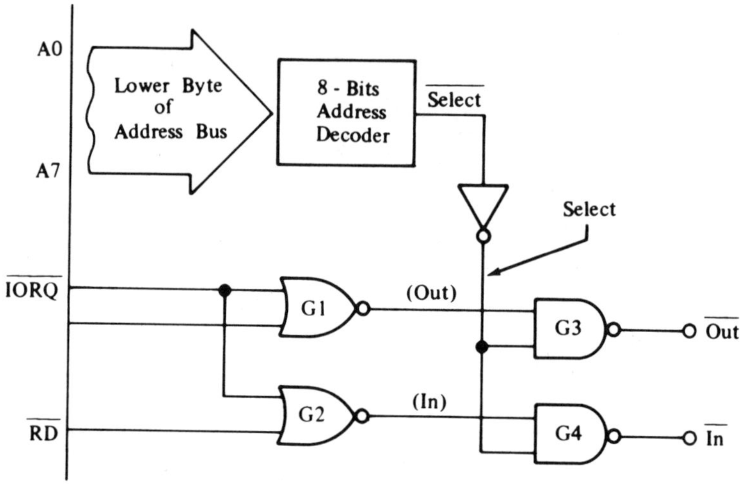

Figure 6-9 Generating NOT-IN and NOT-OUT signals.

the output of G1 (OUT) and that of G2 (IN) in order to indicate that these signals do not account for the port/device address, but only that an output or input operation is taking place, respectively.

Gates G3 and G4 are NAND gates. They produce a HIGH output whenever either input is LOW and a LOW output if and only if both inputs are HIGH. To generate active-LOW output (@OUT) and input (@IN) signals, therefore, we must create a situation in which both inputs are HIGH only at the correct instant. The (OUT) and (IN) signals are connected to one input each for G3 and G4, respectively. The other inputs of G3 and G4 are connected together at the output of the address decoder. If the address decoder output is active-LOW, an inverter is required (as shown in Figure 6-9). If, however, the output of the address decoder is active-HIGH, no inverter is needed. The @OUT and @IN signals generated by the circuit of Figure 6-9 are unique to only one port, that selected by the address decoder.

Figure 6-10 shows a technique that might be used as an economy measure in some designs. Here we are decoding a specific port (port 1) to perform an output operation. Open-collector inverters are used to create an OUT signal at their mutual wire-OR output terminal. If we also wire-OR an output from an address decoder to the same point, however, the OUT signal will be unique to that one port. In this case, we can wire in the output an open-collector noninverting buffer to denote port 1. In the case of port 1, we would require a HIGH on address bus line A0 plus a LOW on both @IORQ and @WR lines in order to create a HIGH at the output of the wired-OR connections.

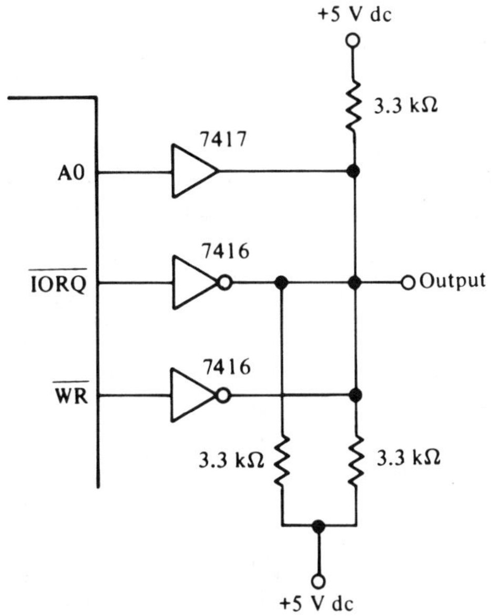

Figure 6-10 Address decoder using open-collector inverters.

There is a limitation to this method. Can you spot it? If we use only 1 bit of the address bus, we must be absolutely certain that the program written for the computer does not call for any other ports than that allowed by the design. In the case of port 1, bit A0 will be HIGH. However, it will *also* be HIGH if the programmer calls up port 3 (0011), port 5 (0101), port 7 (0111), and so forth. As long as we are certain that no ambiguity can be created by the programming of the microprocessor, this method is valid. In some small instrumentation or control systems, therefore, it is a valid technique. In those cases, it is unlikely that someone will want to add a port at some later time. If they do, we can use another bit of the address bus for that port and program accordingly. On a 16-bit address bus, we can accommodate up to 16 different ports if we consider them to be memory-mapped I/O ports. On the Z80, which uses discrete I/O commands in the instruction set, by using the lower 8 bits of the address we can make up to eight discrete I/O ports that are indicated by only 1 bit each. This system is used in many smaller control and instrumentation computers that will never have the full 64K complement of memory. Most such computers use 1K, 2K, 4K, or 8K of memory. A popular tactic is to make a memory-mapped port for a device at the 32K boundary. If there will be no address higher than 32K, we can connect bit A15 of the address bus to denote the device in question. It has been popular in data-converter applications to use A15 to turn on the data device. Bit A15 combined with an OUT signal will turn on a digital-to-analog converter, while bit A15 and an IN signal will turn on an analog-to-digital converter.

Numerous devices on the IC market can be pressed into service to make a device-select signal. It is interesting, however, that certain devices developed years ago tend to be used even in more recent designs. Many engineers are sensitive to using the most modern technology possible for each job, but that is not always the best policy. Remember, the important

	D	C	B	A	
	$\overline{\text{Address Select}}$	$\overline{\text{WR}}$	$\overline{\text{RD}}$	$\overline{\text{IORQ}}$	
In	0	1	0	0	4_{10}
Out	0	0	1	0	2_{10}

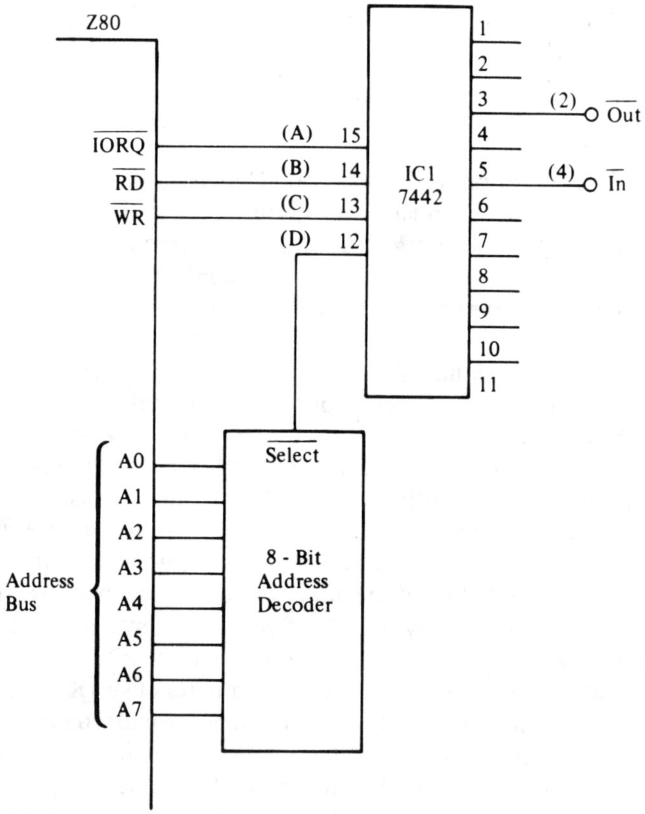

Figure 6-11 Generating NOT-IN and NOT-OUT signals using standard chips.

thing is to get the job done in a timely and *economic* manner, even if this means using devices that have been around for a decade or more. Don't listen to the admonishments of those who think that newest is always best. The NAND and NOR gates described in this chapter are old TTL devices. The use of TTL is not always mandated, however, and NAND/NOR function blocks from newer logic technologies will perform exactly as indicated for the TTL devices.

Another older device that keeps finding its way into new designs is the 7442 described earlier in this chapter. Figure 6-11 shows the use of the 7442 BCD-to-1-of-10 decoder IC as a device-select pulse generator. In the circuit as shown, we can generate both @IN and @OUT pulses.

The basis for this circuit is the @IORQ, @WR, and @RD control signals and the @SELECT signal output from the address decoder. These signals are connected to form a 4-bit binary word at the input of the 7442 device. In this case, we connect @IORQ to the A input, @RD to the B input, @WR to the C input, and @SELECT to the D input (weighting A = 1, B = 2, C = 4, and D = 8). Using this connection scheme, considering that these are active-LOW outputs, we find that the input condition is represented by binary word 0100 (see the table in Figure 6-11) and the output by 0010; these are 4_{10} and 2_{10} in the decimal system, respectively. We can, therefore, use the 4 output (pin 5) for the @IN signal and the 2 output (pin 3) for the @OUT signal. The only time that these respective output terminals on the 7442 are LOW is when the correct port is selected.

MULTIPLE DEVICE-SELECT PULSES

The 7442 and other devices may also be used to select multiple devices in larger systems. We can use the 7442 to select from 10 different devices if the correct connection is used. Similarly, a 74154 device can select up to 16 devices. The 74154 device is a *4-bit-binary-to-1-of-16 decoder,* also called a *4-line-to-16-line decoder* in some manuals and a *data distributor* in others. The latter name is derived from the fact that the data applied to pin 19 (DS) will be transferred to the output line that is selected by the 4-bit word applied to the ABCD inputs. In Figure 6-12 we use this feature to determine whether the outputs are active-HIGH or active-LOW. The level that is applied to DS will be transmitted to the active output. For example, if the switch is open, the level applied to DS will be HIGH. The selected output will then be HIGH and all others will be LOW. If, on the other hand, the switch is LOW, the active output will be LOW and all others will be HIGH.

The @CE terminal (pin 18) is used to turn on the selected output. When this terminal is LOW, the selected terminal will be at the data level that is selected by DS. We use the @IN or @OUT terminals to drive this terminal.

The input word for the 74154 is a 4-bit binary word made up of 4 bits of

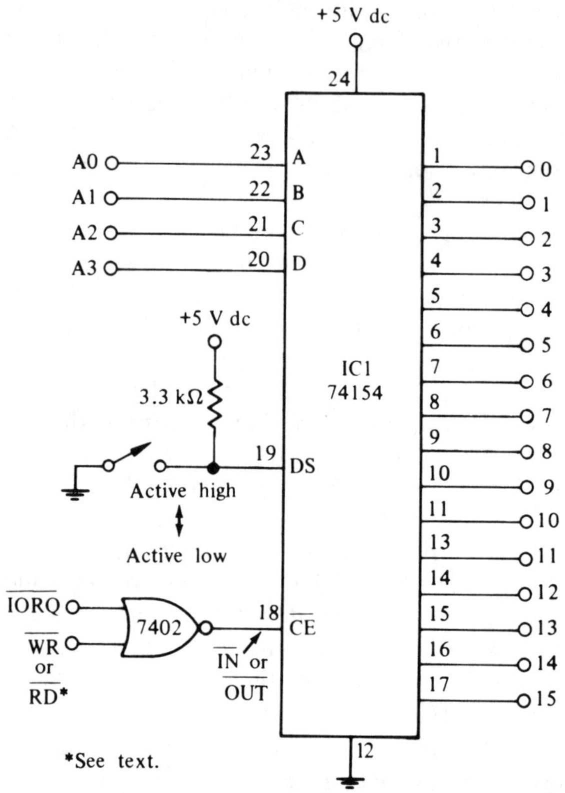

Figure 6-12 Data selector chip used for address decoding.

the address bus. In this case, we have selected bits A0 through A15, so that the first 16 permissible I/O port addresses are the ports addressed. The selection code follows the ordinary binary numbering system. This same system, however, might be used for any four sequential sets of address lines within the system. In some cases, for example, we might wish to use A0, A1, A2, and A15 to locate the ports in the first locations of the upper 32K. In at least one 16-channel analog-to-digital/digital-to-analog (A/D-D/A) converter on the market, the user can program the most significant bit in the address code by connecting it to one particular address bus bit (usually A15 or one of the other high-order bits). The others are connected to the lower-order bits of the address bus, thereby forcing the correct address to be the first 16 sequential addresses from the programmed memory page boundary address.

Both 7442 and 74154 devices can be combined in circuits to provide a large number of discrete device-select signals. In Figure 6-13, for example, we see seventeen 74154 devices connected to form 256 device-select pulses

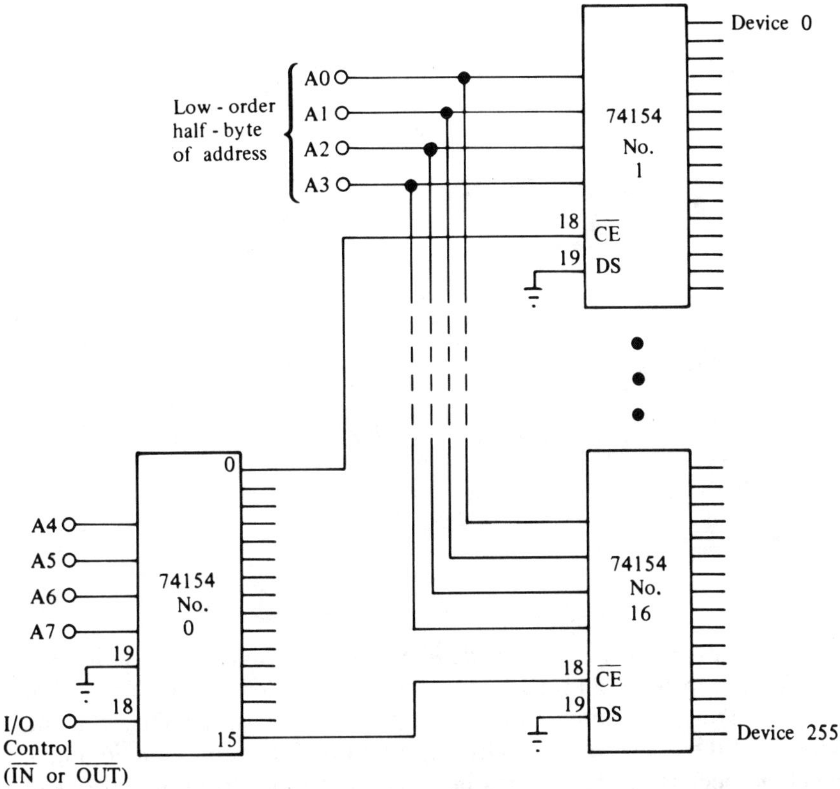

Figure 6-13 Ganging data selector decoders to extend capability.

abeled device 0 (or port 0) up to device 255 (or port 255). In this case, the ower-order 4 bits of the address bus are connected to the 4-bit inputs of 74154s 1 through 16. The next higher-order 4 bits (A4–A7) are connected to the 4-bit input of 74154 number 0. The 16 active-LOW outputs from this 74154 are connected to the @CE lines of the 16 other 74154 devices. A particular 74154 will, therefore, be activated only when its turn comes. Let's examine an example. Suppose we wanted to activate device 24. Decimal 24 is hexadecimal (base 16) 18, so we find that the most significant digit (74154 no. 0) must be 1, while the 4-bit code placed over the A0–A3 lines must be that for 8, or 1000. We must, therefore, transmit the binary word 00011000 over A0–A7. When this happens, the 1 output of 74154 number 0 will drop LOW, thereby enabling 74154 number 2, which will also respond to the 8 on A0–A3 by causing line 24 to drop LOW. This particular scheme can be carried out almost indefinitely, provided you have sufficient address line bits to accommodate the number of devices and the address bus power capacity or drive all the TTL inputs hanging on the line.

7

INPUT/OUTPUT

The topic of input and output devices, components, and circuits is often overlooked in texts and articles on microcomputers because they are not quite as exotic and interesting as some microprocessor chips. But the I/O section of the computer is vitally important to the overall functioning of the machine because it determines how data are transferred in and out of the machine. The utility of a device is often determined, or more often limited, by the structure of the I/O circuitry used. After you design a microprocessor-based instrument and decide to expand its capability, it is almost inevitable that the question of I/O ports will come up: there will probably be too few to support the extra functions that you want to add.

In this chapter we will take a generic look at I/O ports in general, with some emphasis on the types of I/O ports needed to interface the IBM PC series and Apple II series of machines given later in the chapter. The early sections of this chapter pertain to single-board computer and controller machines as might be used at the heart of an electronic instrument.

The input and output functions are operated by the control signals of the microcomputer and may take either of two forms, *direct I/O* or *memory-mapped*. Some microprocessor chips provide for direct I/O in the form of I/O instructions. In the Z80 device, the address of the port will be passed over the lower-order 8 bits (A0–A7) of the address bus, while the data from the accumulator is passed simultaneously over both the data bus (DB0–DB7) and the high-order 8 bits of the address bus (A8–A15). The 8-bit memory address will support up to 256 different I/O ports, which can be numbered 9 through 255. The Z80 device control signals allow for I/O operations and are combined in such a way as to produce unique IN and OUT commands to the I/O devices.

Other microprocessor chips do not provide input and output commands in the instruction set and thus will not have the control signals and capabilities for direct I/O. In those machines, the input and output ports are treated as if they were memory locations; such ports are called *memory-mapped I/O ports*.

Although I/O may not necessarily be the most interesting aspect of microprocessor technology, we must nevertheless study some of these mundane details in order to gain an understanding of how the microcomputer deals with the outside world. To begin this study, we will consider some elementary digital electronics theory and some of the devices used to form I/O ports. With an understanding of these topics, you should be able to progress to designing I/O ports and interfacing techniques.

LOGIC FAMILIES

Digital electronic circuits use assorted *logic blocks* such as gates (AND, OR, NOT, NAND, NOR, XOR, etc.) and flip-flops to perform the various circuit functions. On initial inspection, it seems that digital logic circuit design is made simpler because all the logic blocks are available integrated circuit form and can be simply connected together with seeming impunity. The reason this situation exists is that the IC logic devices are part of various *families* of similar devices. A digital logic family will use standardized input and output circuits that are designed to work with each other, use the same voltage levels for both power supply and logical signals, and generally use the same technology in construction of the devices. Common logic families in current use are TTL, CMOS, NMOS, PMOS, and MOS, with certain subgroups within each of these. Obsolete forms, such as RTL and DTL, although interesting to the owner of older equipment, are of too little interest to justify inclusion here. There are also certain devices that will mix technologies (e.g., an NMOS microprocessor chip that uses TTL input and output circuits) in order to gain some of the advantages of both families.

Transistor-Transistor Logic (TTL)

Transistor-transistor logic (TTL; also called T^2L) is probably the oldest of the currently used IC logic families and is based on bipolar transistor technology. Bipolar transistors are the ordinary PNP and NPN types, as distinguished from the field-effect transistors.

The TTL logic family uses power supply potentials of 0 and +5 V DC, and the +5-V potential must be regulated for proper operation of the device. Most specifications for TTL devices require the voltage to be between

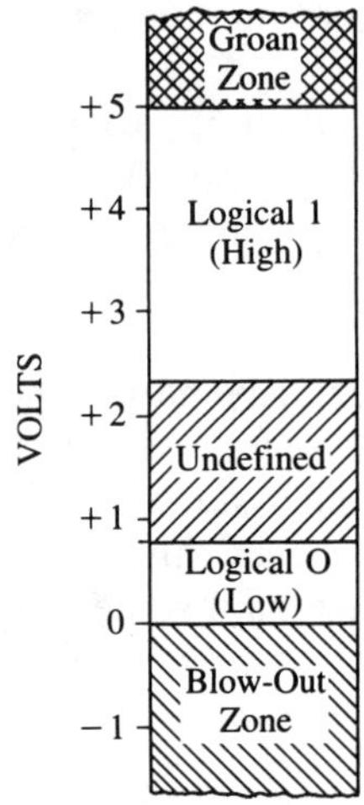

Figure 7-1 TTL logic level diagram.

+4.5 and 5.2 V DC, although there appear to be practical limitations on even these values. Some complex function ICs, for example, will not operate properly at potentials below +4.75 V, despite the manufacturer's statements to the contrary. Also, at potentials above 5 V, even though less than the +5.2-V maximum potential "allowed," there seems to be an excess failure rate that is probably due to the higher internal junction temperatures generated inside the ICs. The best rule of thumb is to keep the potential of the power supply between +4.75 and +5 V; furthermore, the potential must be regulated.

Figure 7-1 shows the voltage levels used in the TTL family of devices to represent logical 1 and logical 0. The logical 1, or HIGH, condition is represented by a potential of +2.4 V or more (+5 V maximum). The device must be capable of recognizing any input potential over +2.4 V as a HIGH condition. The logical 0, or LOW, condition is supposedly 0 V, but most TTL devices define any potential from 0 to 0.8 V as logical 0. The voltage region between +0.8 and +2.4 V is undefined; the operation of a TTL device in this region is not predictable. Care must be exercised to keep the TTL logical signals outside the undefined zone, which can be a source of problems in some circuits that are not properly designed.

The inverter, or NOT gate, is the simplest form of digital logic element and contains all the essential elements required to discuss the characteristics of the family. Figure 7-2A shows the internal circuit of a typical TTL inverter. The output circuit consists of a pair of NPN transistors connected in the "totem pole" configuration in which the transistors form a series circuit across the power supply. The output terminal is taken at the junction between the two transistors.

The HIGH state on the output terminal will find transistor Q4 turned off and Q3 turned on. The output terminal sees a low impedance (approximately 130 Ω) to the +5-V line. In the LOW output state, exactly the opposite situation exists: Q4 is turned on and Q3 is turned off. In that condition, the output terminal sees a very low impedance to ground.

The input terminal of the TTL inverter is a transistor emitter (Al). When the input is LOW, the emitter of Q1 is grounded. The transistor is forward-biased by resistor R1, so the collector of Q1 is made LOW also. This condition causes transistor Q2 to be turned off, so the voltage on its emitter is zero and the voltage on its collector is HIGH. In this situation, we have the conditions required for a HIGH output: Q4 is turned off and Q3 is forward-biased, thereby connecting the output terminal through the 130-Ω resistor to the +5-V DC power supply terminal.

Exactly the opposite situation obtains when the input terminal is HIGH. In that case, we find transistor Q1 turned off and the voltage applied to the base of Q2 is HIGH. Under this condition, the collector voltage of Q2 drops and its emitter voltage rises. Transistor Q4 is turned on, grounding the output terminal, and transistor Q3 is turned off. In other words, a HIGH on the input terminal produces a LOW on the output terminal.

Figure 7-2B shows the current path when two TTL devices are connected together in cascade. The emitter of device A input is connected to the output terminal of device B. The input of a TTL device is a current source that provides 1.6 mA at TTL voltage levels. The output transistors are capable of sinking up to 16 mA. We may conclude, therefore, that for regular TTL devices the output terminal will provide current sinking capability to accommodate up to 10 TTL input loads. Some special buffer devices will accommodate up to 30 TTL input loads.

The input and output capabilities of TTL devices are generally defined in terms of *fan-in* and *fan-out*. The fan-in is standardized in a unit input load rather than specific current and voltage levels. This convention allows us to interconnect TTL devices simply without being concerned with matters such as impedance matching. In interfacing TTL devices it is merely necessary to make sure that the number of TTL input loads does not exceed the fan-out of the driving device. In brief, the fan-in is one unit TTL input load, whereas the fan-out is the output capacity expressed in the number of standard input loads that a device will drive. In the case of the regular TTL devices, the output current capacity is 16 mA, while the standard input load is 1.6 mA; so a fan-out of 16/1.6, or 10, exists.

Asking a TTL device to drive a number of TTL loads in excess of the rated fan-out will result in reduced noise margin and the possibility that the logic levels will be insufficient to reliably drive the inputs connected to the output. Some devices will provide a fan-out margin, but most will not. When it is necessary to drive a large number of TTL loads, it is wise to use a high fan-out buffer.

Multiple TTL inputs are formed by adding extra emitters to the input transistor (see Figure 7-2C). This type of circuit is used on multiple-input

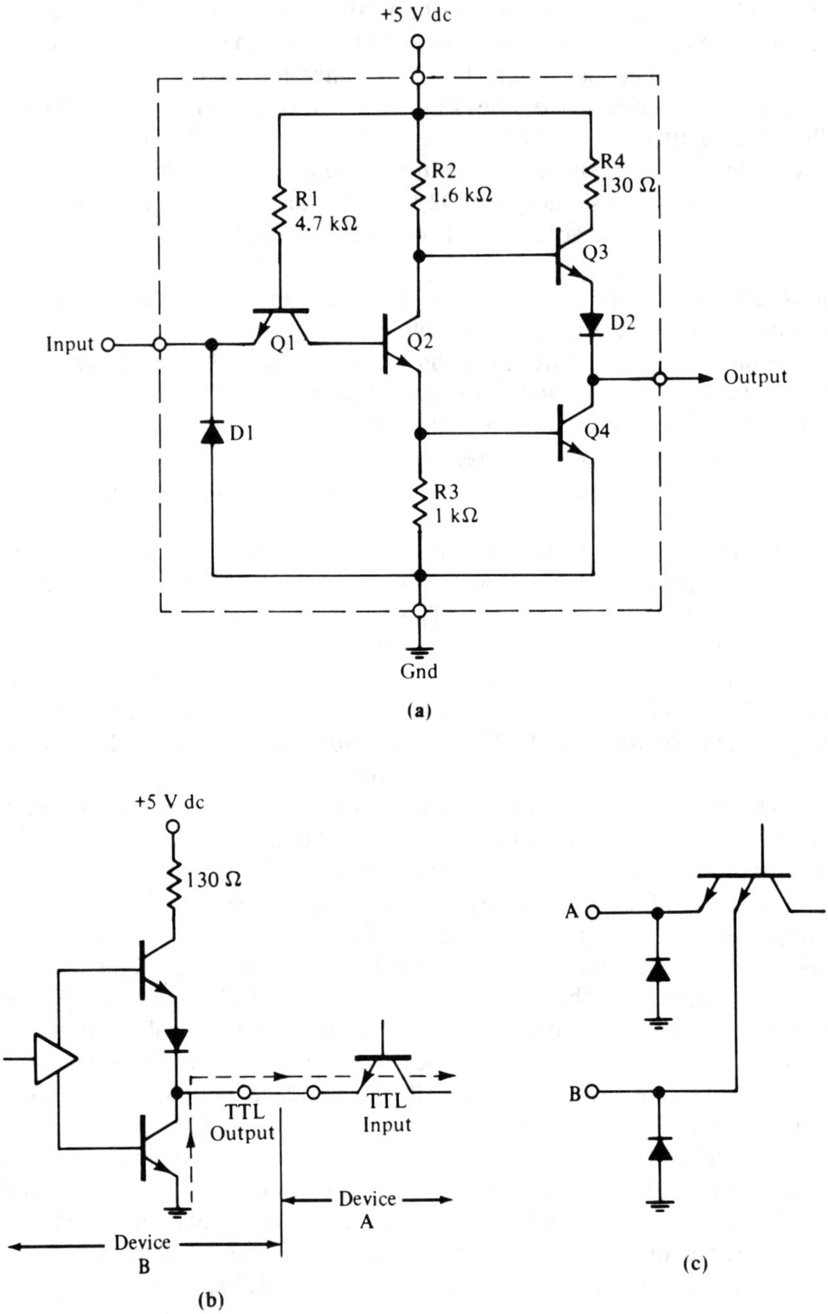

Figure 7-2 A) Typical TTL inverter circuit, B) Interface connection between two TTL chips, C) Multiple emitter inputs.

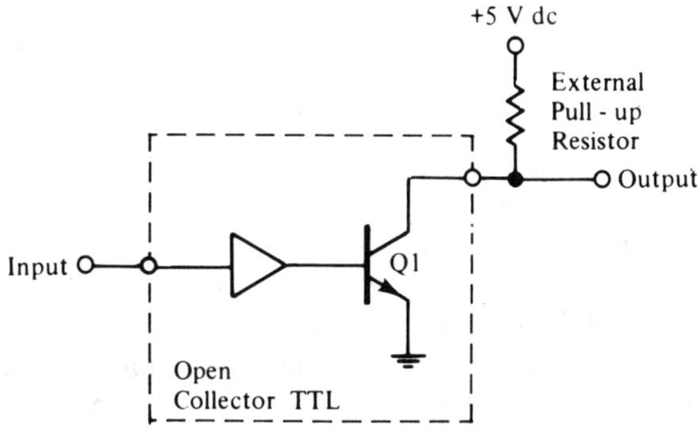

Figure 7-3 Open-collector output circuit.

devices such as NAND gates; each emitter is capable of sourcing 1.6 mA of current and represents a fan-in of one standard TTL load.

Open-collector output. The standard TTL output circuit shown in Figure 7-2A must be connected to a standard TTL input in order to work properly. At times, however, it becomes necessary to interface the TTL device with a device other than a TTL. In some cases, the external load will be at the same voltage level as TTL, but in others the voltage level might be considerably higher than +5 V. The open-collector circuit of Figure 7-3 will accommodate such loads.

Figure 7-3 shows only the output stage of the open-collector device; all the other circuitry will be as in Figure 7-2A. Transistor Q1 is arranged so that its collector is brought out to the output terminal of the device. Since there is no current path to the V+ terminal of the power supply, an external load must be provided for the device to work, as in the case of the situation shown. In the case shown, an external *pull-up resistor* is connected between the output terminal (i.e., Q1 collector) and +5 V DC; for most TTL open-collector devices the value of the pull-up resistor is 2 to 4 Ω. Other loads and higher voltages can be accommodated provided that the DC resistance of the load is sufficient to keep the collector current in Q1 within specified limits.

Speed versus power. The TTL logic family is known for its relatively fast operating speeds. Most devices will operate to 18 to 20 MHz, and some selected devices operate to well over 30 MHz. But the operating speed is not without a trade-off—increased operating power. Unfortunately, higher speed means higher power dissipation. The problem is the internal resistances and capacitances of the devices. The operating speed is set in part by the RC time constants of the internal circuitry. To reduce the time constant and thereby increase the operating speed, it is necessary to reduce the resistances and that will necessarily increase current drain and power consumption.

TTL nomenclature. Each logic family uses a unique series of type numbers for the member devices so that users can identify the technology being used from the number. With very few "house number" exceptions, TTL type numbers will have either four or five digits beginning with the numbers 54 or 74. The normal devices found most commonly are numbered in the 74xx and 74xxx series, while higher-grade military specification devices carry 54xx and 54xxx numbers. The 54 and 74 series retain the same xx or xxx suffix for identical devices. For example, the popular NOR gate will be numbered 7402 in commercial grade components and 5402 in military grade. We can substitute the more reliable 54xx devices for the identical 74xx devices. Parts brokers often sell surplus 54xx devices as 74xx, but they are the same device at higher quality.

TTL subfamilies. Certain specialized TTL devices are used for purposes such as increased operating speed and lower power consumption. These family subgroups include (in addition to regular TTL) low-power (74Lxx), high-speed (74Hxx), Schottky (74Sxx), and low-power Schottky (74LSxx) devices. A principal difference between these groups that must be addressed by the circuit designer or interfacer is the input and output current requirements. In most cases, the levels shown in Table 7-1 apply.

Complementary Metal Oxide Semiconductor (CMOS) Devices

The complementary metal oxide semiconductor, or CMOS, digital IC logic family is based on the metal oxide semiconductor field-effect transistor (MOSFET). In general, CMOS devices are slower in operating speed than TTL devices, but they have one immensely valuable property: low power dissipation. The nature of the CMOS device is such that it presents a high impedance across the DC power supply at all times except when the output

TABLE 7-1

Subfamily	Input Current, mA	Output Current, mA
74XX	1.6	16
74LXX	0.18	3.6
75Hxx	2.0	20
74Sxx	2.0	20
74Sxx	0.4	8.0

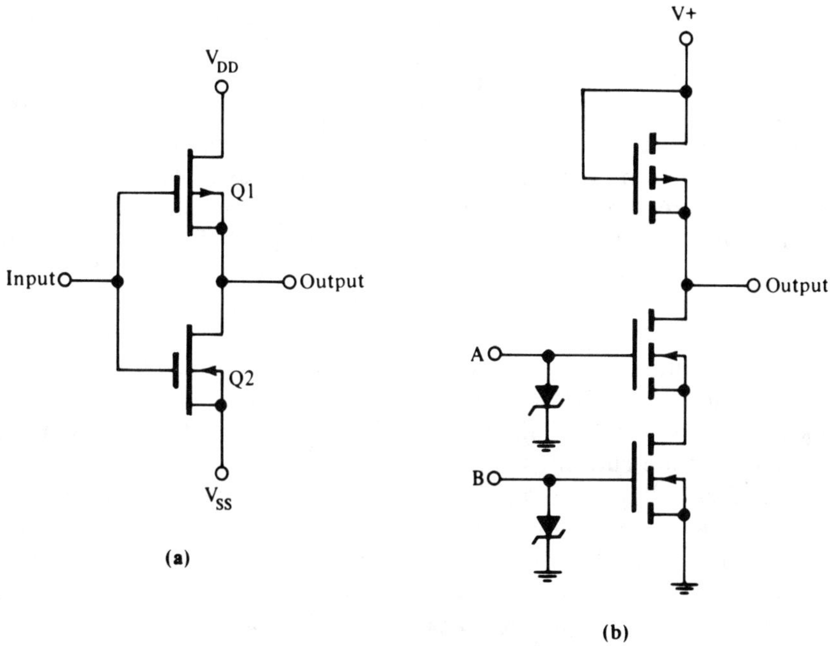

Figure 7-4 A) CMOS inverter circuit, B) CMOS AND gate.

is undergoing transition from one state to the other. At all other times, the CMOS device draws only a few microamperes of electrical current, which makes it an excellent choice for large systems where speed of operation is not the most important specification.

Figure 7-4 shows two CMOS devices that are representative of the larger family of related logic elements. In Figure 7-4A is a simple CMOS inverter. Note that it consists of an N-channel and a P-channel MOSFET connected such that their respective source-drain paths are in series, while the gate terminals are in parallel. This arrangement is reminiscent of push-pull operation, because the N-channel and P-channel devices turn on and off with opposite polarity signals. As a result, one of these two transistors will have a low channel resistance with the input LOW, while the other will offer a very high-resistance (MΩ). When the input is made HIGH, the role of the two transistors is reversed: the one with the low-channel resistance becomes high-resistance, while that with the high resistance goes LOW. This operation has the effect of connecting the output terminal to either V_{dd} or V_{ss} depending upon whether the input is HIGH or LOW. Since, in both cases, one of the series pair is high-resistance, the total resistance across the V_{dd}-V_{ss} power supply is high. Only during the transition period, when both transistors have a medium-range source-drain resistance, will there be any appreciable load in the power supply. The output terminal will not deliver any current, because it will be connected to another CMOS input, which is a very high impedance. As a result, there is never any time when the CMOS IC, operated only in conjunction with other CMOS devices, will draw any appreciable current. An example of the difference between TTL and CMOS current levels is seen by comparing the specs for a common quad two-input NAND gate in both families. The TTL version needs 25 mA, whereas the CMOS device requires only 15 mA.

Figure 7-4B shows a typical CMOS AND gate. The two inputs are connected to independent inputs of a pair of series-connected N-channel MOSFETs. The output of this stage will not change state unless both inputs are active, a result of the series connection.

The operating speed of typical CMOS devices is limited to 4 to 5 MHz, although some 10- to 15-MHz devices are known. The speed is the principal disadvantage to the CMOS line; typical TTL devices operate to 20 MHz but require a lot more current.

Another problem with the CMOS device is sensitivity to static electricity. The typically very thin insulating layer of oxide between the gate element and the channel has a breakdown voltage of 80 to 100 V. Static electricity, on the other hand, can easily reach values of 1000 V or more. Whenever the static is sufficient to cause a biting spark when you touch a grounded object, it is generated by a potential of 1000 V or more. This potential is sufficient to destroy CMOS devices. This problem is especially critical in dry climates or during the low-humidity portions of the year. There are, however, methods of working with a CMOS that allow us to minimize damage to the device. In general, the CMOS working rules require use of a grounded working environment, grounded tools, and avoidance of certain wool or artificial fiber garments. Also, the B series (e.g., CA-4001B) has built-in zener diodes to protect the delicate gate structure by shunting dangerous potentials around the gate.

Tristate Output Devices

Ordinary digital IC logic devices are allowed only two permissible output states: HIGH and LOW, corresponding to TRUE-FALSE logic or 1-0 of the binary number system. In the HIGH state, the output is typically connected through a low impedance to a positive power supply; in the LOW state the output is connected to either a negative power supply or ground. Although this arrangement is sufficient for ordinary digital circuits, there is a problem when two or more outputs are connected together but must operate separately. Such a situation exists in a microcomputer on the data bus. If any one device on the bus stays LOW, then it more or less commands the entire bus: no other changes on any other device will be able to affect the bus, so the result will be chaos. Also, even if we could conspire to make all

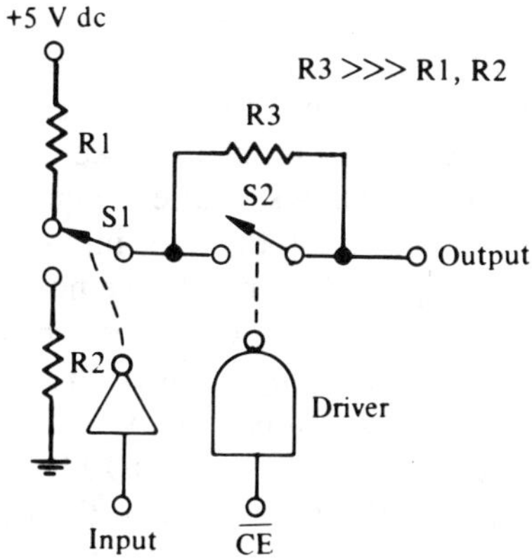

Figure 7-5 Circuit model for tristate output logic devices.

bits HIGH when not in use, there would still be loading factor and also an ambiguity as to which device is turned on at any given time.

The answer to the problem is in *tristate logic,* as shown schematically in Figure 7-5. Tristate devices, as the name implies, have a third permissible output state. This third state effectively disconnects the output terminal from the workings of the IC. In Figure 7-5, switch S1 represents the normal operating modes of the device. When the input is LOW, switch S1 is connected to R1, so the output will be HIGH. Similarly, when the input is HIGH, switch S1 is connected to R2, so the output is LOW. The third state is generated by switch S2. When the active-LOW *chip enable* (@CE) terminal is made LOW, switch S2 is closed and the output terminal is connected to the "output" of S1. When the @CE terminal is HIGH, however, switch S2 is open, so the output floats at a high impedance (represented by R3). Because of this operation, the tristate device can be connected across a data bus line and will not load the line except when @CE is made LOW.

An advantage of tristate digital devices is that the chip enable terminals can be driven by device-select pulses, thus creating a unique connection to the data bus that is not ambiguous to the microcomputer. In other words, the computer will "know" that only the data from the affected input port or device are on the bus whenever that @CE is made LOW.

INTERFACING LOGIC FAMILIES

One of the defining characteristics of a logic family is that the inputs and outputs of the devices within the family can be interconnected with no regard to interfacing. A TTL output can always drive a TTL input and a CMOS output can always drive a CMOS input without any external circuitry other than a conductor. But when we want to interconnect logic elements of different families, some consideration must be given to proper methods. In some cases, it will suffice to simply connect the output of one device to the input of the other; in other cases some external circuitry is needed.

Figure 7-6A shows a series of cascade inverters. The CMOS device is not comfortable driving the TTL input, and the TTL input is not happy with the CMOS output. As a result, we must use a special CMOS input: devices 4049 and 4050. The 4049 device is a hex inverting buffer, while the 4050 is the same in noninverting configuration. The special character of these devices is the bipolar transistor to +5 V DC. The 4049/4050 will operate to potentials up to +15 V, but it is TTL-compatible *only* at a V+ potential of +5 V DC, with the other side of the device power supply grounded. The input of the 4049/4050 is CMOS, so it is compatible with all CMOS outputs.

The TTL input is a current source, so the TTL output depends for proper operation on driving a current source. The CMOS input, however, is a very high impedance because the CMOS family is voltage driven. If we want to interface an ordinary TTL output to a CMOS input (see Figure 7-6B), we must provide a pull-up resistor between the TTL output terminal and the +5-V DC power supply. A value between 2 and 4 kΩ is selected to make the current source mimic a TTL input current level.

The method of Figure 7-6B works well in circuits where both CMOS and TTL devices operate from a +5-V DC power supply. While this is the usual situation in most circuits, there are occasions where the TTL and CMOS devices operate from different potentials, and the correct interfacing method is shown in Figure 7-6C. Here we use an open-collector TTL output with a resistance to the V_{dd} power supply (used by the CMOS device) that is sufficiently high to keep the current flowing in the TTL output at a level within tolerable limits.

We can use a single 4049/4050 device to drive up to two regular TTL inputs (Figure 7-6D), and an ordinary CMOS device will drive a single LS-series TTL input (Figure 7-6D). The 4001 and 4002 CMOS devices are capable of directly driving a single regular TTL input. With the exception of the 4049/4050 device just discussed, these methods depend upon the CMOS and TTL devices operating from a common +5-V DC power supply. If the CMOS devices are operated at higher potentials, we will be forced into using the 4049/4050 method given earlier in order to prevent burnout of the TTL input.

Most microprocessor chips have limited output line capacity; most are limited to one or two TTL inputs load. Most MOS-series microprocessor chips use MOS logic internally but have TTL-compatible output lines. In the case of a two-loads output, the total allowable output current is 3.2 mA. There may be, however, many TTL-compatible inputs connected to the data bus or address bus of the microcomputer. We need a high-current bus driver on each line of the bus in order to accommodate these higher current require-

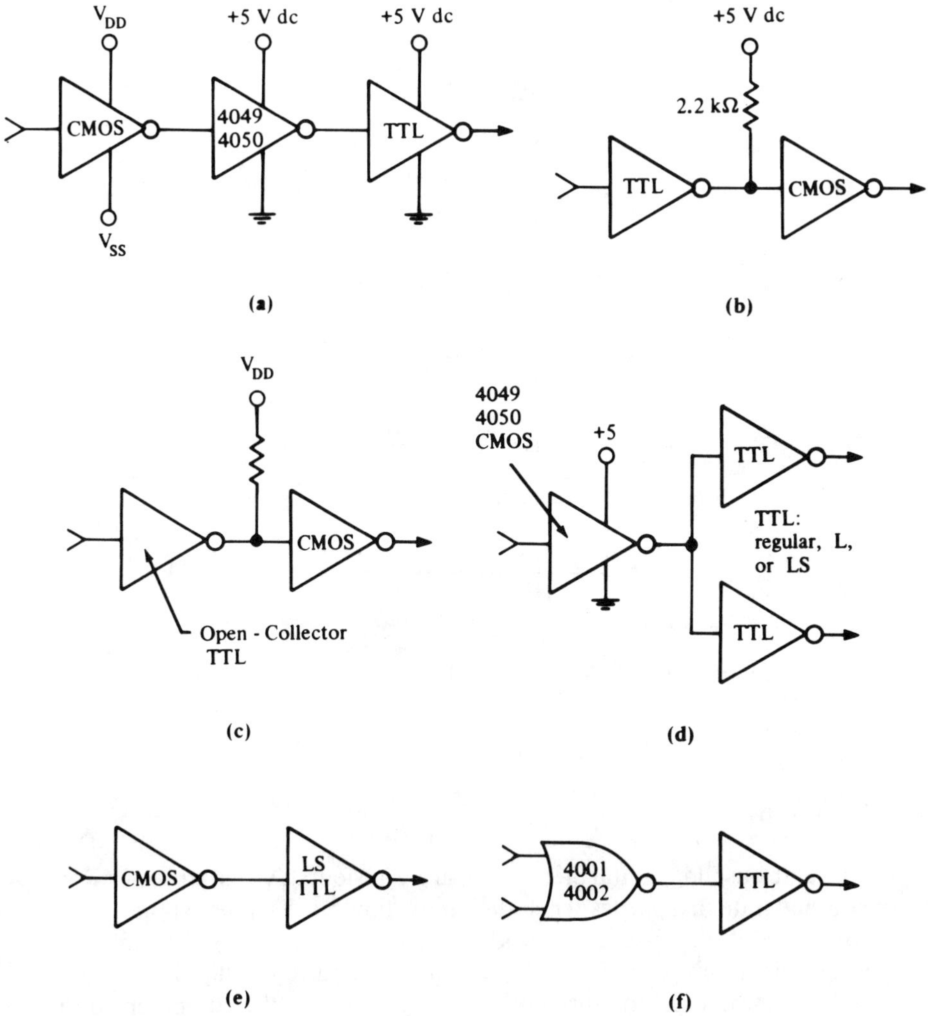

Figure 7-6 I.C. logic element interfacing: A) any-CMOS-to-4049/4050-to-single-TTL; B) TTL-to-CMOS operated from +5 vdc; C) TTL-to −CMOS other than +5 vdc; D) 4049/4050 used to drive TTL; E) CMOS-to-LS/TTL (i.e. 74LS series); F) 4001/4002 CMOS-to-TTL.

ments. Figure 7-7 shows a series of eight noninverting bus drivers interfacing the data bus of a microcomputer (DB0–DB7) with the data bus outputs of the microprocessor chip (B0–B7). This circuit will increase the drive capacity of the microcomputer from a fan-out of 2 to a fan-out of 30 or even 100, depending upon the bus driver selected.

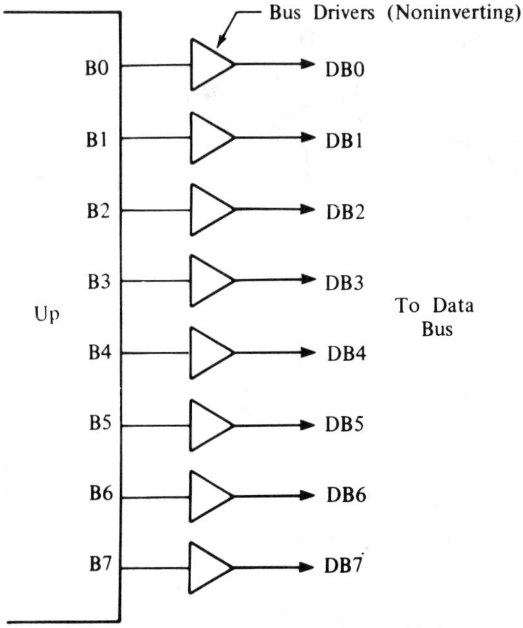

Figure 7-7 Using bus driver chips to improve drive capability.

FLIP-FLOPS

All the gates used in digital electronics are transient devices. In other words the output state disappears when the input stimulus disappears; the gate has no *memory*. A flip-flop, on the other hand, is a circuit that is capable of storing a single bit, one binary digit, of data. An array of flip-flops, called a *register*, can be used to store entire binary words in the computer. In this section, we will examine some of the common flip-flops used in digital circuits. All these circuits can be built with discrete digital gates, even though few modern designers would do so because the various forms of flip-flop are available as discrete units in their own right.

Figure 7-8 shows the basic *reset-set*, or RS, flip-flop. There are two versions, based on the OR and NAND gates, respectively. An RS flip-flop has two inputs, S and R (for set and reset). When the S input is momentarily made active, the output terminals go to the state in which Q = HIGH and NOT-Q = LOW. The R input causes just the opposite reaction: Q = LOW and NOT-Q = HIGH. A rule that must be followed is that these inputs must not be made active simultaneously, or an unpredictable output state will result.

Figure 7-8A shows the RS flip-flop made from a pair of two-input NAND gates. In each case, the output of one gate drives one input of the other; the gates are said to be cross-coupled. The alternate inputs of each gate form the input terminals of the flip-flop.

The inputs of the NAND gate version of the RS flip-flop are active-LOW. This means that a momentary LOW on either input will cause the output action. For this reason, the NAND gate version is sometimes designated as @RS FF (flip-flop), and the inputs are designated @S and @R, respectively.

The NOR gate version of the RS flip-flop is shown in Figure 7-8B. In this circuit, the inputs are active-HIGH, so the output states change by applying a HIGH pulse momentarily. The circuit symbol for the RS flip-flop is shown in Figure 7-8C. In some instances, the NAND version will be indicated by the same circuit, while in others there will be either @R and @S indications for the inputs or circles indicating inversion at each input terminal.

The RS flip-flop operates in an asynchronous manner (i.e., the outputs will change any time an appropriate input signal appears). Synchronous operation, which is required in most computer-oriented circuits, requires that output states change only coincident with a system clock pulse. The circuit in Figure 7-9 is a clocked RS flip-flop. Gates G3 and G4 form a normal NOR-based RS flip-flop. Control via a clock pulse is provided by

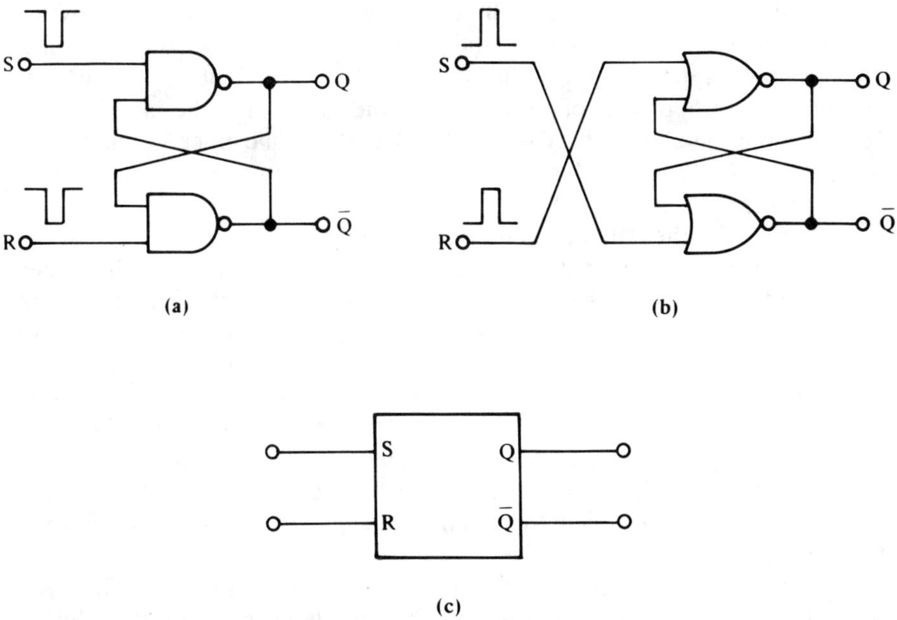

Figure 7-8 R-S flip-flop circuits: A) NAND gate, B) NOR gate, C) Symbol.

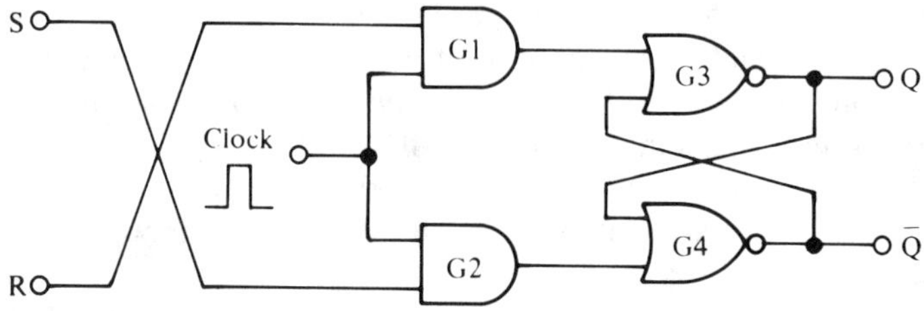

Figure 7-9 Clocked R-S flip-flop.

gates Gl and G2. One input for each is connected to the clock line. These two gates will not pass the R and S pulses unless the clock line is HIGH. The input lines can change all they want between clock pulses, but an output change is affected *only* when the clock pulse is HIGH.

A type-D flip-flop is made by using an inverter to ensure that the S and R inputs of a clocked RS flip-flop are always complementary (Figure 7-10A). The S input of the RS flip-flop and the input of the inverter that drives the R input of the RS FF are connected in parallel. When the S input is made HIGH, therefore, the R input will be LOW. Similarly, a LOW on the S input will place a HIGH on the R input. The circuit symbol for the type-D FF is shown in Figure 7-10B.

The rule for the operation of the type-D flip-flop is as follows: The input data applied to the D terminal will be transferred to the outputs *only* when the clock line is active. Figure 7-10C shows a typical timing diagram for a level-triggered type-D flip-flop that has an active-HIGH clock. The output line of this flip-flop will follow the input line only when the clock line is HIGH. Trace D shows the data at the D input, while trace Q shows the output data; CLK shows the clock line, which is presented with a series of regular pulses.

At time T_0, the data line goes HIGH, but the clock line is LOW, so no change will occur at output Q. At time T_1, however, the clock line goes HIGH and the data line is still HIGH, so the output goes HIGH. Note that the Q output remains HIGH after pulse T1 passes and will continue to remain HIGH even when the data input dips LOW again. In other words, the Q output of the type-D flip-flop will remember the last valid data present on the D input at the time the clock pulse went inactive. At time T_2, we find another clock pulse, but this time the D input is LOW. As a result, the Q output drops LOW. The process continues for times T_3 and T_4. Note that in each case the output terminal follows the data applied to the input *only* when the clock pulse is present.

The example shown is for a level-triggered type-D flip-flop. This type of flip-flop will allow continuous output changes all the while the clock line is HIGH. An edge-triggered type-D flip-flop timing diagram is shown in Figure

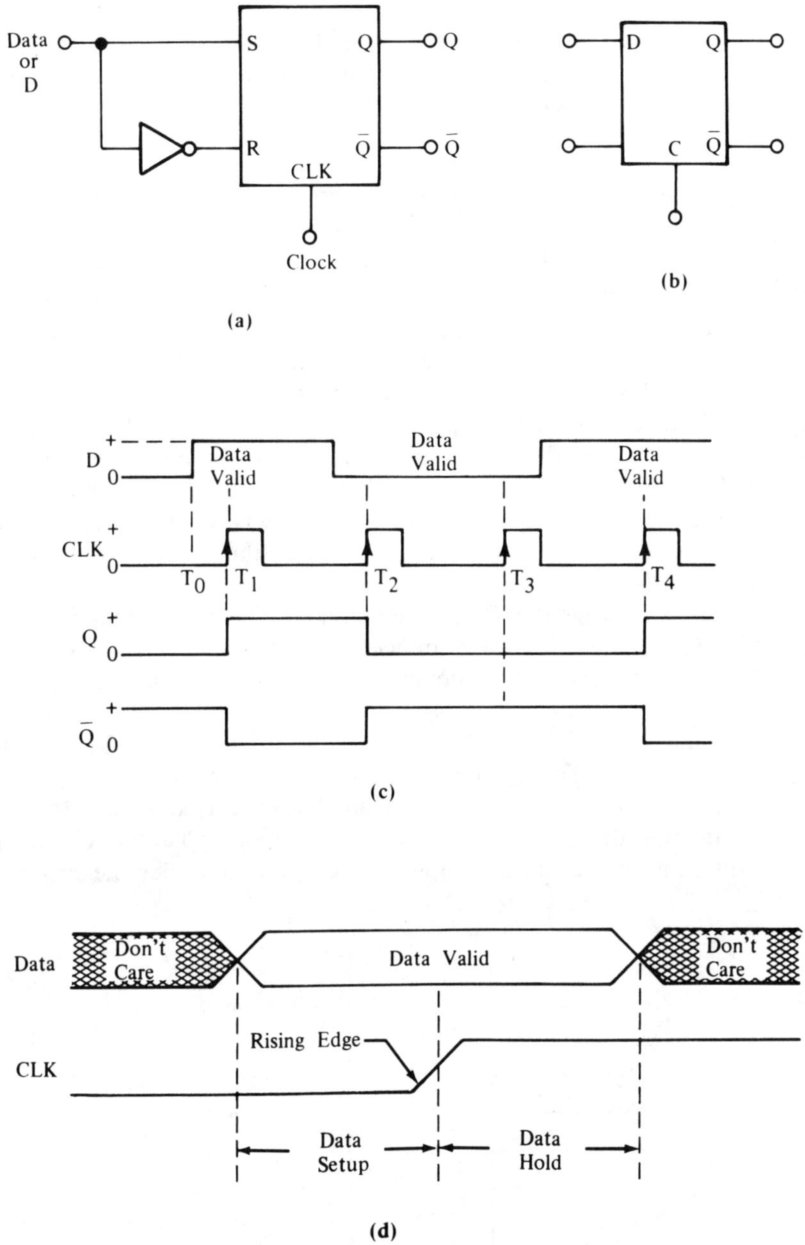

Figure 7-10 Type-D flip-flop: A) Equivalent circuit, B) Symbol, C) Typical timing diagram, D) Use of a data hold circuit.

7-10D. In this case, the data on the outputs will change only during either a rising edge of the clock pulse (positive edge-triggered) or on the falling edge of the clock pulse (negative edge-triggered). The flip-flop will respond only during a very narrow period of time.

I/O PORTS: DEVICES AND COMPONENTS

There are a number of devices on the market that can be used for input and output circuitry in microcomputers. Some devices are merely ordinary TTL or CMOS digital integrated circuits that are adaptable to I/O service. Still others are special-purpose integrated circuits that were intended from their inception as I/O port devices. Most of the microprocessor chip families contain at least one general-purpose I/O companion chip that is specially designed to interface with that particular chip. In this section, we will study some of the more common I/O components. Keep in mind, however, that many alternatives may be better than those shown here. You are advised to keep abreast of the integrated circuits that are available from various manufacturers.

Figure 7-11 shows the TTL 74100 device. This integrated circuit is a dual 4-bit latch circuit. When we connect the latch strobe terminals together (pins 12 and 23), we find that the device is usable as an 8-bit latch. The 74100 device can be used as an *output port*.

The input lines of the 74100 device are connected to bits DB0 through DB7 of the data bus. The Q outputs of the 74100 are being used as the data lines to the external device. The two strobe lines are used to gate data from the data bus onto the Q outputs of the 74100. The data latch (including the 74100) will transfer data at the D inputs to the Q outputs when the strobe line

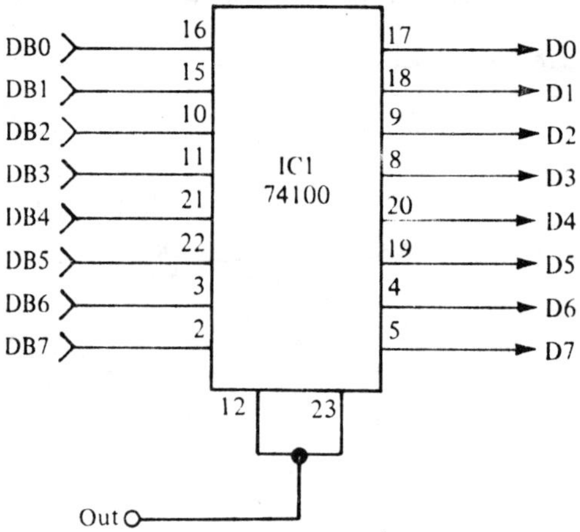

Figure 7-11 74100 used as a data latch.

is HIGH. (Note the similarity to the operation of the type-D flip-flop; the data latch is a special case of the type-D FF in which the *clock* line is labeled *strobe*.) When the OUT signal goes HIGH, therefore, the data on the bus are transferred to the Q outputs of the 74100. The data will remain on the Q outputs even after the OUT signal goes LOW again. This type of output, therefore, is called a *latched output*.

It is not necessary to use a single integrated circuit for the latched output circuit. We could, for example, use a pair of 7475 devices or an array of eight type-D flip-flops (although one wonders why).

Input ports cannot use ordinary two-state output devices because there may be a number of devices sharing the same data bus lines. If any one device, whether active or not, develops a short to ground, that bit will be permanently LOW regardless of what other data are supposed to be on the line. In addition, it is possible that some other device will output a HIGH onto the permanently LOW line and thereby cause a burnout of another IC. Similarly, a short circuit of any given output to the V+ line will place a permanent HIGH on that line. Regardless of the case, placing a permanent data bit onto a given line of the data bus always causes a malfunction of the computer or its resident program. To keep the input ports "floating" harmlessly across the data bus lines, we must use *tristate output* components for the input ports; such components were discussed earlier in this chapter (see Figure 7-5).

A number of 4- and 8-bit tristate devices on the market can be used for input port duty. Figure 7-12A shows the internal block diagram for the 74125

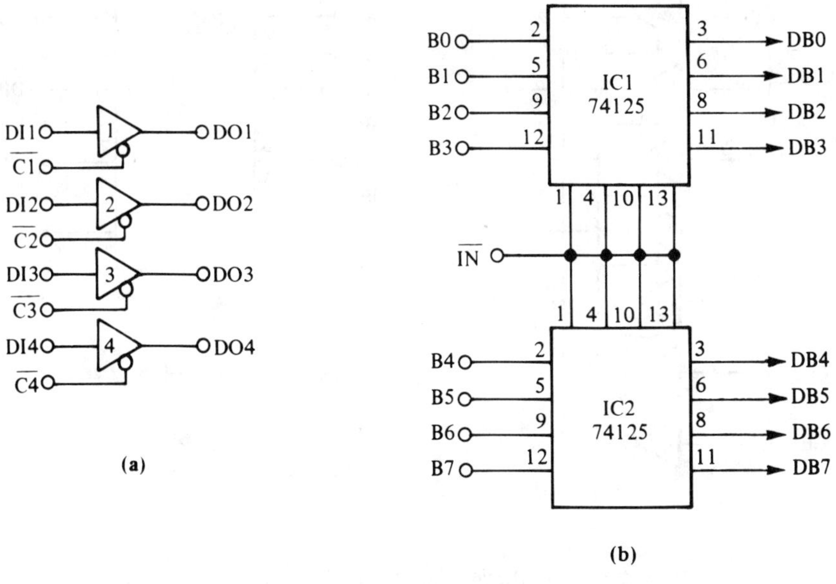

Figure 7-12 A) 74125 chip, B) 74125 used to drive data bus.

TTL device. This device is a quad noninverting buffer with tristate outputs. A companion device (74126) is also useful for input port service if we want or need an inverted data signal. The 74126 device is a quad inverter with tristate outputs. Each stage in the 74125/74126 devices has its own *enable* terminal (@C1–@C4) that is active-LOW. When the enable terminal is made LOW, therefore, the stage will pass input data to the output and operate in the manner normal to TTL devices. If the enable terminal is HIGH, however, the output floats at a high impedance and so will not load the data line to which it is connected.

Figure 7-12B shows a pair of 74125 devices connected to form a single 8-bit input port. The output lines from each 74125 (i.e., pins 3, 6, 8, and 11) are connected to lines DB0 through DB7 of the data bus. The input pins of the 74125 (pins 2, 5, 9, and 12) are used to accept data from the outside world.

The @IN signal generated by the microprocessor chip and the device-select circuits is used to turn on the 74125 devices. Note that all four enable

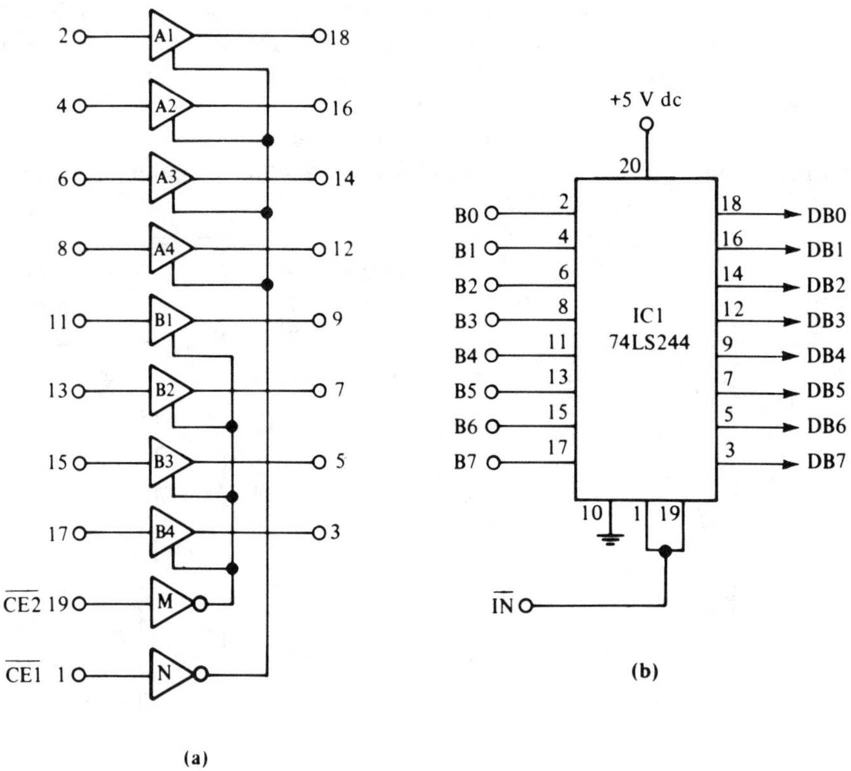

Figure 7-13 A) 74LS244 chip, B) Used as a data bus driver.

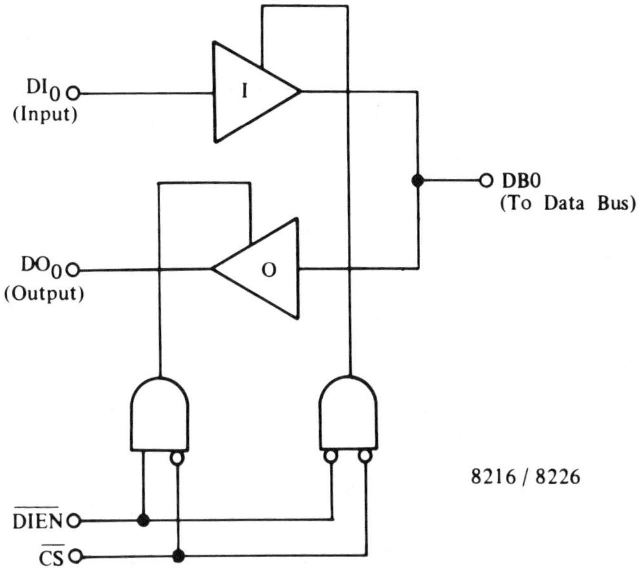

Figure 7-14 Internal circuit for 8216/8226.

lines of each 74125 device are parallel connected so that all stages will turn on at the same time.

The output lines of the input port are *not* latched. The data will, therefore, disappear when the @IN signal becomes inactive, exactly the requirements of an input port on a shared bus.

Another useful input port device is the 74LS244 TTL integrated circuit. Like the 74125 device, the 74LS244 has tristate outputs. The 74LS244 is an array of eight noninverting buffer stages arranged in a *two-by-four* arrangement in which four devices share a common enable terminal. In the case of Figure 7-13A, we find that stages A1 through A4 are driven by chip

8216 / 8226

$\overline{CS}$	$\overline{DIEN}$	State	
0	0	DI	DB
0	1	DB	DO
1	X	High - Z	
1	X	output	

O = Low, 1 = High, X = Either

Figure 7-15 8216/26 truth table.

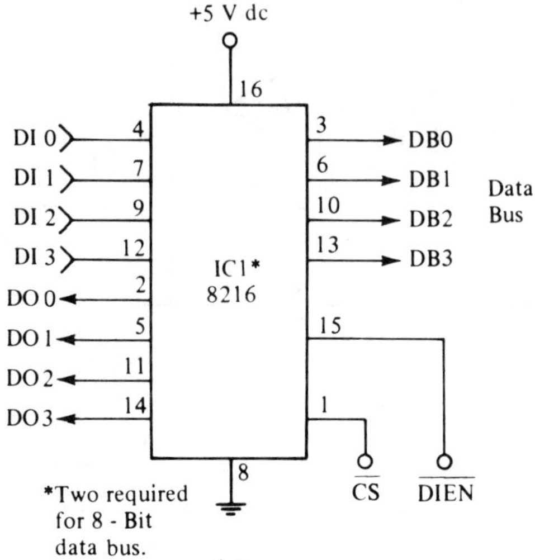

(a)

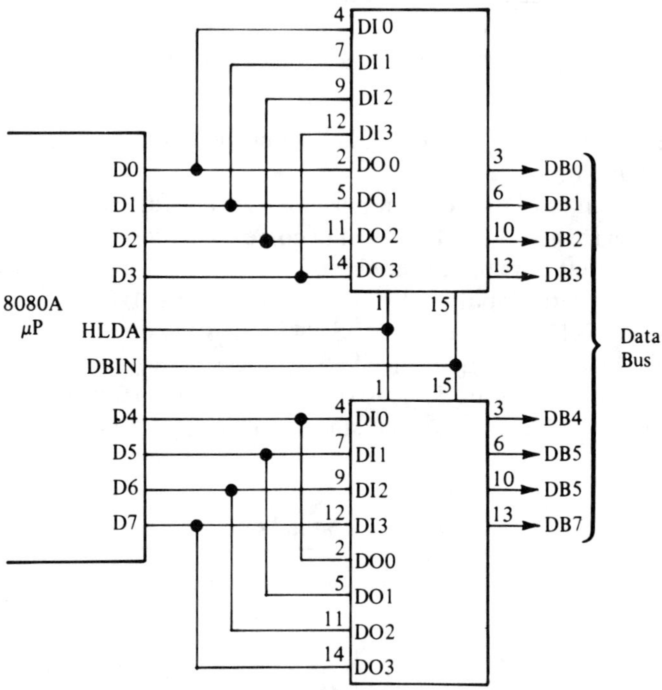

(b)

Figure 7-16 8216/26 circuits: A) Four-bit, B) Eight-bit.

enable input @CE1 (i.e., pin 1), while B1 through B4 are driven by chip enable input @CE2 (i.e., pin 19). In the circuit of Figure 7-13B, we strap the two chip enable terminals together to force the 74LS244 device to operate as a single 8-bit input port. The eight input lines are connected to the respective input terminals of the 74LS244, while the output lines are connected to their respective data bus lines. When the @IN signal becomes active (i.e., LOW), data on B0 through B7 will be gated onto data bus lines DB0 through DB7.

The techniques used thus far in this chapter require separate integrated circuits for input and output functions. While this is often satisfactory, it involves an excessive number of chips for some applications. We can, however, make use of combination chips in which the input and output functions are combined. Several devices on the market are classified as *bidirectional bus drivers*. These devices will pass data in either direction depending upon which device is selected by the control signals. Typical devices used for several years in microcomputer designs are the 4-bit 8216/8226 devices and the 8-bit 8212 device, all from Intel. Originally, these devices were intended for use in the 8080A microprocessor circuit. Even though the 8080A has been long since superseded by newer and more powerful microprocessors, some of the support chips still find wide application.

Figure 7-14 shows the internal structure (simplified) for the 8216 and 8226 devices. The principal difference between the 8216 and the 8226 is that the 8216 uses noninverting stages whereas the 8226 uses inverting stages. Note that the two buffers in each stage are facing in opposite directions with respect to the data bus line (i.e., DB0). In other words, the output line of I is connected to the data bus, so state I can be used as an input port line. Similarly, the input of O is connected to the data bus, thereby allowing us to use O as an output line. The DI and DO lines are for input and output, respectively.

Control of the 8216 and 8226 devices is through the @DIEN and @CS inputs. Figure 7-15 shows the truth table that applies to these chips. The *chip select* line (@CS) is active-LOW, so we find that the output will be in the high impedance state if @CS is made HIGH. The @CS line must be LOW in order for the device to operate. The data direction (@DIEN) line will connect the input lines (DI) to the data bus (DB) when the @DIEN is LOW and connect the data bus lines to the output lines (DO) when @DIEN is HIGH.

Figure 7-16 shows two alternate plans for connecting the 8216 and 8226 devices into actual microprocessor circuits. Figure 7-16A shows the basic connections to make these devices work properly, while Figure 7-16B shows a method for using a pair of 8216 devices with an 8080A microprocessor chip. The control signals from the microprocessor chip are specifically designed for use with the 8216/8226 devices.

8

USING OBSOLETE MICROCOMPUTER BUSES

This chapter is dedicated to two types of reader, one of which the author identifies with very closely. The first is the commercially interested person who is resurrecting old computer equipment for one reason or another. The other, with whom I identify intimately, is the science or engineering graduate student or research associate who is desperately trying to get a thesis-producing experiment working, but who has few funds for the required research. Stumbling around in the professor's equipment room or down in the basement (the only place they let graduate students *spritz* around), she or he comes across a *real computer*. But, darn it all, it's not an IBM AT/386 with math coprocessor—it's old . . . *but it'll do*.

A standard bus is an arrangement of printed wiring board or connector pins that follow a fixed protocol as to pin functions, signal designations, voltage levels, timing, and connector styles. Only a few of the standard buses in the microcomputer market are, in fact, nationally or internationally recognized standards, in the accepted sense of the term. They have instead become "standards" through vigorous promotion by the inventing company. The IBM PC is, perhaps, the best example. Although it started out as a proprietary standard, it is now the current *de facto* industry standard for scientific machines . . . and shares the rest of the market with only the Apple Macintosh as a serious competitor. Hence, a protocol made by, say, the XYZ corporation may well become the "XYZ bus" and be widely imitated. Other companies will note the success of the XYZ bus and will offer either second-source computer boards or peripheral/accessory boards for the XYZ bus.

A disadvantage of such uncontrolled buses comes in the matter of copycat designation of undesignated pins on the connectors. There may be several unreserved pins on any given bus, and some accessory manufacturers may designate them for some special purpose without consulting the originator of the bus. A problem arises when the user attempts to insert two different accessory cards into the computer, each of which designates an unreserved pin for some special purpose. Everything will be all right if no such conflict exists, but it does happen. There are some legendary compatibility problems when after-market IBM PC boards are stuffed into after-market motherboards. Little conflicts such as the memory locations occupied by firmware on the plug-in boards become terribly important, for example. Careful attention to the quirks of the special cards is necessary.

An advantage of the standard bus is the ability to easily interface various elements to form a computer or a computer-based instrument or machine. We can assemble a collection of CPU, memory, and assorted I/O and/or interface cards that are custom configured to perform some specific purpose that may or may not look like a traditional data processing or other computer chore.

Several buses became more or less standards in their time, if not by formal action by some authoritative group, then by common usage. Some are more widely used than others, but one must not assume that popularity denotes either a logical, well-thought-out design or usefulness for any given application. Indeed, engineering opinion generally holds that one of the most popular of the obsolete buses is actually among the least professionally designed and least useful on the market.

The concept of buses refers to situations in which a motherboard contains sockets into which CPU, memory, I/O, interface, and other cards plug. Such a bus will typically have wiring tracks on the motherboard for all bits of the data bus, all bits of the address bus, CPU, and/or system control signals, possibly some I/O lines, and, finally, DC power distribution. All these features are needed to make the plug-in cards work together.

Another type of bus is the I/O-oriented form. One would suppose the 20-mA current loop and RS-232 serial I/O ports could qualify for the designation as "I/O-oriented buses," but that requires the definition to be loose, which it often is. It is better to consider those "buses" as merely serial I/O ports, which they are, and only designate as I/O-oriented buses those that offer certain CPU control signal and addressing capabilities. The IEEE-488 general-purpose interface bus (GPIB) would certainly qualify under this criterion.

Certain problems with microcomputer buses tend to limit or constrain the designer. One such problem is the drive capacity of devices connected to the bus. A typical microprocessor chip output pin (e.g., on the data bus) will drive only two TTL loads (3.2 mA); that is, it has a TTL fan-out of 2. A bus line may represent a much heavier load; in fact, it almost always requires a fan-out much larger than 2. There are at least two reasons why this is true. First, there are *many* TTL inputs connected across each line of the data and address buses in a typical microcomputer, even one that is relatively simple. The load presented by such a situation to each line can be estimated by adding up the total number of devices hanging on each line and then multiplying by 1.6 mA. The second reason is that the multiple parallel bus lines form capacitive loads that the microprocessor chip outputs simply cannot handle. The solution in both cases is to use high-power bus driver or buffer ICs to interface the bus with its drive sources.

Another problem seen occasionally is bus *ringing*. The long bus line represents a complex reactive network of distributed capacitances and inductances, which combine to make the bus act exactly like a high-frequency radio frequency (RF) transmission line. When fast rise-time, high-repetition-rate signals are applied to the line, as they are in digital circuits, the result is exactly as if pulses were applied to a length of coaxial cable transmission line. The pulse will travel the length of the line, where it will be reflected unless the line is properly terminated. The reflected pulses can raise havoc by changing data values or instructions or can cause timing problems that are difficult to deal with.

Improper termination was a severe problem on certain early S-100 bus computers. Very soon after their introduction, however, companies began to offer both active and passive bus line terminator kits to solve the problem. Some had to be wired into the S-100 motherboard from the underside. Others were mounted on a shortened S-100 plug-in card and so could be plugged directly into a motherboard socket (an end socket was selected).

One final constraint is the number of sockets on the bus motherboard. The S-100 motherboard, for example, might have anywhere from five to thirty 100-pin sockets. Obviously, if expansion or a large number of optional cards is needed, one must select a motherboard with sufficient sockets to do the job. In this same vein, some thought had to be given to later expansion; rare is the computer system that does not expand as the owner becomes more enthusiastic or proficient, or both!

The type of bus required, as well as its size, depends much on the type of applications. A quite different machine is required for a number-crunching data-processing chore than for a controller of small scientific experiments.

When a standard-bus microcomputer is used, we can often obtain a computer-based instrument by designing only an interface board. Let's consider a simple example, an "evoked potentials computer" for studying the human electroencephalograph (EEG) response to specified stimuli, such as audio clicks or a flash of light.

Evoked potentials are a means of recording the minute component of the EEG signal. The EEG is a record of the brain's minute electrical activity as acquired from a set of differential scalp electrodes. Normally, the scalp-surface EEG signal is the summation of many dozens of signals from throughout the brain, only one component of which is due to the specified stimulus. In other words, the surface EEG potential is the algebraic sum of many time-varying signals. The analogy would be the situation of trying to discern one voice from the crowd by lowering a microphone 10 feet inside the Houston Astrodome during a football game. The problem is one of too much signal to discern the minute contribution of one small voice.

The solution to the problem used in evoked potentials work is to repeat the stimulus many times and then coherently average the EEG signal in a digital computer. *Coherent averaging* means that many samples are taken following the stimulus, but they are only compared with the signal taken at the same poststimulus time as previous trials. For example, all signals taken 100 ms after the stimulus will be averaged together, and all signals taken 101 ms after the stimulus are averaged together independently. If this is done properly, the component due to the stimulus will be enhanced, while the rest of the signals will tend to zero because of randomness. The result is that the signal remaining will represent the brainwave component caused by the stimulus.

Typical stimulii tested by this method have included light (the most common form of stimulus because of the ease of acquiring the relevant EEG potentials), sound, touch, and smell. Coherent signal averaging requires either that the computer be synchronized to the stimulator or that it synchronize the stimulus; both methods are used.

There is a simple solution to the problem. Select a computer system that has (in addition to a CPU) enough read-only memory (ROM) to contain the program, plus 25 to 50 percent reserve capacity, enough random-access memory (RAM) for the data points plus expansion reserve (2K to 4K are probably sufficient), and I/O capability. For the I/O function, you might use either an I/O card (or existing ports) or build I/O ports onto each interface card.

A suitable interface card is needed that will plug into one standard slot of a bus-organized microcomputer. The card may be acquired under any of several options. First, the computer manufacturer may build suitable blank interface cards or I/O cards as one of the standard system accessories (available at extra cost, of course). Second, you could design a suitable printed circuit wiring board from scratch. Such a card would have to adhere to the computer maker's specifications regarding pin-outs, card-edge connectors, size, shape, and voltage levels if it is to be successful. Finally, you could purchase a *prototyping card*. These cards are of the correct size and shape and contain an appropriate card-edge connector suitable for the computer on hand. Vector still offers S-100 prototyping cards, although few if any others do. The card may also contain power distribution tracks and printed circuit

IC pads (usually for DIP ICs). Otherwise, however, they are blank. The user adds components and point-to-point wiring (which may be either solder or wire-wrap).

In the simplest case, the interface card simply plugs into a motherboard slot, and analog signals are brought via connectors to the interface card from the outside world. We can, therefore, build other computer-based instruments by changing only the interface card and panel connectors and/or designations.

STANDARD (IF OBSOLETE) BUSES

There are several "standard" microcomputer buses on the market. Some are very good, others are not; some became very popular, while others disappeared rapidly; some offer extensive optional accessories, while others have none. Oddly, popularity does not guarantee either proper or even acceptable design.

In the remainder of this chapter we will discuss several previously popular microcomputer buses. Space limits both the depth and scope of the discussions. Many perfectly valid systems are not covered for lack of space. Do not construe either endorsement or condemnation of any bus from its exclusion or inclusion here.

THE S-100 BUS

The S-100 bus is probably the oldest popular microcomputer bus. It was introduced in 1975 by MITS, Inc., in their Altair microcomputer. The original Altair was based on the Intel 8080a microprocessor chip, although many later S-100 computers used the more powerful Z80 device instead. The Z80 was designed by Zilog, Inc., and is more or less compatible with the 8080a system. The two chips are not pin-for-pin compatible, and the Z80 contains a larger instruction set. Most programs designed for 8080a will also run on the Z80 machine, except for those that are dependent upon certain timing relationships.

There were once more than 350 different plug-in boards available from a variety of manufacturers that are allegedly compatible with the S-100 bus. There were literally dozens of memory boards, I/O boards (serial, parallel, 20 mA, RS-232, etc.), video boards, disk controllers, cassette tape controllers, speech synthesizers, and other special accessories. Unfortunately, many companies took it upon themselves to designate certain of the unreserved lines on the S-100 bus for some special purpose of their own. If two different boards try to use these lines for their own purpose, then a conflict will arise that will lead to strange results.

The S-100 connector consists of two rows of 50 pins (100 total) spaced 0.125 in. apart. The S-100 standard configuration calls for the lower 50 (1 through 50) pins to be on the component side, while the upper 50 are on the foil side of the board. The connector is offset along the 10-in. side of the 5.3 × 10 in. S-100 printed circuit board so that it is impossible to insert the card incorrectly. Otherwise, catastrophic failure will result. Pins 55 to 67 and 12 to 17 are unreserved in the standard connector, but they are often designated for special purposes by S-100 computer manufacturers or accessory makers.

Power supply for the S-100 bus microcomputer consists of three lines: an 8-V unregulated line with high current capacity and positive 15-V and negative 15-V low-current lines. The circuits on each card usually want to see +5 V regulated for the digital circuitry and either +/−12 or +/−15 V for the analog circuitry and some digital devices such as EPROMs (the 8080a microprocessor required −12 V at one pin for proper operation). In the S-100 scheme, each circuit card contains its own voltage regulators. There are some cards available with three independent +5-V (1 or 3 A) voltage regulators. This *distributed voltage regulation* scheme accomplishes several goals. For one thing, it all but eliminates the problem of voltage drop in the high-current lines. Ordinarily, a high-current regulated voltage line will see a voltage drop. In some computers, there will be a *sense* line from the voltage regulator that is attached to the printed circuit board at the point where the regulated voltage value must be precise. This approach does not work well in some digital equipment where there are plug-in printed circuit cards. After all, at which card can you accept a voltage error? The second advantage of distributed voltage regulation is that a failure in the regulator will not wipe out the entire computer, but only that fraction of the circuit severed by the smaller regulator. In most S-100 cards, the voltage regulation is provided by three-terminal IC voltage regulators in TO-3 or TO-220 power transistor packages; examples are the LM-309K, LM-340K-05, and 7805.

The S-100 pin-outs and signals are given in Table 8-1. Note that not all pin-outs are found in all systems. The originally unreserved pins are specified in this table.

RADIO SHACK TRS-80 BUS

The TRS-80 bus was developed by the Tandy Corporation for use in their Z80-based Radio Shack computers. With the possible exception of the Apple II bus, the TRS-80 bus became the most popular with the general public. This popularity is not only because of the Radio Shack marketing organization, which consists of many thousands of local company-owned and franchise stores, but also because the machine was easy for the nonexpert to learn to operate.

TABLE 8-1 S-100 Bus

Pin No.	Designation	Function
1	+8 V	Main power line for digital circuits; +8 V is supplied to the +5-V regulators on the cards.
2	+16 V	Unregulated voltage to the +12-V regulators on the circuit cards.
3	@XRDY	External ready.
4	V10	Vectored interrupt 0. Main vectored interrupt.
5	VI1	
6	VI2	
7	VI3	
8	VI4	Lower-priority vectored interrupts.
8	VI5	
10	VI6	
11	VI7	
12–17	Undesignated	
18	@STATDSB	Status disable. Active-LOW input that will force SINP, SMI, SMEMR, SOUT, SHLTA, and SSTACK into the high-impedance output state (i.e., disconnected).
19	@CCDSB	Command and control disable. Active-LOW input that will force PDBIN, PSYNC, PHLDA, PINTE, Not-PWR, and PWAIR into the high-impedance output state (disconnected).
20	UNPRO	Unprotect line. An active-HIGH input that will reset the protect flip-flop on the board addressed so that board can become active. See also pins 69 and 70.
21	SS or SSO	Single-step operation. Used by the front panel in S-100 computers equipped with such a panel. Will disable the bus buffer so that the front panel can drive the bus instead. Active-HIGH.
22	@ADDDSB or @ADRDSB	Address bus disable. Active-LOW input that forces the 16-bit address bus into the high-impedance output state (disabled).
23	@DODSB or @DBDSB	Data bus disable. Active-LOW input that forces 8-bit data bus into the high-impedance output state (disabled).
24	@-2	Phase-2 clock.
25	@-1	Phase-1 clock (not used on Z80 systems).
26	PHLDA	HALT acknowledgment. Active-HIGH output that indicates a HOLD signal is received. The CPU will go into the dormant high-impedance output (disabled) state after execution of the current instruction cycle.
27	PWAIT	Wait. Active-HIGH output that indicates the CPU is in the wait state.

TABLE 8-1 S-100 Bus

Pin No.	Designation	Function
28	PINTE	Interrupt enable flag. An active-HIGH output that indicates the CPU will respond to interrupts (i.e., when PINTE is HIGH). The internal interrupt flip-flop that controls this signal is under program control, not hardware control.
29	A5	Address bus bit 5.
30	A4	Address bus bit 4.
31	A3	Address bus bit 3.
32	A15	Address bus bit 15.
33	A12	Address bus bit 12.
34	A9	Address bus bit 9.
35	DO1	Data bus bit 1 output.
36	DO0	Data bus bit 0 output.
37	A10	Address bus bit 10.
38	DO4	Data bus bit 4 output.
39	DO5	Data bus bit 5 output.
40	DO6	Data bus bit 6 output.
41	DI2	Data bus bit 2 input.
42	DI3	Data bus bit 3 input.
43	DI7	Data bus bit 7 input.
44	SM1	Machine cycle 1. Active-HIGH output that indicates CPU is in instruction op-code fetch cycle.
45	SOUT	Active-HIGH output that, in conjunction with @PWR, indicates when the address bus data is valid for an output device.
46	SINP	Input status line. Active-HIGH output that, in conjunction with PDBIN, indicates when the address bus data are valid for an input device. Tells a peripheral when data should be put on the data bus.
47	SMEMR	Active-HIGH output indicating a memory read operation.
48	SHLTA	Active-HIGH output signal that indicates execution of a HALT command.
49	@CLOCK	Complement of @-2 main system clock.
50	GND	Ground for digital signals and power supply.
51	+8 V	Same as +8 V on pin 1 and occupies the space on the other side of the PC board, opposite pin 1, so that pins 1 and 51 can be connected in parallel to the +8-V unregulated power supply.
52	−16 V	Unregulated negative 16 V used to supply −12-V regulators.
53	@SSWDSB	Sense switch disable. An active-LOW input that permits sense switches on front panel to input data.

TABLE 8-1 S-100 Bus

Pin No.	Designation	Function
54	@EXTCLR	Signal from front panel that resets all I/O devices.
55	RTC	Real-time clock line, if used.
56–67	Undesignated	
68	MWRITE	Active-HIGH output that indicates contents of the data out bus (DO0–DO7) are being written into memory at the location specified by A0–A15 of the address bus.
69	@PS	Protect status line. Active-LOW output indicates the status of the protect flip-flop on the plug-in card.
70	PROT	Active-HIGH line resets protection flip-flop (see pin 20 for both pins 69 and 70).
71	RUN	Active-HIGH signal that indicates that the front panel RUN/STOP flip-flop is in the *set* condition.
72	@PRDY	Ready.
73	@PINT	Active-LOW interrupt request input that will cause the CPU to recognize the interrupt line unless the interrupt enable flip-flop is reset (see pin 28) or if the CPU is in HOLD status.
74	@PHOLD	Active-LOW input that requests the CPU to go to the HOLD state at the end of the current instruction execution.
75	@PRESET	Active-LOW reset line. This line is essentially a "hardware jump to location 00000000" instruction, and it will reset (clear) the contents of the program counter when made LOW.
76	PSYNC	Active-HIGH output that indicates the onset of a machine cycle.
77	@PWR	Active-LOW write output indicates that a write operation to either memory or an output port is taking place.
78	PDBIN	Active-HIGH read output signal indicates a request from the CPU to either memory or input to place data on the data in (DI0—DI7) bus.
79	A0	Address bus bit 0.
80	A2	Address bus bit 1.
81	A2	Address bus bit 2.
82	A6	Address bus bit 6.
83	A7	Address bus bit 7.
84	A8	Address bus bit 8.
85	A13	Address bus bit 13.
86	A14	Address bus bit 14.
87	A11	Address bus bit 11.
88	DO2	Data bus bit 2 output.

TABLE 8-1 S-100 Bus

Pin No.	Designation	Function
89	DO3	Data bus bit 3 output.
90	DO7	Data bus bit 7 output.
91	DI4	Data bus bit 4 input.
92	DI5	Data bus bit 5 input.
93	DI6	Data bus bit 6 input.
94	DI1	Data bus bit 1 input.
95	DI0	Data bus bit 0 input.
96	SINTA	Active-HIGH output that indicates acknowledgment of an interrupt request. See pin 73.
97	@SWO	Active-LOW signal indicating data transfer from CPU to either memory or I/O.
98	SSTACK	Active-HIGH output signal indicating that the address of a push-down stack is on the address bus.
99	POC	Hardware jump to location 00000000 instruction that clears the program counter on either power-on or PRESET.
100	GND	Ground. Opposite side of board from pin 50 so that grounds may be connected together.

Unlike the S-100 and Apple II buses, the TRS-80 bus does not permit plug-in accessories within the mainframe. There is, however, an interface bus that may be used to good advantage. The pin-outs and signal definitions are given in Table 8-2.

TABLE 8-2 TRS-80 Bus

Pin No.	Designation	Function
1	@RAS	Active-LOW row address select. This signal is used in the refresh operation of dynamic RAM devices used in the TRS-80. See also pins 3 and 16.
2	@SYSRES	Active-LOW reset output goes LOW when the main system reset line is operated internal to the computer. This output may be used to reset peripherals and other devices interfaced with the TRS-80. The @SYSRES line will go LOW whenever the reset button is pressed or when power is applied to the TRS-80 (i.e., the power-on reset circuit is operated). The effect of the reset is to load the program counter register of the Z80 with 00H, so the reset is basically a hardware JUMP to 00H instruction.
3	@CAS	Active-LOW column address select signal

TABLE 8-2 *continued*

Pin No.	Designation	Function
		used in the refresh operation of dynamic RAM used in the TRS-80. See pins 1 and 16 also.
4	A10	Address bus bit 10.
5	A12	Address bus bit 12.
6	A13	Address bus bit 13.
7	A15	Address bus bit 15.
8	GND	Ground.
9	A11	Address bus bit 11.
10	A14	Address bus bit 14.
11	A8	Address bus bit 8.
12	@OUT	System output. This active-LOW output denotes an output operation being executed by the Z80 microprocessor used in the TRS-80. This device-select pulse is generated by applying the @IORQ and @WR control signals of the Z80 to an OR gate.
13	@WR	Active-LOW output that goes LOW during a memory write operation. This signal is not to be confused with the Z80 @WR signal, even though it is the product of ORING @WR and @MREQ Z80 control signals. The Z80 write signal is also used in output operations (see pin 12).
14	@INTAK	Interrupt acknowledge. This active-LOW signal tells the outside world that an interrupt subroutine is beginning. In modes 0 and 2, this signal can be used to signal the peripheral to place the interrupt vector address onto the data bus.
15	@RD	Active-LOW output that tells the outside world that a memory read operation is taking place. This signal is not to be confused with the Z80 control signal using the same mnemonic. The @RD signal on the TRS-80 bus indicates only memory read, while the @RD Z80 control signal is also used for input operations. In the TRS-80, @RD is generated by ORING @MREQ and @RD. See also pin 13.
16	MUX	Active-HIGH signal that controls internal data multiplexers used in memory refresh operations.
17	A9	Address bus bit 9.
18	D4	Data bus bit 4.
19	@IN	Active-LOW output that tells the outside world that an input operation is taking place. See also pin 12.
20	D7	Data bus bit 7.

TABLE 8-2 *continued*

Pin No.	Designation	Function
21	@INT	Active-LOW interrupt request line. May be used by external devices to interrupt the CPU.
22	D1	Data bus bit 1.
23	@TEST	Active-LOW input that will allow external devices interfaced with the TRS-80 to gain control of the data bus.
24	D6	Data bus bit 6.
25	A0	Address bus bit 0.
26	D3	Data bus bit 3.
27	A1	Address bus bit 1.
28	D5	Data bus bit 5.
29	GND	Ground.
30	D0	Data bus bit 0.
31	A4	Address bus bit 4.
32	D2	Data bus bit 2.
33	@WAIT	Active-LOW input that forces the Z80 microprocessor used in the TRS-80 to go into the wait state. This signal is used to allow the Z80 to work with slow memory and peripherals and will continue to insert wait states into the CPU cycles until @WAIT is HIGH again.
34	A3	Address bus bit 3.
35	A5	Address bus bit 5.
36	A7	Address bus bit 7.
37	GND	Ground.
38	A6	Address bus bit 6.
39	(see text)	This pin is designated +5 V DC on level I TRS-80 machines and GND on level II machines.
40	A2	Address bus bit 2.

The Radio Shack TRS-80 microcomputers are a wise selection for many users, especially since so many of them are still around (and many still in service). The interface connector pin-outs described in Table 8-2 allow wide latitude for interfacing chores, despite the lack of plug-in capability inside the machine. The TRS-80 has a wide variety of software written for it, both by Radio Shack (Tandy Corporation) and independent vendors.

KIM-1, SYM-1, AND AIM-65 BUS

The KIM-1 microcomputer was a single-board trainer that was introduced by MOS Technology, Inc., of Norristown, Pennsylvania, the originator of the 6502 microprocessor chip. It was apparently intended as a means of introducing the world of microprocessing to engineers who would incorporate the 6502 into their instrument and computer designs. The KIM-1 computer, however, developed into a popular starter computer as well as a trainer. Many current computer experts began their careers with a KIM-1 device.

The KIM-1 was a single-board computer that contained 1K of 8-bit memory, a 6522 versatile interface adapter (VIA), a 20-mA TTY current loop for making hard copies, and a cassette (audio) interface to allow storage of programs on ordinary audiotape. One feature of the KIM-1 tape interface not found on others of the era is the ability to search for programs on the tape by a designator applied to the beginning of the program on the cassette.

The SYM-1 (Fig. 8-1) is a more recent single-board trainer computer that uses the KIM-1 bus. The SYM-1, however, is still easily obtained and contains more features than the original KIM-1. For the *aficionado* of the KIM-1, the SYM-1 is a good substitute.

The AIM-65 (Rockwell Microelectronics, Inc.) is a more advanced microcomputer based on the KIM-1 bus. The AIM-65 computer uses a standard ASCII typewriter keyboard instead of the hexadecimal pad of the KIM-1. It also has a 20-character 5 × 7 dot matrix LED display and a 20-column 5 × 7 dot matrix thermal printer instead of the standard seven-segment LED readouts of the KIM-1 (which require some training to read hexadecimal digits above 9). The printer uses standard calculator printer paper, which is available at stationery stores.

The AIM-65 also has a sophisticated monitor program stored in ROM and the ability to incorporate BASIC and a 6502 assembler into other on-board ROMs. In contrast, the KIM-1 originally used a relatively simple monitor. To write and input programs, one had to "fingerbone" instructions into the computer on a step-by-step basis. The AIM-65 comes with a text editor. Also, the AIM-65 can be configured with either 1K or 4K of memory, and external memory to 48K can be added if desired.

Two interfacing connectors are etched onto the boards of the KIM-1, SYM-1, and AIM-65 computers. The *applications connector* is basically an I/O connector, while the *expansion connector* is more similar to a genuine bus connector. Both are of primary interest to microprocessor users who must interface the computer with an external device. Pinouts are described in Table 8-3 for the applications connector and in Table 8-4 for the expansion connector.

Numbered connector pins are on the top or component side of the printed wiring board; alphabetic pins are on the bottom or foil side of the board.

The KIM-1 and related computers use the 6522 VIA device. The 6522 contains two 8-bit I/O ports, designated ports A and B. These ports are represented by bits PA0 and PA7 and PB0 and PB7. Both ports can be configured under software control for either input or output port service on a

TABLE 8-3 KIM-1/SYM-1/AIM-65 Application Connector

Pin No.	Designation	Function
1	GND	Ground.
2	PA3	Port A, bit 3.
3	PA2	Port A, bit 2.
4	PA1	Port A, bit 1.
5	PA4	Port A, bit 4.
6	PA5	Port A, bit 5
7	PA6	Port A, bit 6.
8	PA7	Port A, bit 7.
9	PB0	Port B, bit 0.
10	PB1	Port B, bit 1.
11	PB2	Port B, bit 2.
12	PB3	Port B, bit 3.
13	PB4	Port B, bit 4.
14	PA0	Port A, bit 0.
15	PB7	Port B, bit 7.
16	PB5	Port B, bit 5.
17	KB RO	Keyboard row 0.
18	KB CF	Keyboard column F.
19	KB CB	Keyboard column B.
20	KB CE	Keyboard column E.
21	KB CA	Keyboard column A.
22	KB CD	Keyboard column D.
A	+5 V	+5-V DC from main board power supply.
B	K0	
C	K1	
D	K2	
E	K3	Memory-bank select signals (active-LOW).
F	K4	
H	K5	
J	K7	
K	DECODE	Memory decode signal. Used to increase memory capacity with off-board memory devices.
L	AUD IN	Audio input from cassette.
M	AUDOUTL	Low-level audio output to cassette with "microphone" input.
N	+12 V	+12-V DC power from main board.
P	AUDOUTH	High-level audio output to cassette player with "line" input.
R	TTYKBD+	Positive terminal of 20-mA teletype keyboard loop.
S	TTYPNT+	Positive terminal of 20-mA teletypewriter printer loop.
T	TTYKBD−	Negative terminal of 20-mA teletypewriter keyboard loop.
U	TTYPNT−	Negative terminal of 20-mA teletypewriter printer loop.
V	KB R3	Keyboard row 3.
W	KB CG	Keyboard column G.
X	KB R2	Keyboard row 2.
Y	KB CC	Keyboard column C.
Z	KB R1	Keyboard row 1.

TABLE 8-4 KIM-1/SYM-1/AIM-65 Expansion Connector

Pin No.	Designation	Function
1	SYNC	Active-HIGH output line that goes HIGH during the phase-1 clock signal during instruction fetch operations. This line is used to allow the 6502 to operate with slow memory, dynamic memory, or in the direct-memory access mode.
2	@RDY	Has the effect of inserting a wait state into the CPU operating cycle. See similar description for same signal in Apple II discussion.
3	01	Phase-1 clock signal.
4	@IRQ	Maskable interrupt request line. Active-LOW.
5	RO	Reset overflow input. A negative-edge-triggered input that will reset the overflow flip-flop in the CPU.
6	@NMI	Active-LOW nonmaskable interrupt input line. This interrupt line cannot be masked by the internal interrupt flip-flop.
7	@RST	In parallel with the reset line on the 6502 and on the microcomputer. When brought LOW, this line will cause the program counter inside of the 6502 to be loaded with 00H. The effect of this line is to form a hardware "JUMP to 00H" instruction.
8	DB7	Data bus bit 7.
9	DB6	Data bus bit 6.
10	DB5	Data bus bit 5.
11	DB4	Data bus bit 4.
12	DB3	Data bus bit 3.
13	DB2	Data bus bit 2.
14	DB1	Data bus bit 1.
15	DB0	Data bus bit 0.
16	K6	Address decoder output that goes HIGH whenever the CPU addresses a location from 1800H to 1BFFH.
17	SSTOUT	Single-step output.
18	(NC)	No connection.
19	(NC)	No connection.
20	(NC)	No connection.
21	+5 V	+5-V DC power supply from main board.
22	GND	Ground.
A	AB0	Address bus bit 0.
B	AB1	Address bus bit 1.
C	AB2	Address bus bit 2.
D	AB3	Address bus bit 3.
E	AB4	Address bus bit 4.
F	AB5	Address bus bit 5.
H	AB6	Address bus bit 6.
J	AB7	Address bus bit 7.
K	AB8	Address bus bit 8.

TABLE 8-4 *continued*

Pin No.	Designation	Function
L	AB9	Address bus bit 9.
M	AB10	Address bus bit 10.
N	AB11	Address bus bit 11.
P	AB12	Address bus bit 12.
R	AB13	Address bus bit 13.
S	AB14	Address bus bit 14.
T	AB15	Address bus bit 15.
U	02	Phase-2 clock signal.
V	R/@W	Read/write line is HIGH during read operations and LOW during write operations.
W	W/@R	Complement of the R/@W line. This line is LOW for read operations and HIGH for write operations.
X	PLLTST	Phase-locked-loop test. This line is used in testing and adjusting the PLL that operates the audiocassette tape recorder audio FM signal.
Y	@02	Complement of phase-2 clock signal.
Z	RAMRW	Line turns on RAM during read/write operations during phase-2 clock periods.

bit-by-bit basis. In other words, PA0 might be an input bit, while PA1 is an output port bit. Or we can configure all 8 bits of either or both ports as either input or output.

CONCLUSION

The use of the older, now obsolete, microcomputer bus is not generally recommended. But when there is a money crunch, and the older machine is available, then there is no reason why it should not be used. A friend of mine used a KIM-based computer, and then a Z-80 machine, to build a product that has sold over $1,000,000 worth of instrumentation . . . all based on "obsolete" machines. Those 6502, Z80, and 8080a single-board computers and controllers are still sold, and in surprising numbers, so perhaps there is even a current market for them in applications where the expense of using the 80386 is not justified.

9

INTERFACING STANDARD AND CUSTOM PERIPHERALS

Rarely will a digital computer stand alone for long. It is almost certain that the owner will soon want to add capability by incorporating peripherals into the system. There is an almost endless variety of peripherals that perform a large assortment of different jobs. The strong desire for hardcopy readouts instead of volatile video prints will likely cause the microcomputer owner to buy a printer or teletypewriter. Other peripherals include remote instruments and assorted forms of display.

Figure 9-1 shows a typical wide-carriage printer. These peripherals come in a variety of printing mechanisms, with two standard input formats (parallel and serial) and two different standard paper carriages (80 column and 130 column). They are also available in three different paper-handling formats: single sheet, continous "track" feed, and tray feed (which is used on laser printers).

SERIAL VS. PARALLEL COMMUNICATIONS

Most microcomputer internal data buses are configured in a parallel format in which all 8 or 16 bits are transmitted simultaneously. This method is an example of *parallel* data transmission and is used internally because it is the fastest method available. The disadvantage of parallel data transmission, however, is that it requires one or more separate lines for each bit of the data word. This presents little problem inside the computer where these parallel data lines are merely a few centimeters of copper track laid side by side on a

Figure 9-1 Computer printer.

printed circuit board. But when we go outside the computer, the situation changes. For short runs, say across a table or within an equipment rack, parallel transmission is still economical, but the situation changes rapidly as distance increases. Even a few yards of wire might provide enough problems to make it worthwhile to convert to serial transmission, where a single pair of conductors carries all data. If the data have to be transmitted through media other than wire, say telephone lines or radio channels, it is extremely costly to use parallel transmission. For sending data across the country we would need the equivalent of eight telephone circuits in order to send an 8-bit data word asynchronously. But if we use serial transmission, the number of channels reduces to one.

The key to serial data transmission is a serial input/output port for the computer. There are several ways to accomplish a serial I/O port: discrete hardware, IC UART (defined below), and software UART. The discrete hardware method requires a shift register with a length equal to the number of bits in the data word. The proper form of shift register is *parallel in, serial out* (PISO). Data from the parallel data bus are input to the PISO shift register and then clocked out through the serial output one by one. This requires at least one clock operation for each bit of the shift register and so will take a fair amount of time. The IC UART, or universal asynchronous receiver/transmitter, is a special device that makes *parallel-to-serial* conversions for the transmitter section and *serial-to-parallel* conversions for the receiver section. Finally, we have the software implementation. In this type of UART the data word to be transmitted is loaded into the accumulator of the CPU, then shifted one place right or left, and output one time for each bit of the word. All three methods are used in various computers, although the discrete hardware method is probably used least today.

SERIAL DATA COMMUNICATIONS

The two major serial data communication standards are the 20-mA current loop and RS-232. The 20-mA current loop uses an electrical current to carry the data; the RS-232 uses voltage levels. The RS-232 is a standard of the Electronic Industries Association (EIA) and is extensively used throughout the computer industry. There are two extant RS-232 versions, the older RS-232B and the more recent (and current) RS–232C. Because RS-232B has been obsolete for many years, we will not cover it in detail here.

The RS-232C Standard

The RS-232C standard is issued by the EIA in an attempt to make it possible to interface equipment made by a wide variety of manufacturers without the need for special engineering for each case. The idea is to use the same electrical connector (i.e., the DB-25 family of D-shell connectors) wired in the same manner all the time, and to use the same voltage levels for the binary digits 1 and 0 all the time. Supposedly, if everyone interprets the standard the same way, it should be possible to connect together any two devices with RS-232 ports without any problem, and it usually does work that way. Modems (modulator/demodulators), CRT video terminals, printers, and other devices all come with RS-232 connectors, at least as options. Some computers provide RS-232 serial output ports, and almost all have RS-232 capability available as an option from either the original computer manufacturer or a specialty after-market house.

It is relatively easy to design an RS-232 port, especially if the application will operate asynchronously at a low data rate such as 110 to 1200 baud. RS-232C receiver and transmitter integrated circuits are available from various manufacturers (e.g., Motorola). The RS-232 transmitter/buffer chip will accept TTL-level inputs and convert them to the appropriate RS-232C level for transmission. The RS-232 receiver works exactly the opposite: it will accept RS-232 input signals and convert them to equivalent TTL levels.

The RS-232 is a very old standard and predates TTL standards. As a result, the RS-232 standards use what appear to younger eyes as very odd voltage levels to recognize logical 0 and logical 1 levels. Besides voltage

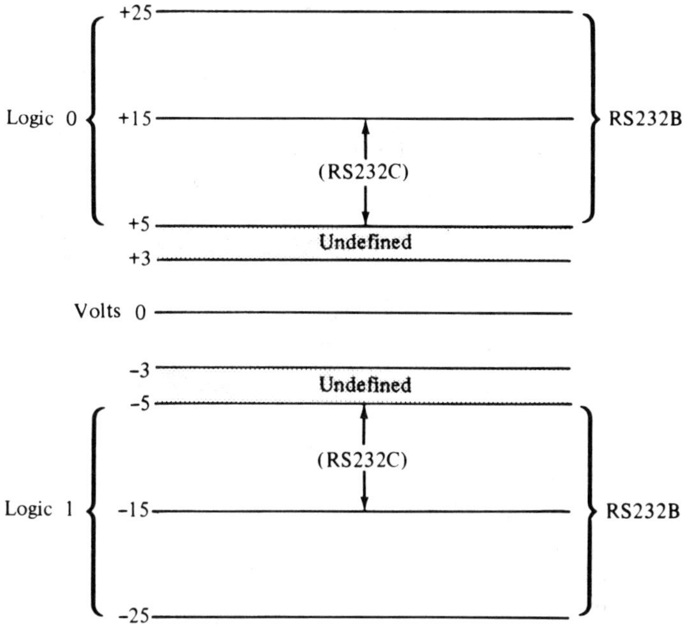

Figure 9-2 RS-232 logic protocols.

levels, the standard also fixes load impedances presented to the bus by receivers and the output impedances of transmitter/drives.

The two RS-232 standards (RS-232B and RS-232C) are both depicted in Figure 9-2. In the older version, RS-232B, logical 1 is any potential in the range from −5 to −25 V, while logical 0 is anything from +5 to +25 volts. The voltages in the −3 to +3 volt range are a transition state, while the ranges from +/−3 to +/−5 V are undefined and will produce unpredictable results if used (a situation that can occur in poorly designed systems).

The RS-232C standard uses narrower limits between logical 0 and logical 1 in order to make the data transmission speedier. The upper limits for the logical 0 and logical 1 levels are +/−15 V, rather than +/−25 V as in the RS-232B standard. In addition to narrowing the voltage ranges, the newer RS-232C standard fixes the load impedance to 3000 to 7000 ohms, and the driver output impedance is lower than previously. Also, the driver must provide a slew rate of 30 volts per microsecond (30 V/μS). The Motorola MC1488 and MC1489 meet these specifications.

The RS-232 standard specifies a standard connector so that all products will be compatible. The DB-25 (i.e., the male DBM-25 and female DBF-25) D-shell connector is used for this purpose and is the identifying feature of an RS-232 equipped piece of equipment. Figure 9-3 shows the pin-out designations for the RS-232C connector, while pin definitions are listed below.

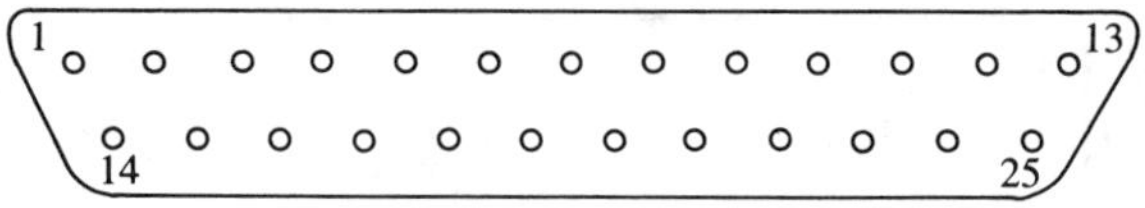

Figure 9-3 DB-25 connector used in RS-232 serial communications.

Pin No.	RS232 Name	Function
1	AA	Chassis ground
2	BA	Data from terminal
3	BB	Data received from modem
4	CA	Request to send
5	CB	Clear to send
6	CC	Data set ready
7	AB	Signal ground
8	CF	Carrier detection
9	undef	
10	undef	
11	undef	
12	undef	
13	undef	
14	undef	
15	DB	Transmitted bit clock, internal
16	undef	
17	DD	Received bit clock
18	undef	
19	undef	
20	CD	Data terminal ready
21	undef	
22	CE	Ring indicator
23	undef	
24	DA	Transmitted bit clock, external
25	undef	

The electrical requirements for the RS-232C standard are as follows:

1. The *mark* (logical 1) level shall be −3 to −15 V; the *space* (logical 0) level shall be +3 to +15 V.
2. Load impedances on the receiver side of the circuit shall be greater than 3000 ohms but less than 7000 ohms.
3. The maximum data rate is 20 kilobits/second.
4. Inputs for receivers must have a capacitance of less than 2500 picofarads (pF).
5. There should be not more than 50 feet of hardwire between inputs and outputs without an audio modem at each end of the circuit.
6. Standard data rates are 50, 75, 110, 150, 300, 600, 1200, 2400, 4800, 9600, and 19,200 baud.

9-Pin RS-232 Connector

The standard 25-pin DB-series connector is large and clumsy, and it contains a large number of unused pins. After microcomputers became widespread, a new RS-232 connector evolved that used a smaller 9-pin connector.

Current-Loop Transmission

The current-loop form of data communications systems was derived from the standard method used on teletypewriter equipment. These electromechanical typewriters were popularized in the late 1930s with equipment from companies such as Kleinschmitt and The Teletype Corporation (Skokie, Illinois). The word Teletype is a registered trademark of The Teletype Corporation, even though it is frequently used erroneously as a generic term for all teletypewriter equipment. Such use is improper, however, unless the machine being discussed was manufactured by The Teletype Corporation. There are actually two different current-loop standards: 60-mA and 20-mA. The 60-mA standard is now obsolete.

The reason current loops became popular for teletypewriters is that the characters are selected by five or seven electrical solenoids that activate the mechanical selector bars inside the machine. These solenoids are connected in series banks of differing numbers depending upon the character being formed. As a result, voltage transmission is not as effective as current-loop transmission.

There are at least three different types of current-loop device: *printer only, keyboard only,* and *keyboard/printer combinations*. The printer-only type contains the solenoids and typing mechanism and will print the characters transmitted over the current loop. There is no method for sending data back from the printer-only machine. The keyboard-only device is exactly the opposite: it contains the encoder and keyboard but is incapable of printing. Such machines are now quite rare but were once used for remote entry of data. The keyboard/printer combination machine contains both the receiver and sender sections in one cabinet. Figure 9-4 shows the circuit for a 20-mA current-loop keyboard/printer teletypewriter unit. The transmitter consists of a keyboard and an encoder (that forms the data word) that can be modeled as a simple electrical switch. When the switch is closed, the circuit passes current down the 20-mA loop and to its own printer. Similarly, when the 20-mA loop is active from the other end, the current will flow in the solenoid, causing the remote print operation. Since the keyboard switch is in series with the circuit, some means must be provided to close the printer circuit during receive operations. This function is provided by switch S1, labeled in Figure 9–4 as "send-receive."

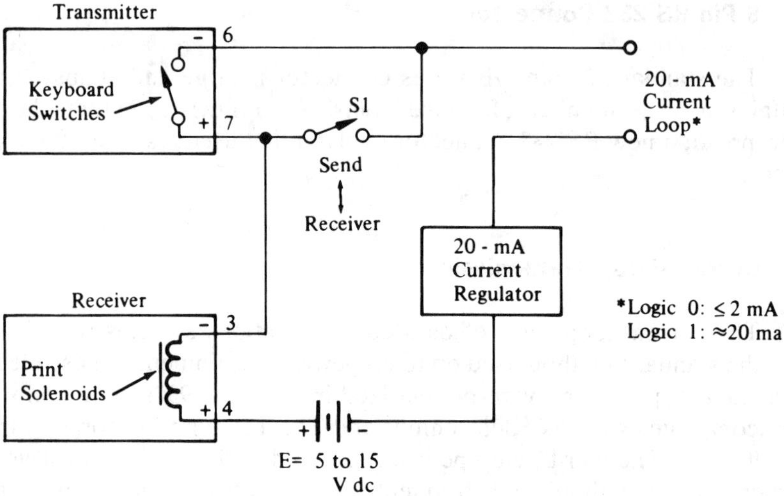

Figure 9-4 20-mA current loop for local printer/instrument control.

Three modes of communication are used with current-loop systems: *simplex, half-duplex,* and *full-duplex*. The simplex system allows communication in one direction only. There will be a dedicated receiver and a dedicated transmitter that never change roles; data flow is always in one direction only. The half-duplex system allows two-way communications, but only in one direction at a time. We can send data from, say, A to B one time, and then reverse the situation and transmit from B to A; we may not, however, simultaneously send data from A to B and from B to A. Simplex and half-duplex current loops require only one pair of wires, while full-duplex normally requires two pairs of wires (one for each direction). Note, however, that full-duplex operation over a single pair of wires is possible if the data levels are first converted to audio tones. In that case, the system will use different pairs of tones to represent *mark* and *space*, one pair for each direction.

60-mA current loops. The 60-mA current loop is now obsolete and is only found in old installations or where old surplus equipment is used. The 60-mA system shown in Figure 9-5 shows how one of these teletypewriters can be interfaced to the TTL-compatible serial output port. In many cases, one bit of a parallel output port will be configured as a serial port through either software or hardware implementation. In Figure 9-5, the least significant bit (LSB) of the parallel output port is designated as the serial output.

The TTL level from the serial output port in Figure 9-5A drives the base terminal of a high-voltage, NPN power transistor (a Motorola MJE-340 or equivalent). The collector-emitter path of the transistor is connected in series with the 60-mA current loop and so acts as a switch. When the TTL level is HIGH, transistor Q1 is turned on and current flows in the loop. However, when the TTL bit is LOW, the transistor is turned off, so no current will flow in the circuit. The transistor therefore provides a *mark* (i.e., logical 1) when the TTL bit is HIGH, and a *space* (i.e., logical 0) when LOW.

The current loop is powered from a 120- to 140-V DC power supply that has a series-connected rheostat (R2), which is used to set the approxi-

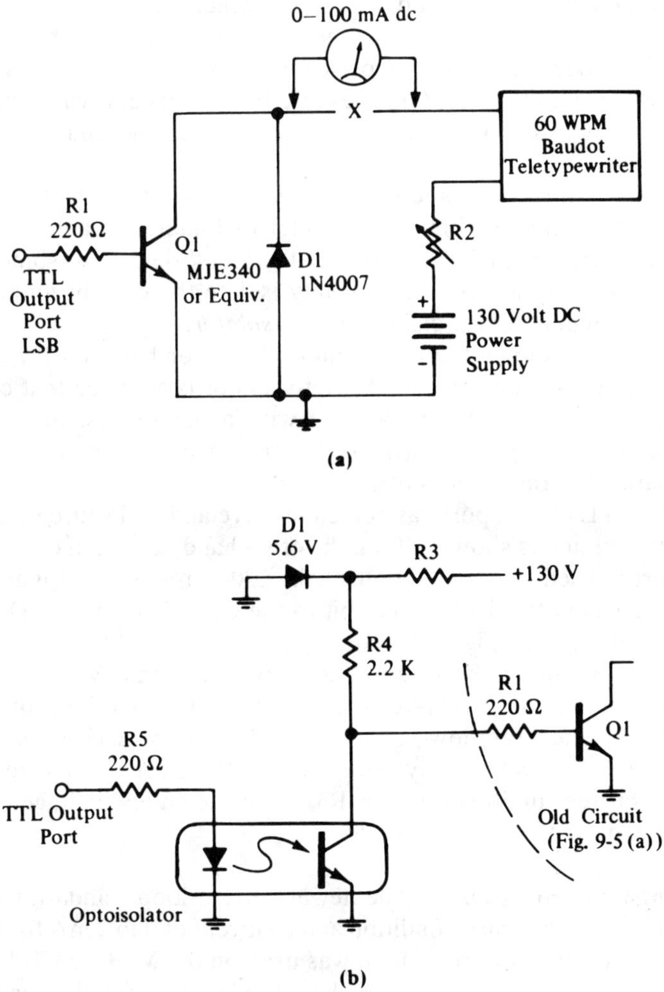

Figure 9-5 A) Obsolete 6-mA loop (all current models use 20-mA loop or another protocol, but some 60-mA machines may still be around); B) Isolated circuit.

mate current level. The current level in the 60-mA current loop is adjusted by breaking the loop and inserting a 0 to 100 milliammeter into the circuit. A key should be pressed on the keyboard, or in the case of Figure 9-5A, a HIGH must be written to the serial output port. That action will turn on the loop, allowing current to flow. Potentiometer R2, which is rheostat connected, is then adjusted for approximately 60-mA current flow.

The circuit in Figure 9-5A suffers from a problem. When the 60-mA current flows in the solenoids, a magnetic field is built up around each coil. Abruptly interrupting the current (i.e., going to a space bit) will cause the field to collapse, giving rise to a high-voltage spike created by the *inductive kick* phenomenon. As a result of this spike, which can damage semiconductor devices, we must provide some means of suppression. This function is provided by diode D1, a 1000-V peak inverse voltage (PIV), 1-A rectifier-type diode. Diode D1 is normally reverse-biased, except when the spike is present, so it will clip off the spike before it has the chance to do any damage.

There is a further problem: *isolation*. High-voltage, high-current circuits can cause "glitches" in the computer that alter data and interrupt the process. In fact, this problem is one of the worst defects in some types of computer. The solution is to completely isolate the current loop from the computer through a device called an *optoisolator*.

Figure 9-5B shows the use of an optoisolator between the computer and the current-loop peripheral. An optoisolator is a device that contains a light-emitting diode (LED) juxtaposed with a phototransistor. When the LED illuminates the phototransistor, the transistor is turned on; when the LED is dark, the transistor is off.

If the TTL output port has sufficient drive and will source current, we may use the circuit as shown. If, on the other hand, a normal open-collector output port is used, we must connect the 220-Ω resistor in Figure 9-5B to +5 V and connect the TTL output bit to the cathode of the LED (which is shown as grounded in Figure 9-5B).

The transistor in the optoisolator will not normally operate from a 130-V DC source, so a lower-voltage power supply must be provided. We could provide a separate low-voltage power supply or derive a low voltage from the +130-V power supply. In Figure 9-5B we use a 5.6-V zener diode (D1) and a current-limiting resistor (R3) to provide a low voltage consistent with the needs of the phototransistor.

20-mA current loops. The newer current-loop standard uses a current of 20 mA for the mark condition and a current of 0 to 2 mA for the space condition. The 20-mA current loop was used on the Model 33 Teletype and all subsequent models. There was also a code change with these models. The older 60-mA machines used the 5-bit Baudot Code, while 20-mA machines most frequently use the modern ASCII (American Standard Code for

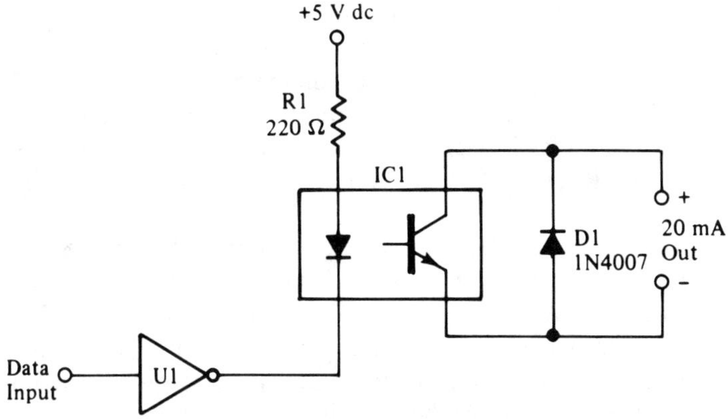

Figure 9-6 TTL-to-20-mA converter.

Information Interchange) code. Keep that in mind when using older machines, because computers today support only ASCII unless code conversion software is provided.

Figure 9-6 shows a simple method for using an optoisolator to interface a computer serial output port with a 20-mA current loop. The operation of this circuit is exactly like that of Figure 9-5B. When the data input is HIGH, indicating a mark or logical 1 condition, the output of the open-collector inverter (Ul) will go LOW, thereby grounding the cathode of the LED. This will turn on the transistor, which allows current to flow in the circuit. Again, a reverse-biased diode is used to prevent damage and other troubles caused by the inductive spike generated when the solenoids are de-energized.

A transmitting version is shown in Figure 9-7. Here we have a keyboard or 20-mA transmitter sending data to a computer that has a TTL-

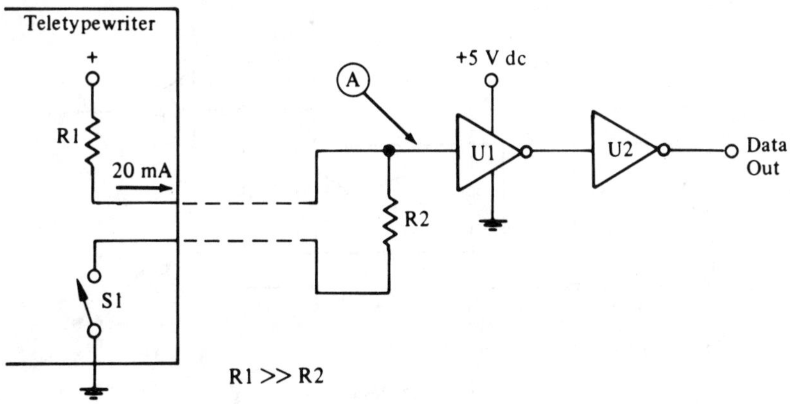

Figure 9-7 20-mA-to-TTL converter.

compatible input. A DC power supply (+5 to +15 V) and a pair of series resistors is selected to provide a current of 20 mA so that when switch S1 is open, there is no current flow, so the voltage at point A will be HIGH. A double inverter sequence makes the output of the circuit HIGH also. If, however, the switch is closed, current flows in the circuit, so a voltage drop is created across Rl and R2. If R1 is much greater than R2, the input of U1 sees a LOW condition.

A somewhat more satisfying circuit is shown in Figure 9-8. This circuit uses an optoisolator with the LED in series with the current loop to interface with the computer. When the current flows in the loop, the LED is turned on, so the phototransistor is illuminated and turned on. When the transistor is on, the resistance from collector to emitter is very low, so the input of the inverter (IC2) sees a LOW. If the LED is off, indicating a space, the transistor is off. This condition makes the collector-emitter resistance high, so the voltage applied to the input is also high; the input of IC2 is HIGH. The values of resistor Rl and the supply voltage (shown in Figure 9-8 as +5 V) can be varied to other values if TTL-compatibility is not needed. Of course, IC2 cannot be a TTL inverter in that case; a CMOS 4049 or 4050 is recommended.

Figure 9-9A shows the connections to the popular Model 33 Teletype. The terminal strip shown in this figure is normally found on the right-rear panel (viewed from the operator's seat) under a cover. Be careful to unplug the teletypewriter from the 110-V AC when accessing this terminal strip, because that potential is found on pins 1 and 2 of the strip. Isolated versions of the receive and transmit circuits are shown in Figures 9-9B and 9-9C, respectively.

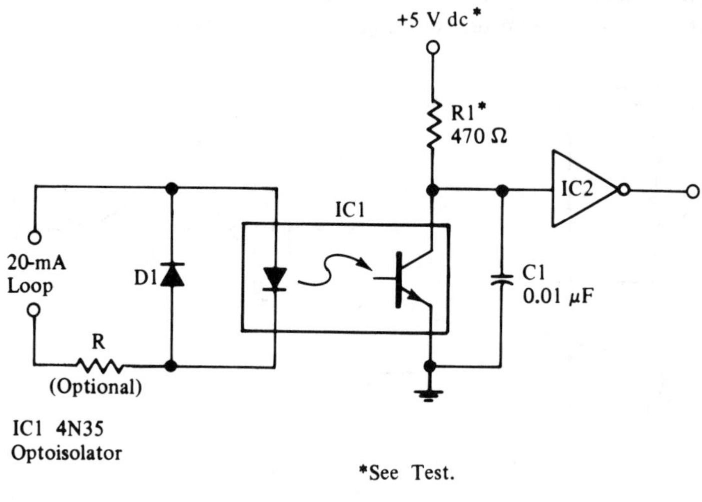

Figure 9-8 Isolated 20-mA-to-TTL converter.

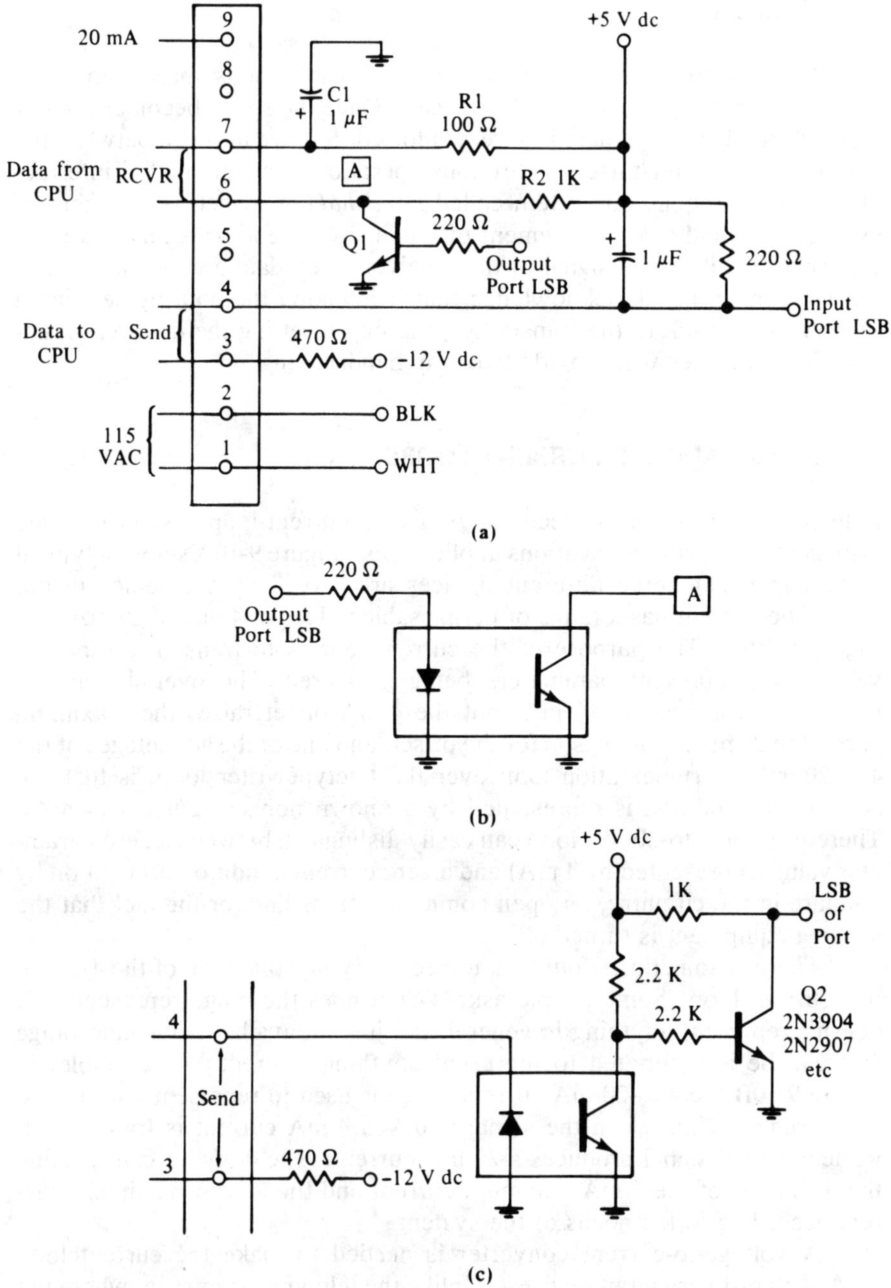

Figure 9-9 A) Proper connections between computer (TTL output) and a Model 33 Teletype machine, B) Optical isolation for receiver, C), Send optical isolation.

Handshaking

Most peripherals can operate in an asynchronous manner only at slower data rates (e.g., 300 baud and less). Once the speed becomes greater than 300 baud or when there is a tremendous difference in speed between the two devices, we must use synchronous operation. This method of interfacing generally requires a system called *handshaking*, which is a system of interrogation and acknowledgment of readiness to send or receive data. A sender unit will send a signal to the receiver when data are available. The receiver, in turn, will acknowledge that it is taking the data by sending a second signal back to the transmitter, thereby resetting the data ready signal. Some devices will provide two-way handshaking.

4-to-20-MILLIAMPERE CURRENT LOOPS

Industrial process control technology uses a current-loop system in some instrumentation communications applications. Figure 9-10A shows a typical system in which three different devices are served by the same current loop. The current has a range of permissable values of 4 mA to 20 mA (see Figure 9-10B). The purpose of the current loop is to transmit a range of values that represent parameters being measured. The overall dynamic range of the current is 16 mA, but the 4-mA offset raises the maximum current to 20 mA. The reason for the offset, and one of the advantages of the 4-to-20-mA instrumentation loop over the teletypewriter loop, is that the zero input condition is represented by a known nonzero current (4 mA). Therefore, the 4-to-20-mA loop can easily distinguish between a zero parameter value (represented by 4 mA) and a zero current condition brought on by a failure in the circuitry, an open communications line, or the fact that the sending equipment is turned off.

There is sometimes found an unnecessary mystification of the 4-to-20-mA current loop. Some people ask, "What does the range represent?" It doesn't represent anything in general, it's just an available dynamic range that can be appropriated to represent anything desired. For example, in Figure 9-10B the 4-to-20-mA current range is used to represent a 0 to 10 V signal range. Thus when the signal is 0 V a 4-mA current is transmitted, while a +10-V signal produces a 20-mA current. The designer can redefine the meanings of the 4-mA minimum current and the 20-mA maximum current according to the needs of the system.

A voltage-to-current converter is needed to make the current loop work with ordinary input devices. Unlike the teletypewriter case, where the current was either on or off depending upon whether a 1 or 0 was being transmitted, the analog 4-to-20-mA current loop requires a circuit that will produce an output load current that is proportional to the input voltage.

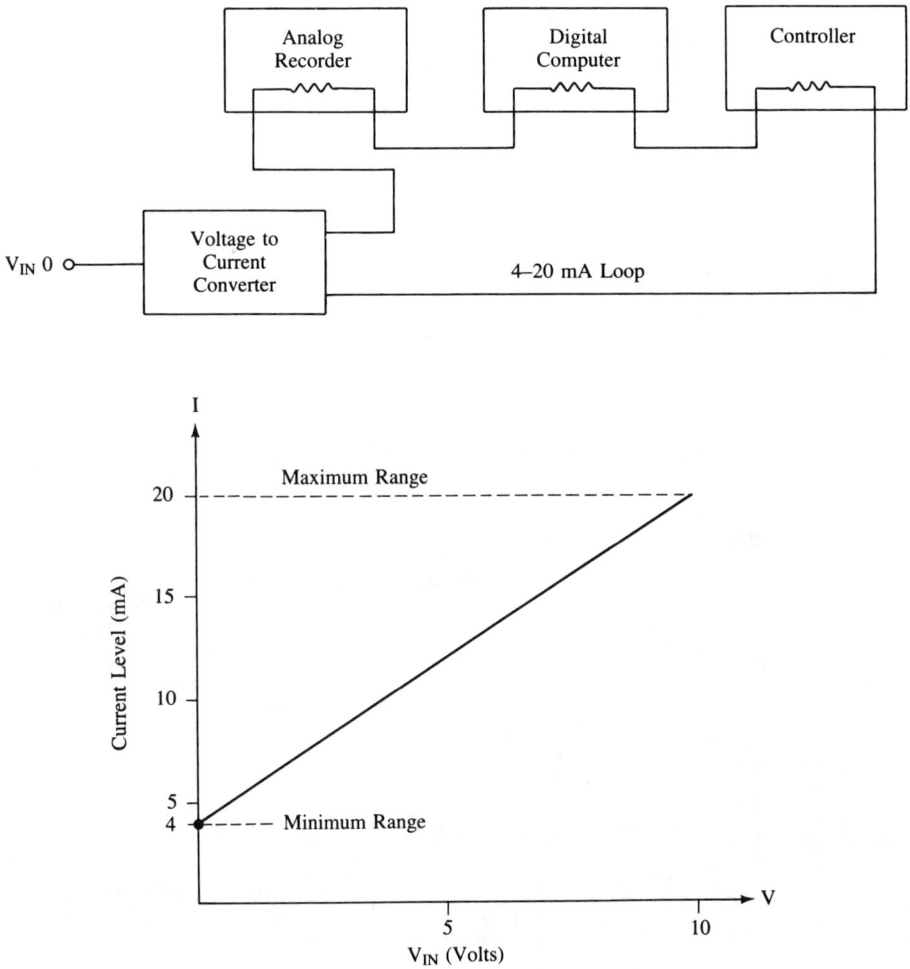

Figure 9-10 4-to-20-mA instrumentation current loop: A) Circuit, B) Transfer function.

Figure 9-11 shows one such circuit. The current loop, here represented by load resistor R_L, is the feedback current of a noninverting follower operational amplifier circuit. The current is proportional to the input voltage according to the following rule:

$$I_L = \frac{V_0 - V_{in}}{R_L} \tag{9.1}$$

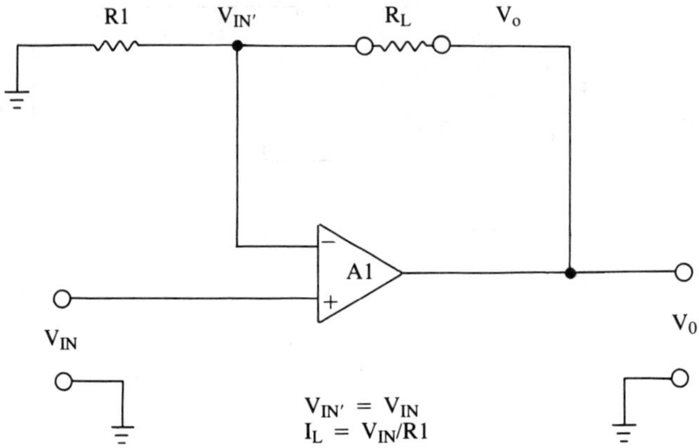

Figure 9-11 Simple current generator for floating loads.

Output voltage V_0 is found as follows:

$$V_0 = \frac{V_{in}R_L}{R_{in}} + V_{in} \tag{9.2}$$

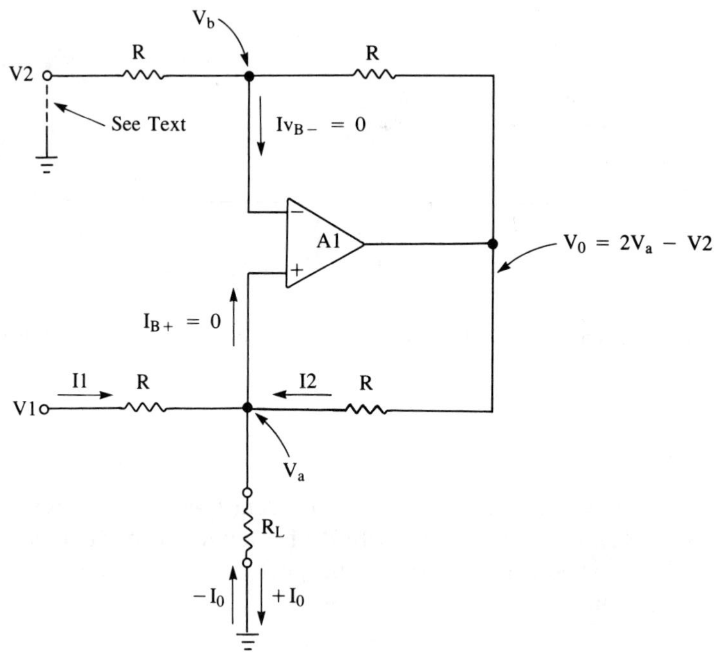

Figure 9-12 A) Current generator for grounded load.

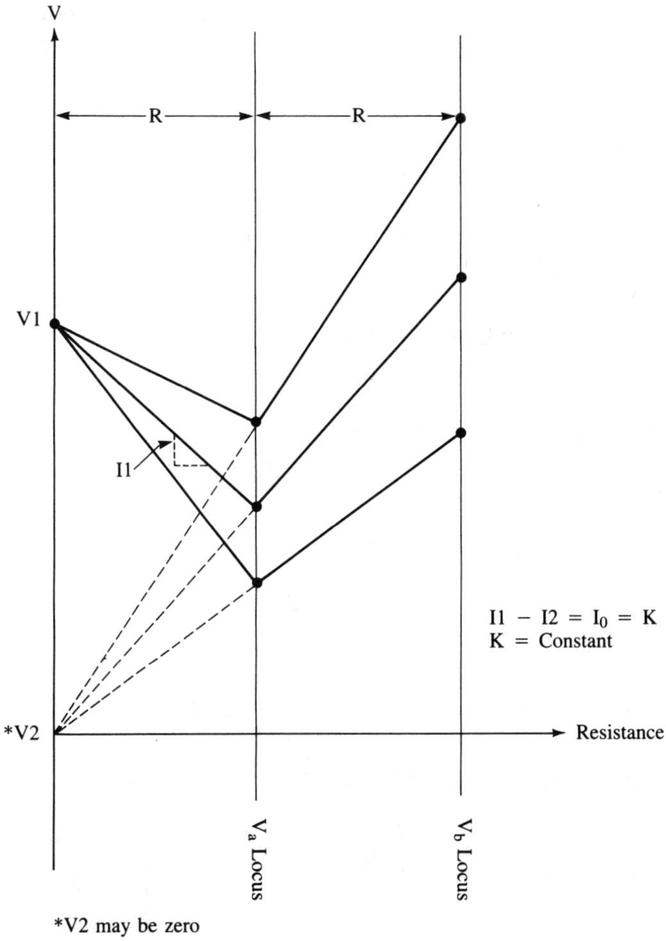

Figure 9-12 B) Transfer function.

By algebra one can demonstrate that substituting Equation (9.2) into Equation (9.1) produces

$$I_L = \frac{V_{in}}{R_L} \tag{9.3}$$

The circuit of Figure 9-11 operates over a wide range of load resistance values but suffers from the fact that the load must be floating with respect to ground. A grounded load circuit is shown in Figure 9-12A. This circuit is called the Howland current pump after B. Howland of the MIT Lincoln Laboratories. The uniqueness of the Howland current pump is that the load current is independent of load resistance. The current is proportional to the difference between voltages V1 and V2. In the generalized case of the Howland current pump V2 is not zero, although in many practical applications V2 is zero and the associated input resistor is grounded. Figure 9-7 is a graphical solution of the circuit evaluation for the case V2 = 0.

Another distinction of the Howland current pump is that it can either sink or source current depending upon the relationship of V1 and V2. If $I_L > 0$, then the current flows out of the circuit (i.e., "down" in Figure 9-12B), but if $I_L < 0$ the current flows into the circuit (i.e. "up" in Figure 9-12B). For the sake of simplicity the case where V2 = 0 is evaluated.

There are two "output" voltages in this circuit. Voltage V_a is the load voltage and appears across the load resistance, as follows:

$$V_a = I_l R_l \tag{9.4}$$

According to Kirchhoff's current law,

$$I_l = I1 + I2 \tag{9.5}$$

By Ohm's law,

$$I1 = \frac{V1 - V_a}{R} \tag{9.6}$$

and

$$I2 = \frac{V_0 - V_a}{R} \tag{9.7}$$

Substituting Equations 9.6 and 9.7 into Equation 9.5 results in the following:

$$I_L = \frac{V1 - V_a}{R} + \frac{V_0 - V_a}{R} \tag{9.8}$$

which reduces to

$$I_L R = V1 + V_0 - 2V_a \tag{9.9}$$

Therefore, we may write

$$V_a = \frac{V1 + V_0 - I_L R}{2} \tag{9.10}$$

For voltage V_a the operational amplifier is connected in the noninverting follower configuration. Because the two feedback network resistors are equal, the gain of the circuit is two. Therefore

$$V_0 = 2V_a \tag{9.11}$$

or,

$$V_0 = V1 + V_0 - I_L R \tag{9.12}$$

which can be combined as follows:

$$0 = V1 - I_L R \tag{9.13}$$

Equation (9.14) informs us that output load current I_L is proportional to input voltage V_{in} and inversely proportional to the feedback resistance (assuming all four resistors are equal).

In the more general case where V2 is nonzero, Equation (9.14) becomes

$$I_L = \frac{V1 - V2}{R} \tag{9.15}$$

To make the Howland current pump produce the 4-mA current that represents the 0-V signal level requires that V2 be set to a value that will force Equation (9.15) to evaluate to 4 mA when V1 = 0.

GENERAL-PURPOSE INTERFACE BUS (IEEE-488)

The general-purpose interface bus (GPIB) is a system that allows interconnection of up to 15 electronic instruments or devices so that they can interact with each other. There are three categories of device on the GPIB: talkers, listeners, and controllers. The system is programmable and so can form the basis of automatic test equipment systems.

Automatic test equipment (ATE) is now one of the leading methods for testing electronic equipment in factory production and troubleshooting situations; it has been used for this purpose for years. The data acquisition system can also be designed around standard ATE modules and equipment.

The usual method is to use a programmable digital computer to control a bank of test instruments. The program, often in BASIC language, turns on and off the various instruments and then evaluates the results as measured by other instruments.

The bank of equipment (see Figure 9-13 for an example) can be configured for a special purpose or for general use. For example, we could select a particular lineup of equipment needed to test, say, a broadcast audio console, and provide a computer program to make the various measurements: gain, frequency response, total harmonic distortion, and so forth. Alternatively, we could also make a generalized test set. This is the method selected by a number of organizations that have large amounts of different electronic devices to test. There will be a main bank of electronic test equipment, adapters to make the devices under test interconnect with the system, and a special program for each type of equipment. Such an approach makes for a cost-effective system of test equipment.

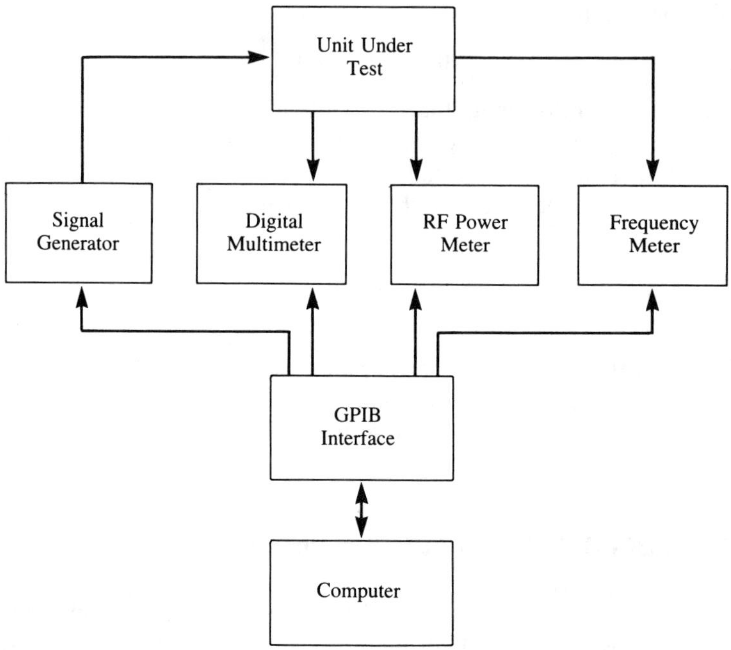

Figure 9-13 Instrument bank using GPIB system.

Previously, the main problem in attempting to make ATE was that it was impossible to use unmodified off-the-shelf commercial test instruments. Thus many of the best and most useful pieces of test equipment could not be used at all or had to be extensively modified by the ATE maker before they could be used. The problem was programmability: How could one make a signal generator respond to the computer commands?

In 1978, the Institute of Electrical and Electronic Engineers (IEEE) released their specification titled "IEEE Standard Digital Interface for Programmable Instrumentation," or "IEEE-488" as it is called in the trade. This specification provides details for a standard interface between a computer and instruments. It also calls out ASCII codes and mnemonics for program instructions. The IEEE-488 bus is also called the General Purpose Interface Bus (GPIB). The Hewlett-Packard Interface Bus (HPIB) is a proprietary version of the IEEE-488 bus. The main purpose for the IEEE-488/GPIB is use with ATE, both generalized and specific.

Test instrumentation intended for GPIB service will have a 24-pin "blueline" connector on the rear panel. This connector is one of the Amphenol blueline series not unlike the 36-pin connector used for parallel printer interface on microcomputers. There will also be a GPIB address DIP switch on the rear panel, usually near the connector. The purpose of the switch is to set the 5-bit binary address where the instrument is located in the system, determine whether or not the device is a listener only or a talker only, and determine certain other details.

GPIB BASICS

The IEEE-488/GPIB specification provides technical details of the standard bus. The logic levels on the bus are generally similar to TTL: a LOW is less than or equal to 0.8 V, while a HIGH is greater than 2 V. The logic signals can be connected to the instruments through a multiconductor cable up to 20

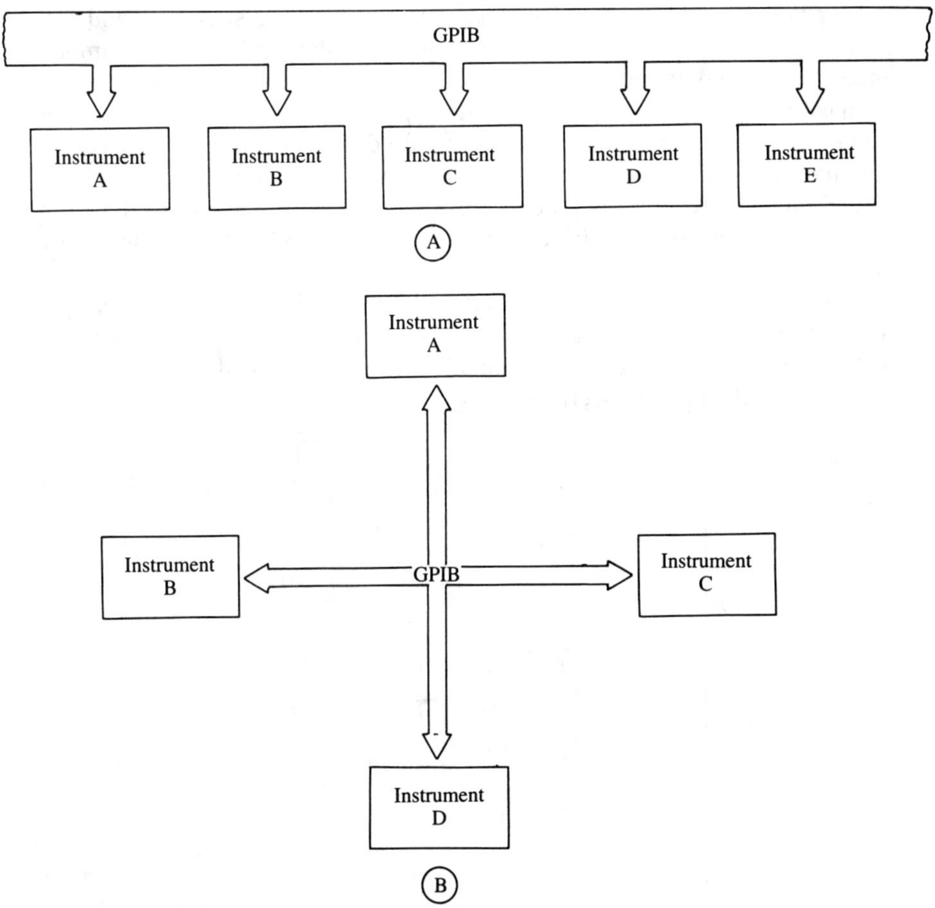

Figure 9-14 A) Serial or sequential GPIB configuration, B) Star or parallel GPIB configuration.

meters (66 feet) in length, provided that an instrument load is placed every 2 meters. This specification works out to a cable length in meters twice the number of instruments in the system. Most IEEE-488/GPIB systems operate unrestricted to 250 kilobytes per second, or faster with certain specified restrictions.

There are two basic configurations for the IEEE-488/GPIB system (Figure 9-14): linear and star. These configurations are created with the cable connections between the instruments and the computer. The linear configuration is a daisy-chain method, in which the tap-off to the next instrument is taken from the previous one in the series. In the star configuration, the instruments are connected from a central point.

There are three major buses in the IEEE-488/GPIB system. Each line in each bus has a circuit similar to that shown in Figure 9-15. Besides the shunt protection diode and stray capacitance, there are also pull-up and pull-down resistors that effectively determine the standardized input impedance. Connected to the bus line are receiver and driver circuits. These similar-to-TTL logic elements provide input or output to the instrument. The driver is an output and will be a tristate device. That is, it is inert until commanded to turn on. A tristate output will float at high impedance until turned on. The receiver is a noninverting buffer with a high-impedance input. This arrangement of drivers and receivers provides low loading to the bus.

Figure 9-16 shows the structure of the IEEE-488/GPIB. There are three buses and four types of devices. The devices are: controllers, talkers only, listeners only, and talker/listeners. These buses and devices are described in the following sections.

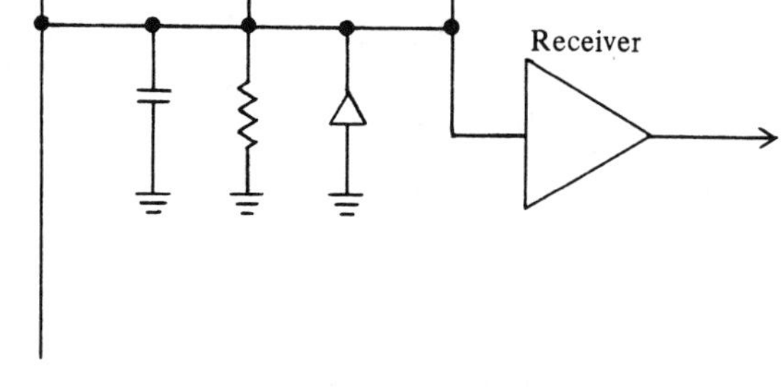

Figure 9-15 Typical internal bus circuit for GPIB instrument.

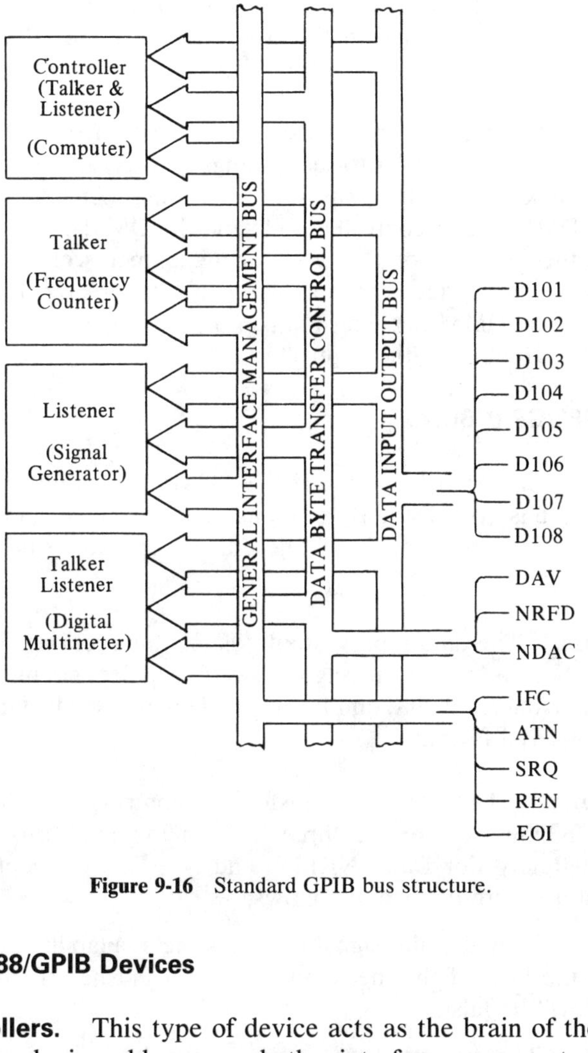

Figure 9-16 Standard GPIB bus structure.

IEEE-488/GPIB Devices

Controllers. This type of device acts as the brain of the system: it communicates device addresses and other interface messages to instruments in the system. Most controllers are programmable digital computers. Both Hewlett-Packard and Tektronix offer computers that serve this function, and certain other companies produce hardware and software that permit other computers to act as IEEE-488/GPIB controllers.

Listeners. A device capable of listening will receive commands from another instrument, usually the controller, when the correct address is placed on the bus. The listener acts on the message received but does not send back any data to the controller. As shown in Figure 9-16, an example of a listener is the signal generator.

Talkers. The talker responds to the message sent to it by the controller and then sends data back to the controller over the DIO data bus. A frequency counter is an example of a talker.

Talker/listeners. There is also a combination device that accepts commands from the controller to set up ranges, tasks, and the like, and then returns data back over the DIO bus to the controller. An example is a digital multimeter (DMM). The controller will send the DMM commands that determine whether it is AC or DC, volts or milliamperes or ohms, and what specific range—the device is thus acting as a listener. When the measurement is made, the DMM becomes a talker and transmits the data measured back over the DIO bus to the controller.

IEEE-488/GPIB Buses

There are three major buses in the IEEE-488/GPIB system: the Data Input Output (DIO) bus, the Data Byte Transfer (DBT) bus, and the General Interface Management (GIM) bus. These buses operate as described in the following paragraphs.

DIO bus. The data input output bus is a bidirectional 8-bit data bus that carries data, interface messages, and device-dependent messages between the controller, talkers, and listeners. This bus sends data asynchronously in byte-serial format.

DBT bus. The data byte transfer bus controls the sending of data along the DIO bus. There are three lines in the DBT bus: DAta Valid (DAV), Not Ready For Data (NRFD), and Not Data ACcepted (NDAC). These signal lines are defined as follows:

DAV. The data valid signal indicates the availability and validity of the data on the line. If the measurement is not finished, for example, the DAV signal will be false.

NRFD. The not ready for data signal lets the controller know whether or not the specific device addressed is in a condition to receive data.

NDAC. The not data accepted signal line is used to indicate to the controller whether or not the device accepted the data sent to it over the DIO bus.

GIM bus. The general interface management bus coordinates the system and ensures an orderly flow of data over the DIO bus; it has the following signals: InterFace Clear (IFC), ATtentioN (ATN), Service ReQuest (SRQ), Remote ENable (REN), and End Or Identify (EOI). These signals are defined as follows:

ATN. The attention signal is used by the controller or computer to let the system know how data on the DIO bus lines is to be interpreted, and which device is to respond to the data.

IFC. The interface clear signal is used by the controller to place all devices in a predefined quiescent or standby condition.

SRQ. The service request signal is used by a device on the system to ask the controller for attention. This signal is essentially an interrupt request.

REN. The remote enable signal is used by the controller to select from between two alternate sources of device programming data.

EOI. The end or identify signal is used by talkers for two purposes. It will follow the end of a multiple-byte sequence of data in order to indicate that the data is now finished, and it is also used in conjunction with the ATN signal for polling the system.

The 7-bit binary signals used in the IEEE-488/GPIB system for ASCII and GPIB message codes are shown in Figure 9-17. The signals are implemented as conductors in a system interface cable. Each IEEE-488/GPIB-compatible instrument will have a female 36-pin Amphenol-style connector on the rear panel. The pin-out definitions are as follows:

Pin No.	Signal Line
1	DIO1
2	DIO2
3	DIO3
4	DIO4
5	EOI
6	DAV
7	NRFD
8	NDAC
9	IFC
10	SRQ
11	ATN
12	shield
13	DIO5
14	DIO6
15	DIO7
16	DIO8
17	REN
18	ground (6)
19	ground (7)
20	ground (8)
21	ground (9)
22	ground (10)
23	ground (11)
24	logic ground

ASCII & GPIB CODE CHART

<table>
<tr><th>B7 B6 B5</th><th>ø ø ø</th><th>ø ø 1</th><th>ø 1 ø</th><th>ø 1 1</th><th>1 ø ø</th><th>1 ø 1</th><th>1 1 ø</th><th>1 1 1</th></tr>
<tr><th>BITS
B4 B3 B2 B1</th><th colspan="2">CONTROL</th><th colspan="2">NUMBERS SYMBOLS</th><th colspan="2">UPPER CASE</th><th colspan="2">LOWER CASE</th></tr>
<tr><td>ø ø ø ø</td><td>0
NUL
0 0</td><td>20
DLE
10 16</td><td>40 0
SP
20 32</td><td>60 16
0
30 48</td><td>100 0
@
40 64</td><td>120 16
P
50 80</td><td>140 0
`
60 96</td><td>160 16
p
70 112</td></tr>
<tr><td>ø ø ø 1</td><td>1 GTL
SOH
1 1</td><td>21 LLO
DC1
11 17</td><td>41 1
!
21 33</td><td>61 17
1
31 49</td><td>101 1
A
41 65</td><td>121 17
Q
51 81</td><td>141 1
a
61 97</td><td>161 17
q
71 113</td></tr>
<tr><td>ø ø 1 ø</td><td>2
STX
2 2</td><td>22
DC2
12 18</td><td>42 2
"
22 34</td><td>62 18
2
32 50</td><td>102 2
B
42 66</td><td>122 18
R
52 82</td><td>142 2
b
62 98</td><td>162 18
r
72 114</td></tr>
<tr><td>ø ø 1 1</td><td>3
ETX
3 3</td><td>23
DC3
13 19</td><td>43 3
#
23 35</td><td>63 19
3
33 51</td><td>103 3
C
43 67</td><td>123 19
S
53 83</td><td>143 3
c
63 99</td><td>163 19
s
73 115</td></tr>
<tr><td>ø 1 ø ø</td><td>4 SDC
EOT
4 4</td><td>24 DCL
DC4
14 20</td><td>44 4
$
24 36</td><td>64 20
4
34 52</td><td>104 4
D
44 68</td><td>124 20
T
54 84</td><td>144 4
d
64 100</td><td>164 20
t
74 116</td></tr>
<tr><td>ø 1 ø 1</td><td>5 PPC
ENQ
5 5</td><td>25 PPU
NAK
15 21</td><td>45 5
%
25 37</td><td>65 21
5
35 53</td><td>105 5
E
45 69</td><td>125 21
U
55 85</td><td>145 5
e
65 101</td><td>165 21
u
75 117</td></tr>
<tr><td>ø 1 1 ø</td><td>6
ACK
6 6</td><td>26
SYN
16 22</td><td>46 6
&
26 38</td><td>66 22
6
36 54</td><td>106 6
F
46 70</td><td>126 22
V
56 86</td><td>146 6
f
66 102</td><td>166 22
v
76 118</td></tr>
<tr><td>ø 1 1 1</td><td>7
BEL
7 7</td><td>27
ETB
17 23</td><td>47 7
'
27 39</td><td>67 23
7
37 55</td><td>107 7
G
47 71</td><td>127 23
W
57 87</td><td>147 7
g
67 103</td><td>167 23
w
77 119</td></tr>
<tr><td>1 ø ø ø</td><td>10 GET
BS
8 8</td><td>30 SPE
CAN
18 24</td><td>50 8
(
28 40</td><td>70 24
8
38 56</td><td>110 8
H
48 72</td><td>130 24
X
58 88</td><td>150 8
h
68 104</td><td>170 24
x
78 120</td></tr>
<tr><td>1 ø ø 1</td><td>11 TCT
HT
9 9</td><td>31 SPD
EM
19 25</td><td>51 9
)
29 41</td><td>71 25
9
39 57</td><td>111 9
I
49 73</td><td>131 25
Y
59 89</td><td>151 9
i
69 105</td><td>171 25
y
79 121</td></tr>
<tr><td>1 ø 1 ø</td><td>12
LF
A 10</td><td>32
SUB
1A 26</td><td>52 10
*
2A 42</td><td>72 26
:
3A 58</td><td>112 10
J
4A 74</td><td>132 26
Z
5A 90</td><td>152 10
j
6A 106</td><td>172 26
z
7A 122</td></tr>
<tr><td>1 ø 1 1</td><td>13
VT
B 11</td><td>33
ESC
1B 27</td><td>53 11
+
2B 43</td><td>73 27
;
3B 59</td><td>113 11
K
4B 75</td><td>133 27
[
5B 91</td><td>153 11
k
6B 107</td><td>173 27
{
7B 123</td></tr>
<tr><td>1 1 ø ø</td><td>14
FF
C 12</td><td>34
FS
1C 28</td><td>54 12
,
2C 44</td><td>74 28
<
3C 60</td><td>114 12
L
4C 76</td><td>134 28
\
5C 92</td><td>154 12
l
6C 108</td><td>174 28
|
7C 124</td></tr>
<tr><td>1 1 ø 1</td><td>15
CR
D 13</td><td>35
GS
1D 29</td><td>55 13
–
2D 45</td><td>75 29
=
3D 61</td><td>115 13
M
4D 77</td><td>135 29
]
5D 93</td><td>155 13
m
6D 109</td><td>175 29
}
7D 125</td></tr>
<tr><td>1 1 1 ø</td><td>16
SO
E 14</td><td>36
RS
1E 30</td><td>56 14
.
2E 46</td><td>76 30
>
3E 62</td><td>116 14
N
4E 78</td><td>136 30
^
5E 94</td><td>156 14
n
6E 110</td><td>176 30
~
7E 126</td></tr>
<tr><td>1 1 1 1</td><td>17
SI
F 15</td><td>37
US
1F 31</td><td>57 15
/
2F 47</td><td>77 UNL
?
3F 63</td><td>117 15
O
4F 79</td><td>137 UNT
_
5F 95</td><td>157 15
o
6F 111</td><td>177
DEL (RUBOUT)
7F 127</td></tr>
<tr><td></td><td>ADDRESSED COMMANDS</td><td>UNIVERSAL COMMANDS</td><td colspan="2">LISTEN ADDRESSES</td><td colspan="2">TALK ADDRESSES</td><td colspan="2">SECONDARY ADDRESSES OR COMMANDS</td></tr>
</table>

KEY

<table>
<tr><td>octal</td><td>25 PPU</td><td>GPIB code</td></tr>
<tr><td></td><td>NAK</td><td>ASCII character</td></tr>
<tr><td>hex</td><td>15 21</td><td>decimal</td></tr>
</table>

Tektronix®
COMMITTED TO EXCELLENCE

REF: ANSI STD X3. 4-1977
IEEE STD 488-1978
ISO STD 646-1973

Figure 9-17 ASCII/GPIB codes (Courtesy Tektronix, Inc.).

The IEEE-488 GPIB can be used to marry various pieces of standard commercial test equipment, computers, and customized equipment into a system that is capable of providing specified tests and measurements under the control of a computer program, often in BASIC. The GPIB makes either special-purpose or generalized automatic test equipment possible with a minimum of effort.

KEYBOARDS AND CUSTOM PANELS

The ASCII keyboard is probably the most common method for talking to a microcomputer. On simpler computers we might find a calculatorlike keyboard that will allow us to input hexadecimal numbers for direct machine language programming or data entry. In both cases we will need to know the methods for interfacing the keyboard and the programming needed to accommodate the keyboard (the computer does not "naturally" understand the keyboard).

Figure 9-18 shows the circuit for a typical non-IBM-PC keyboard. This particular type is for the ASCII code, but the circuit is similar for hexadecimal keyboards (which would use a smaller keyboard X-Y matrix and probably a different IC).

Device IC1 is a keyboard integrated circuit. It is a read-only memory in which the binary codes for the various ASCII codes are stored in address locations that are uniquely accessed by single contact closures on the X-Y matrix.

The actual keyboard is a crosspoint switch in which depressing a given key will short together a unique combination of X and Y terminals on IC1. When a key is depressed, then, the binary code (representing an ASCII character) is output to lines B1 through B7.

The keyboard contains two additional switches, S1 and S2. These switches are for the control (CNTL) and shift (SHFT) functions, respectively.

A strobe pulse or signal is used to let the outside world know when the data is new and valid. In some designs, there will be a trashy signal on the output lines except when a push button is pressed. The computer must know when this data is valid ASCII characters and when it represents nothing more than random noise. In still other cases, the output will be latched, so the computer needs to know when the data is truly new and when it is the old data that has already been input. Two types of strobe signal are used, transient and level (see Figure 9-19). The transient form of strobe pulse is a short-duration pulse that is issued once per keystroke. When the data on the keyboard output lines becomes stable and valid, then a brief pulse is generated. The time between the closure of the switch and the production of the output pulse may be only nanoseconds.

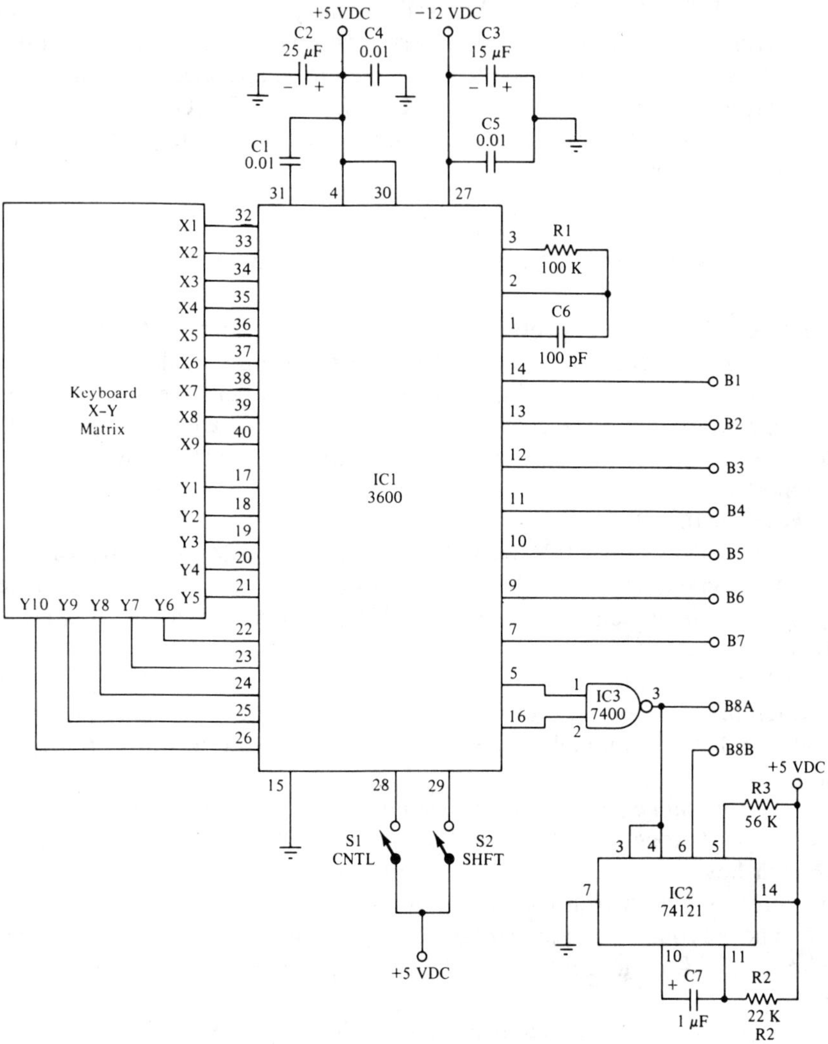

Figure 9-18 X-Y matrix keyboard (non-IBM) interface.

The level form of strobe signal is shown in Figure 9-19A. This signal is a constant voltage level that remains as long as the keyboard operator is depressing a key. The level type of strobe signal is seen in low-cost keyboards and in those for certain special-purpose applications. In Figure 9-18, the level type of strobe signal is available at the output of the NAND gate (IC3), while the transient form is available at the output of IC2 (a one-shot). In an 8-bit microcomputer, we can use the lower-order 7 bits of the data bus

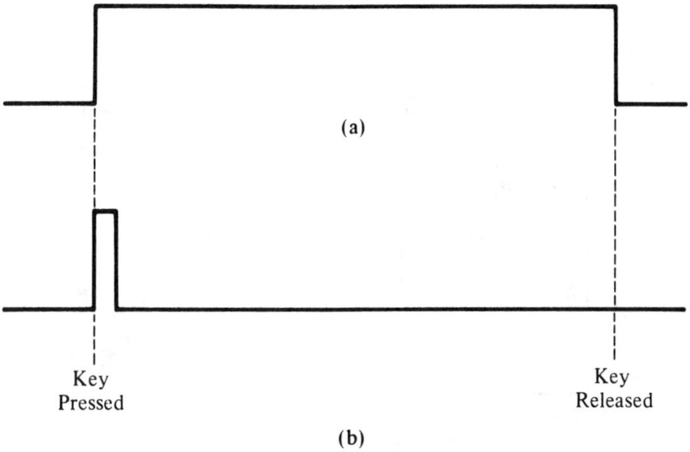

Figure 9-19 A) Continuous strobe, B) Transient strobe.

or input port for the ASCII data and the highest-order bit to carry the strobe signal. In Figure 9-18, the level strobe signal is available as bit B8A, while the transient form is at bit B7B. Of course, in any given application, only one of these will be used.

INTERFACING KEYBOARDS TO I/O PORTS

Most currently available microcomputer keyboards have TTL-compatible output lines. Similarly, most modern microcomputers have TTL-compatible input ports. We can, therefore, directly connect the output lines of the keyboard to the input lines of the computer. Furthermore, the 7 data bits plus strobe bits are easily compatible with the 8-bit format of the typical microcomputer input port. It is merely necessary to connect the lines of the output to the computer inputs.

We must, however, write a simple program to service the keyboard. It is the usual practice of designers to have the microcomputer loop endlessly while looking for a valid strobe bit. An example is shown in Figure 9-20. Here we are inputting the binary word from the input port (i.e., the keyboard data). Bit B7 of the input port (B8 of the keyboard—they are often numbered slightly differently) will be logical 1 (HIGH) when the data is valid. We will, therefore, want to mask all bits but B7 and then test for zero or nonzero. For example, we can AND the input word with 10000000, which will set all bits to zero except B7 when B7 of the input signal is also 1. In a typical program, the computer will branch back to the input instruction on result = 0 and to the program for storing the input data on result = 1.

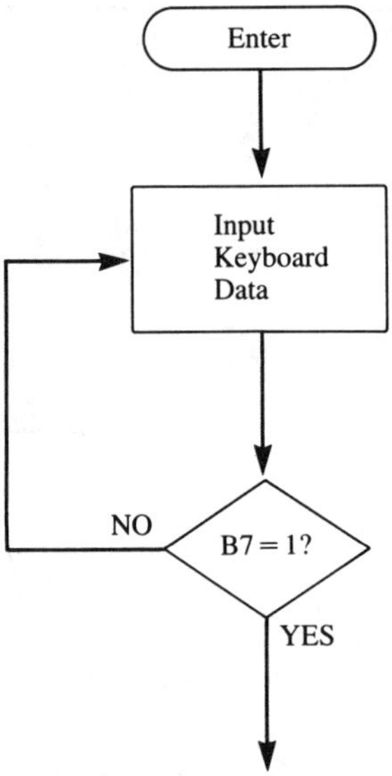

Figure 9-20 Data input software logic.

INTERFACING KEYBOARDS TO THE DATA BUS

The data bus in a microcomputer is used for many different purposes. The use of the bus changes virtually every time a new instruction is either fetched or executed—in other words once every cycle. We must not, therefore, allow any one device to command the data bus all of the time. In the case of a keyboard or similar peripheral we can use a circuit similar to Figure 9-21 to interface to the data bus.

The device used to connect the keyboard to the data bus is a tristate driver. This IC has a third possible output state that exists when pin number 1 is HIGH: high impedance to both +5 V and ground. In this third state the internal circuitry is effectively disconnected from the output pins. Pin number 1 controls the operation of the IC. When pin number 1 is HIGH, then the device floats across the data bus at high impedance, so it cannot affect the bus. If, on the other hand, the level applied to pin number 1 is LOW, then the outputs of the internal buffer amplifiers are connected to the data bus. If the IN signal drops LOW when the keyboard data is valid, then data will be input to the computer.

Notice that we show two alternate methods for connecting the strobe signal to the computer. In one case, we would apply the strobe to the highest-order bit of the 8212 input port device. In that case, we would write a loop program (as described earlier) and hope that it works properly. But there is also another method that is not so wasteful of CPU time. We would allow the microcomputer to perform other chores unless the strobe signal causes an interrupt. This is accomplished by connecting the strobe output of

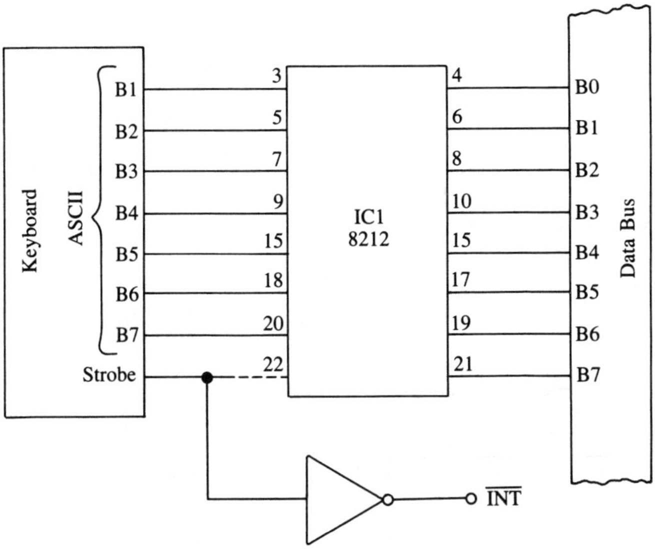

Figure 9-21 Keyboard interfacing via 8212 tristate logic chip.

the keyboard to the interrupt line of the computer (inverting polarity is sometimes needed). When the computer senses that the interrupt line is LOW, it will stop executing the program that is in progress and jump to a subroutine specified by the interrupt. In this case, the subroutine would have to be a keyboard input program.

In an earlier chapter we discussed the generation of device-select pulses and decoder circuits. Let's review how the IN and OUT signals are generated. Interfacing the keyboard to the data bus required an IN signal, i.e., a signal that will drop LOW when the computer wants to input keyboard data. This signal is shown for the Z80-based machines in Figure 9-22. For the Z80 there are three conditions that must be met before an input operation takes place: (1) the I/O request line (IORQ) goes LOW, (2) the read line (RD) goes LOW, and (3) the correct input port address is present on the lower 8 bits of the address bus.

In the circuit of Figure 9-22 the 7430 device is used as the address decoder; its output will drop LOW only when the correct address (in this example, 11010011) is present on the address bus. A 7442 BCD-to-1-of-10 decoder is used to detect when all three conditions are simultaneously present. This device will examine a BCD word applied to its inputs and then cause the correct decimal output to drop LOW. The way we have the 7442 connected requires that the binary word 0100 be applied to the BCD inputs before pin number 5 (i.e., the "4" output) goes LOW. When that occurs we have our IN signal.

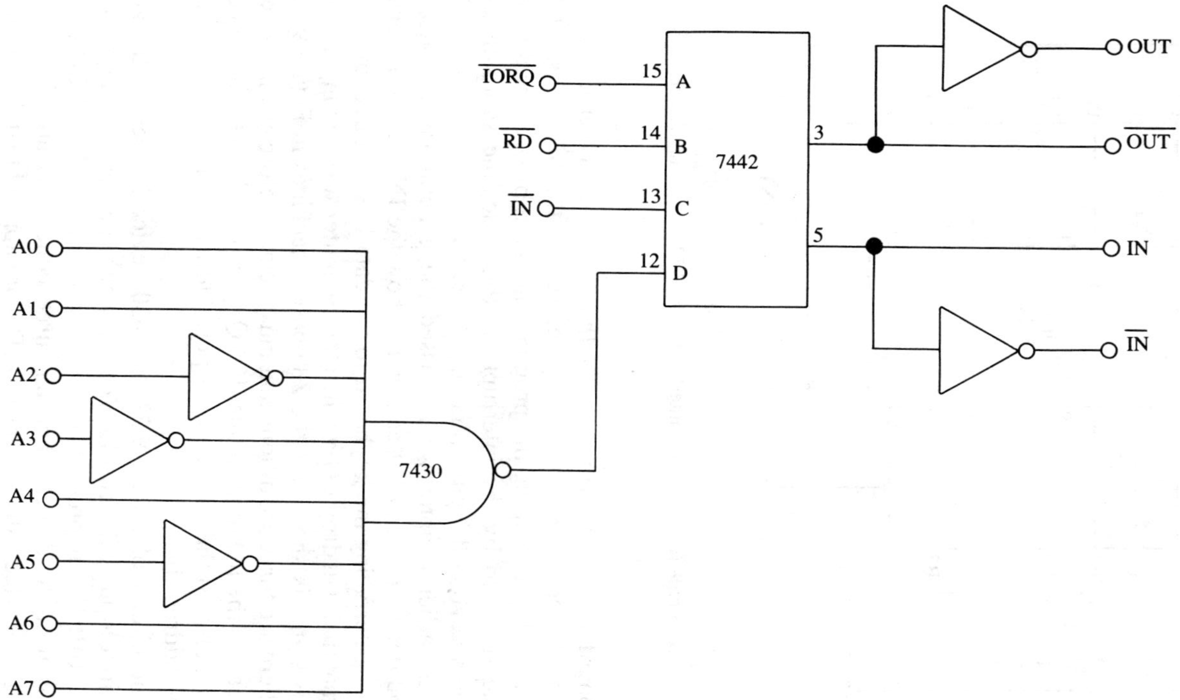

Figure 9-22 7430/7442 address/operation decoder circuit.

INTERFACING PUSH BUTTONS TO THE MICROCOMPUTER

Push buttons might be interfaced to a microcomputer or microprocessor for any of several different reasons. Among them might be a special-purpose or limited-use keyboard. We find several different kinds of electrical switch in electronic instruments. We know the different forms (SPST, SPDT, DPST, DPDT, and so forth). We also find alternate action switches. These will perform one operation on the first press and then the opposite operation on the second press. For example, we might find an SPST "AA" switch that will close on the first press and then open on the second press. All of these switches can be simulated using simple switches and some software in a microcomputer.

Consider a simple situation such as that shown in Figure 9-23. The actual switch is a simple SPST type and may be either a toggle switch or a push button (it can also be an electronic switch). There is a pull-up resistor between the junction of the switch and port bit B0 and the +5-V line. This means that B0 will be HIGH when S1 is open and LOW when S1 is closed. Let's say that the desired action is to have the computer write a HIGH to bit B0 of output port 1 whenever the switch is closed (i.e., B0 of input port 1 is LOW). We would write a short program that would continuously monitor port 1 and determine when B0 is LOW. It would then write a data word with B0 HIGH (i.e., 01H, 07H, FFH, etc.) to output port 1. Of course, it is not necessary that the program perform this specific operation. Just as often, the computer would use the data provided for some other purpose. It is common on microprocessor-controlled instruments to use one input port to inform the computer of certain desired conditions. In that case, all 8 bits of the selected input port might have a switch connected. The operator could then program the computer for whatever conditions are existing or, possibly, desired.

Figure 9-24 shows two methods for interfacing switches to the data bus of the computer without the need for existing input ports. Of course, this operation requires us to make an impromptu input port for the switch involved.

The circuit in Figure 9-24A shows how to interface with a single switch. We require an input buffer (either inverting or noninverting will suffice, but changes will be required in the software) that has a tristate output terminal; the 74125 and 74126 are examples. In addition, the chip enable (CE) terminals must be independent of other sections of the chip. When the CE terminal is brought LOW by the generation of an IN1 device-select pulse, the data at the input of the tristate buffer is transferred to the output, hence to the data bus of the microcomputer. We could use either of two approaches shown for connecting a switch to the input of the buffer. In the case of S1, we use the same sort of circuit as in Figure 9-23; the pull-up resistor is needed for S1 but not if the alternate method (using the flip-flop) is

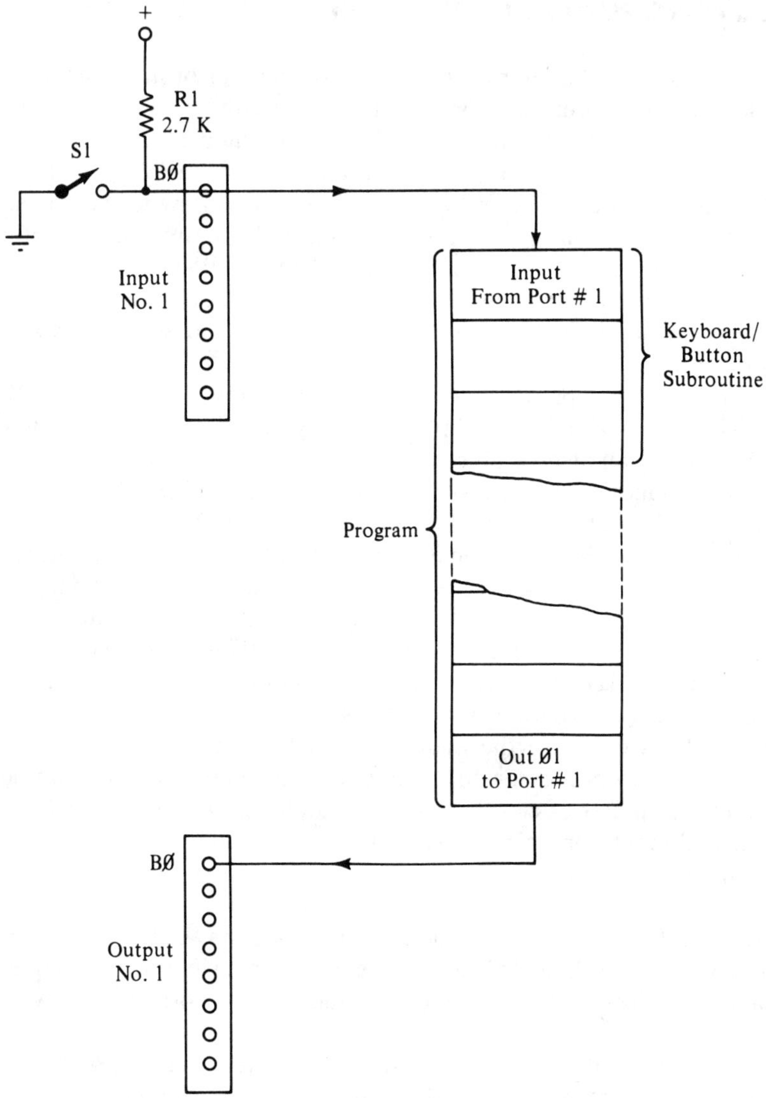

Figure 9-23 Custom switch and program operation.

used. Alternatively, we could use a type-D flip-flop to hold the data after the operator releases the button. We would want to use this method when (a) the computer does not cycle back to the input port containing the switch fast enough to not miss an operation, or (b) when the computer will periodically interrogate the switch to find if some preset condition still exists. In some cases, for example, the intent will be to have the computer continuously

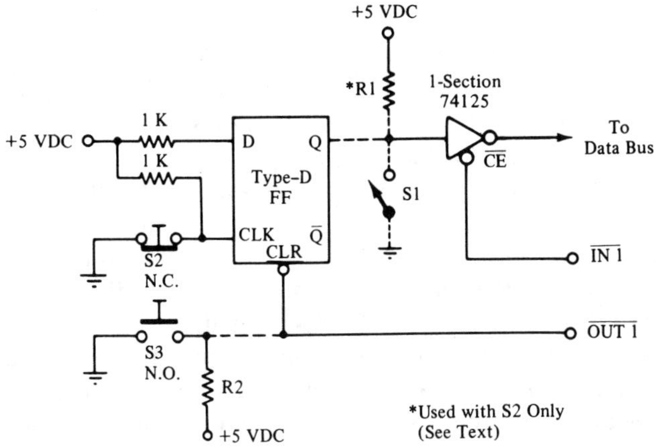

(a)

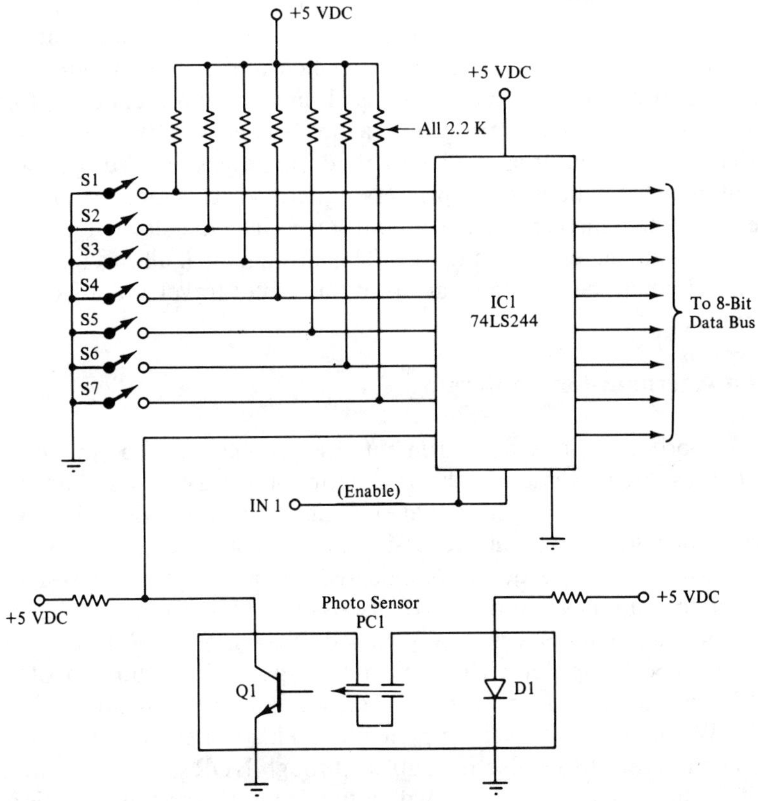

(b)

Figure 9-24 A) Switch circuit, B) Interfacing switches to data bus.

execute the program until someone resets the output of the flip-flop. In the case shown, we can reset the flip-flop either by a software command (i.e., execute a dummy output to generate an OUT1 signal) or by depressing a hardware push button (S3 in Figure 9-11).

The other circuit is shown in Figure 9-24B. In this case, we are doing essentially the same thing as in Figure 9-23, except that we are required to create our own input port. The switches are connected to the input of an 8-bit noninverting buffer that has tristate outputs. When the two strobe lines are brought HIGH by generation of the IN1 signal, the data on the respective input lines are transferred to the output lines, hence to the 8-bit data bus of the microcomputer. The software of the computer will then sort out the respective meanings of the various open and closed conditions for the switches.

Note that this circuit also includes an electronic photo switch. These devices are often used in mechanical instruments to indicate some external condition. The photo switch consists of an LED and a phototransistor. These components are mounted inside of a light-tight housing that has an optical path that turns on the transistor when the LED is illuminated. In some, the entire photo switch is enclosed, and these are called optoisolators. In still other cases, there will be a slot between the transistor and the LED to all or some external device to blind the transistor. We see this type of circuit in instruments such as printers. There will be a metal or plastic tab on the print-head carriage that will be inserted into the slot when the head carriage reaches the end of its travel. This arrangement allows us to generate a HIGH when the head carriage is at the end of travel and a LOW at all other times.

An Alternate-Action Circuit

Microprocessor-based instruments sometimes require an alternate-action (AA) switch. Sometime this is a matter of necessity; sometimes it is a matter of convenience or front panel design. An AA switch is one that performs opposite actions on successive closures of the contacts. Mechanical AA switches can be quite complicated when compared with a simple SPST switch. In this section we will discuss a method for making an SPDT AA switch using a simple, normally closed (NC) SPST push button switch and a dual type-D flip-flop. The typical TL type-D FF requires a LOW on the CLR line for the FF to be reset (i.e., make Q = LOW and NOT-Q = HIGH). When the power is first applied to the instrument, a power-on reset pulse is generated; this pulse is coupled through NOR gate G1 to the CLR lines of the flip-flops. This action will initialize the FFs to the condition that makes output A = LOW and output *A* = HIGH.

Recall the action of the type-D flip-flop. When the clock input is made HIGH, the data on the D input will be transferred to the Q output. In the circuit of Figure 9-25A, the D input of FF1 is connected permanently HIGH. The normally closed push button is connected to the CLK input, along with a pull-up resistor to +5 V DC. When the button is pressed, the switch contacts open, and this brings the CLK input HIGH, thereby trans-

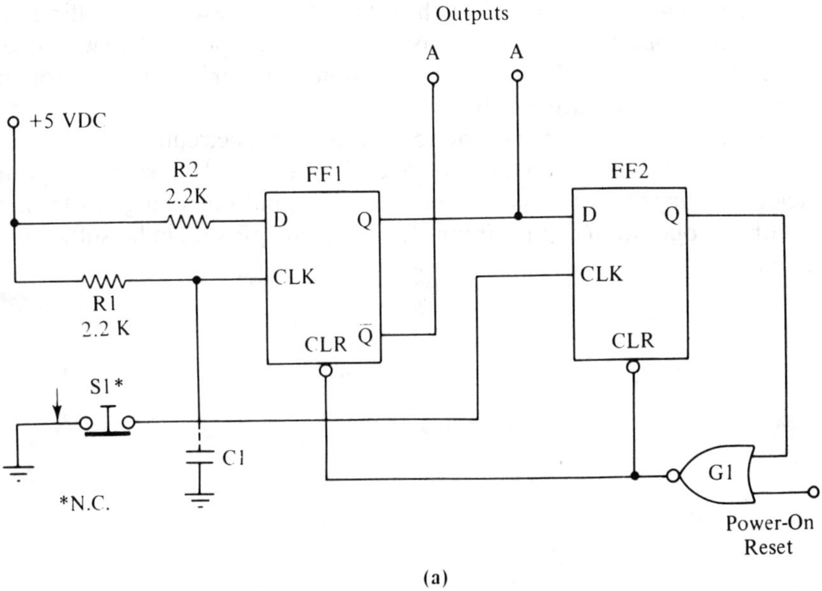

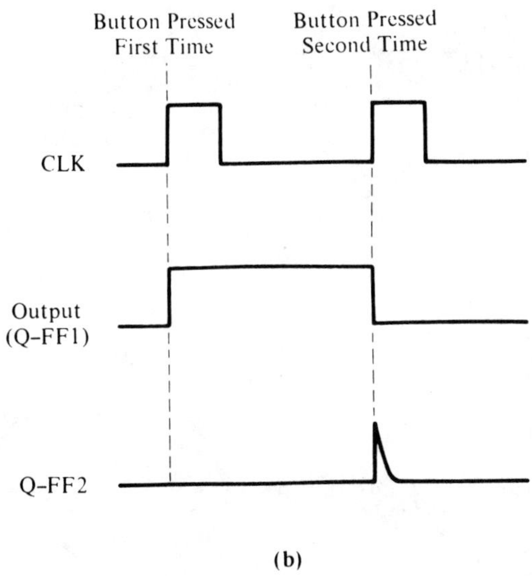

Figure 9-25 Alternate action switch A) Circuit, B) Timing.

ferring the HIGH on the D input to the Q output. At this time, the A output of the circuit is made HIGH and the *A* output of the circuit is LOW. When the button is pressed a second time, we find that the D input of FF2 is now HIGH (it was LOW before), so the Q output of FF2 goes HIGH. This level is coupled back to the CLR line of FF1 and FF2 through gate G1 (the same as for the power-on reset pulse). This has the effect of resetting the flip-flops, making both Q outputs LOW. At this time, the A output of the switch goes LOW and the *A* goes HIGH; alternate action! The timing diagram for this circuit is shown in Figure 9-25B.

Notice that an optional capacitor is used in this circuit. Its purpose is to suppress push button contact bounce. Capacitor C1 must have a value sufficient to suppress the contact bounce pulses, but not so high as to interfere with the operation of the circuit; less than 500 pF should be sufficient in most cases.

10

INTERFACE SOFTWARE METHODS

The art of interfacing with microcomputers and microprocessors is essentially an exercise in input/output strategies. Although memory interfacing chores have largely the same requirements as do other types of interfacing, most microcomputers come with either a full set of memory or the provision to easily add extra memory. We discussed I/O hardware earlier, so we will now confine our discussion to I/O software methods.

Two different situations may be faced by various readers. In one case, it will be necessary to generate a device-select pulse to select an I/O port or some peripheral device that connects to the data bus. In other cases, it might be necessary to execute a simple I/O operation and then do something with the data besides leaving it in the accumulator.

We find that there are three different strategies for handling interface input. In all three, we assume that an outside device is either turned on or off by the computer or wants to send data to or receive data from the computer. One method requires the microcomputer to continuously poll an I/O port looking for new data. The second method calls for the microcomputer to periodically poll an input port for new data. During the rest of the time it is free to perform other chores. Finally, we have the interrupt method. In this type of operation, the CPU executes the main program until an external device activates the CPU interrupt line. After the interrupt signal is received, the CPU will complete the operation currently being executed and then jump to an interrupt subroutine. Later in this chapter we will examine interrupt functions and hardware to facilitate the interrupt capabilities of the microprocessor; the Z80 will be used as the example, but other microprocessors have similar functions.

GENERATING TIMING LOOPS

Unless a system has a built-in hardware timer (many do), it may sometimes be necessary to generate timing loops in software. There are several instructions in both 6502 and Z80 machines that will facilitate this type of operation. But before examining actual microprocessor instructions, let's consider the overall software strategy.

Figures 10-1 and 10-2 give flow diagrams for typical timing loop subroutines. It is assumed that the microprocessor contains X and Y index registers, although the technique will work on any register or memory location that can be either decremented or incremented (the former is preferred) by software instructions. In fact, there is one instruction in the Z80 repertoire that makes it desirable on that chip to use the B register for timing subroutines.

The subroutines in Figures 10-1 and 10-2 both depend upon the system clock to set the time duration. Every subroutine requires a certain number of clock cycles to execute. A typical subroutine for Figure 10-1 in 6502 language may require five clock cycles. At a system clock rate of 1 MHz, each cycle will require one microsecond (1 μs), so the time to complete each loop is 5 μs.

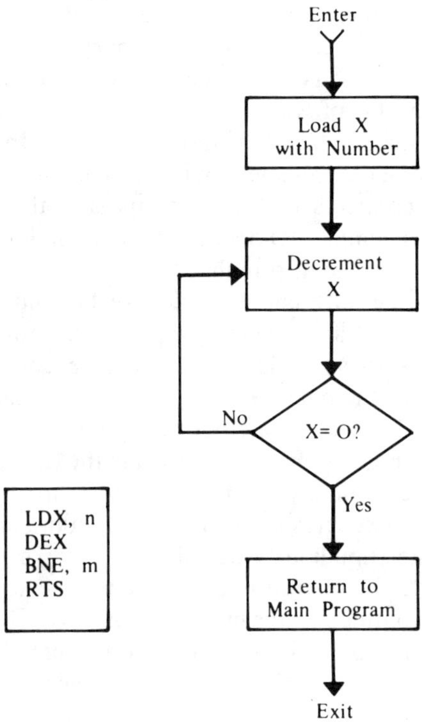

Figure 10-1 Program flowchart.

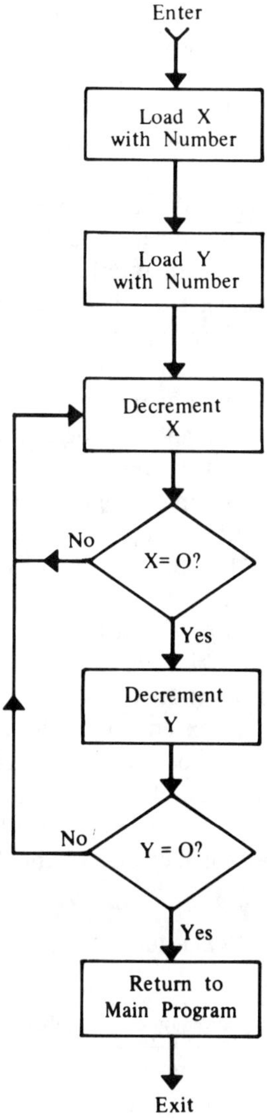

Figure 10-2 Testing subroutine.

The basic technique is to load index register X (or whichever location is selected) with the number of times the loop must be exercised, less the time required to enter and exit the subroutine (i.e., JSP, LDX, and RTS instructions) to form the desired duration. The X register is then decremented and tested for the condition X = 0. As long as X > 0, the program will branch backward and decrement X one more time. When the required number of iterations has expired, X will be zero and the program can exit the loop to return to the main program.

Using the 6502 example, we know that each loop requires 5 μs. In addition, there is the overhead of 8 μs internally and 6 μs for the JSR instruction. Suppose we want to generate a 0.5-ms (i.e., 500-μs) time delay. We have an 8-μs overhead, so the loop time is (500 − 8) or 492 μs. To find the number of loops required, we must divide the total loop time by the time required to execute each loop: (492 μs/5μs) = 98_{10} executions of the loop. Since 98_{10} = 52H (hexadecimal), we will load 52H into the X register.

The instructions required for execution of the timing loop in 6502 language are LDX,n (load X with number n), DEX (decrement X), BNE,aa (branch forward or backward by displacement as on result equal to nonzero), and, of course, RTS (return from subroutine).

Let's look at a typical 6502 program to generate a 500-μs time delay. We will locate the subroutine at memory location 0F00H. To call this subroutine, thereby creating our 500-μs delay, we would use the JSR 0F00H (jump to subroutine at location 0F00H) instruction.

Table 10-1 shows a sample program in 6502 language This subroutine can be changed for any time delay from 13 to 1275 μs (i.e., 1.275 ms) by changing the index stored in the X register to a hexadecimal number from 01H to FFH, respectively.

A Z80 implementation of Figure 10-1 may well use the B register, because there is a single 3-bit instruction that handles all of step 2: DJNZ (decrement B and jump on nonzero).

Figure 10-2 shows an extension of the concept of Figure 10-1. In this example, both X and Y registers are used such that the loop containing X is nested within the loop containing Y. Like the previous example, a minimum time delay is associated with the overhead. In this case, the internal overhead is 20 μs, with a 6-μs offset due to the JSR instruction (which affects

TABLE 10-1 Program Listing

Step No.	Memory Location	Mnemonic	Code Number	Cycles	Remarks
1	0F00H	LDX,52H	A2	2	
	0F01H	(data)	62	—	
2	0F02	DEX	CA	2	Decrement X register
	0F03	BNE	D0	2	Branch to 0F03 on X not equal 0
	0F04	(data)	FD	—	Two's complement of −3
3	0F05	RST	60	6	Return to main program

both Figures 10-1 and 10-2), for a total overhead of 26 μs. The minimum duration occurs with 01H loaded into both X and Y registers. As with the previous example, the minimum resolution is 5 μs (the one least significant bit—1 LSB-value of X). Thus we can generate timing delays of 26 μs (X = Y = 01H) to 325,125 μs (X = Y = FFH).

GENERATING SOFTWARE PERIPHERAL/DEVICE-SELECT PULSES

In Chapter 6 we discussed hardware-generated device-select pulses. These pulses are generated by combining certain control and address signals to turn on I/O ports. But we also sometimes use bits of existing I/O ports to provide device selection, which is the subject of this section.

Figure 10-3A shows a hypothetical situation in which up to eight peripherals can be signaled without the need for decoding. A single output port, which we see here as memory-mapped at location A003, serves the signaling function. All bits of port 2 are normally kept LOW (i.e., 00H), so the bit line connected to the desired device is brought HIGH for a time T when that device is commanded by the computer to turn on. In the example shown, peripheral 4 is connected to bit B3 of part 2, so to turn on only that one device we would write 00001000_2 (or 08H) to location A003.

Note that adding 8-bit decoding to each peripheral will permit us to uniquely address up to 255 devices with one all-off state, or 256 devices if there is no all-off state. If we designate 00H as the all-off signal, the 255 codes from 01H to FFH will each be available to turn on one device. Such an application, incidentally, requires all 8 bits of the output port to be connected to all peripherals and probably necessitates using a high-power 8-bit line driver or buffer for the output port.

A typical program flow for this application is shown in Figure 10-3B. We are assuming here that output port 2 is memory-mapped to location A003, and its companion input port 2 is at A004. The input port receives the data that the peripheral wants to deliver. If there is no need for data from the peripheral, input port 2 is not used. Such an application might make use of the computer to turn on lights or certain other devices.

The program also assumes that a time-delay subroutine (Table 10-1) is stored at location 0F00H. Peripheral 3 requires a turn-on pulse of not less than 250 μs. If the time-delay routine loads the X register with 32H, the time delay will be 264 μs, providing a margin of error. (Note: 32H = 50_{10}; $50_{10} \times 5$ μs = 250 μs; 250 μs + 6 μs for JSR + 8 μs internal delay yields 264 μs.)

When executing the main program, the computer comes across the device-select segment at location 0300H. The first instruction (LDA #08H) loads into the accumulator the binary number 00001000_2, or 08H. This number will form the bit pattern on part 2 that will make B3 HIGH (turning on device 4) and all others LOW. The following instruction (STA A003) stores #08H from the accumulator in location A003, which is the memory location allocated to port 2. The program will then jump to the 264-μs subroutine located at 0F00H. Since port 2 is a latching-type circuit, bit B3 remains HIGH for 264 μs. When program control returns to the main program at location 0306H, the instruction sequence requires #00H to be loaded into port 2 at A003. This sequence returns all bits of port 2 to LOW.

If the external device is to input data to the computer, a simple instruction sequence must be followed. In our example of Figure 10-3B, we will input data from port 2 at A004 and then store it at location 0500H.

In some cases, the program will, like the example, operate asynchronously. This protocol assumes that the data at the peripheral will be ready and valid at the end of the 264-μs period of the device-select pulse. If this is

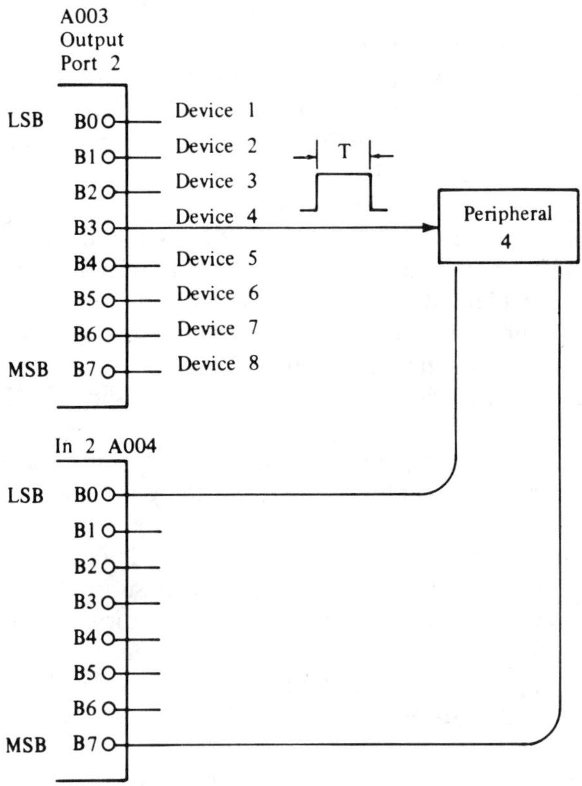

(a)

Figure 10-3 A) Interface to peripheral.

Main Program

Memory Location	Instruction	Comment
0300	LDA,n	
0301	#08H	
0302	STA (mm)	
0303	03H	
0304	A0H	
0305	JSR (mm)	
0306	00H	
0307	0FH	
0308	LDA,n	
0309	#00H	
030A	STA (nn)	
030B	03H	
030C	A0H	
030D	LDA, (nn)	
030E	04H	
030F	A0H	
0310	STA (00)	
0311	00H	
0312	00H	
0313	*	
0314	*	

Location	Intruction
0F00	8*
0F01	*
0F02	*
0F03	*
0F04	RTS

(b)

B) Program flow.

not the case, some sort of scheme must be provided to have the CPU wait until the peripheral indicates that it is ready to transmit data. Such schemes are sometimes called *handshaking routines*. In the simplest case, the computer just loops, doing nothing until a data ready signal is received. The computer will then input the data and, sometimes, send a data received signal back to the peripheral.

SAMPLE KEYBOARD ROUTINE

Most microcomputers use ASCII encoded keyboards, which use 7 bits to represent 128 different alphanumeric symbols and control signals. We can apply the seven parallel ASCII lines to the lower 7 bits of a computer input port.

The 8th bit of the input port is reserved for the strobe signal generated by the keyboard. This bit is a data ready signal that tells the computer that the data on B0 to B6 are valid. Prior to an active strobe being received, the keyboard data would be either 00H, 7FH, or trash (the usual case), depending upon design.

In the example of Figure 10-4, it is assumed that the strobe signal is active-HIGH. The program inputs data from the port located at A004 and then stores it in memory (this is done to save the data). We must then test the data (which are still in the accumulator) to determine if the most significant bit (B7) is 1 or 0.

The strategy for ascertaining the state of B7 is to perform a logical-AND operation between the data in the accumulator and a binary number

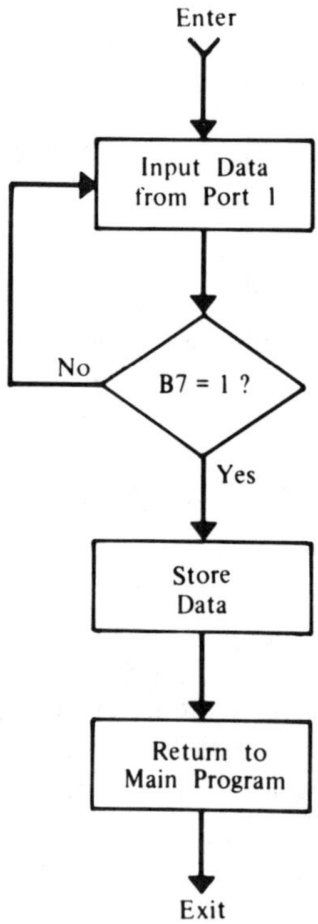


```
LDA (A004)
STA 0050H
AND # 80H
BEQ
RTS
```

Figure 10-4 Input routine.

that will yield a 1 only when the tested bit is also a 1. Let's review the rules for logical AND:

0 AND 0 = 0
1 AND 0 = 0
0 AND 1 = 0
1 AND 1 = 1

If we wish to mask an 8-bit word stored in the accumulator to find B7 = 1, we would AND the accumulator with 10000000_2 (i.e., 80H). If B7 is LOW (i.e., 0), the AND instruction will yield 00000000_2 (00H), no matter what state the other bits take on. If the result is 00H, therefore, we will branch back to the instruction that inputs data from A004. The program will loop through this program until a valid word is received, as indicated by B7 = 1. In that case, the program will exit the keyboard subroutine and return to the main program with the keyboard data stored in 0050H (or whatever other location that the programmer selected; there is nothing special about 0050H).

INTERRUPTS

One very useful feature of most computers is interrupt capability. An interrupt permits the CPU to occupy itself with other more profitable chores than looping while some sluggish peripheral makes up its mind to transmit data. The interrupt capability may also be used for alarms and other applications. In other words, an interrupt is a process in which a computer stops executing the main program and begins executing another program located somewhere else in memory. This is not a mere "jump" or "call" operation, but a response to an external stimulus.

There are several reasons an interrupt capability may be required. One of these is the case of an alarm condition. We could, for example, use a computer in an environmental control system and use the interrupt capability to allow response to alarm situations (e.g., smoke detector, liquid level, burglar alarm, overtemperature). The computer would ordinarily go about some other chore, perhaps the business of controlling the system. But once during the execution of each instruction of the program, the CPU will interrogate the interrupt system. It is thus monitoring the alarm status while executing some unrelated program. When an interrupt is received, indicating an alarm status, the computer would jump immediately to the program that services the interrupt—it would ring a bell, call the fire department, turn on a light, or the like.

Another application is to input data that occur only occasionally or whose periodicity is so long as to force the computer to do nothing for an inordinate amount of time. A real-time clock, or timer, for example, might want to update its input to the computer only once per second or once per minute. An analog-to-digital converter (ADC) might have a 20-ms conversion time. Even the now obsolete slower version of the Z80 CPU chip can perform hundreds of thousands of operations while waiting for the ADC to complete its conversion—waiting for those data would be a tremendous waste of CPU time.

Another use is to input or output data to or from a peripheral device such as a line printer, teletypewriter, keyboard, or terminal. These electromechanical devices are notoriously slow to operate. Even so-called "high-speed" line printers are considerably slower than the CPU.

There are at least two ways to handle this situation, and both involve having the peripheral device signal the CPU when it is ready to accept another character. This is done by using a strobe pulse from the peripheral, issued when it is ready to receive (or deliver) another data byte. One way to handle this problem is to have the programmer write a periodic poll of the peripheral. The strobe pulse is applied to 1 bit of an input port. A program is written that periodically examines that bit to see if it is HIGH. If it is found to be HIGH, the program control will jump to a subroutine that services the peripheral. But this approach is still wasteful of CPU time and places undue constraint on the programmer's freedom.

A superior method is to use the computer's interrupt capability. The peripheral strobe pulse becomes an interrupt request. When the CPU recognizes the interrupt request, it transfers control to an interrupt service subroutine (i.e., a program that performs some function required for the operation of the peripheral that generates the interrupt). When the service program is completed, control is transferred back to the main program at the point where it left off. Note that the CPU does not recognize an interrupt request until after it has finished executing the current instruction in the main program that would have been executed had no interrupt occurred.

Types of Interrupt

There are two types of interrupt recognized by the CPU: nonmaskable and maskable. The nonmaskable interrupt is executed next in sequence regardless of any other considerations. Maskable interrupts, however, depend upon the condition of an interrupt flip-flop inside of the microprocessor. If the programmer wishes to mask (i.e., ignore) an interrupt, the appropriate flip-flop is turned off. There are three distinct forms of maskable interrupt in the CPU, and these take the designations mode 0, mode 1, and mode 2.

There are two interrupt input terminals on the Z80 chip. The @NMI (pin 17) is for the nonmaskable interrupt, while the @INT is for the maskable interrupt.

The nonmaskable interrupt (@NMI) is much like a restart instruction, except that it automatically causes program control to jump to memory location 00 66 (hex), instead of to one of the eight standard restart addresses. Location 00 66 (hex) must be reserved by the programmer for some instruction in the interrupt service program, very often an unconditional jump to some other location higher in memory.

The mode 0 maskable interrupt causes the Z80 to pretend that it is an 8080A, preserving some of the software compatibility between the two CPUs. During a mode 0 interrupt, the interrupting device places any valid instruction on the CPU data bus and the CPU executes this instruction. The time of execution will be the normal time period for that type of instruction, plus two clock pulses. In most cases, the interrupting device will place a restart instruction on the data bus, because all of these are 1-byte instructions. The restart instructions transfer program control to one of eight page 0 (i.e., locations from 00 00 hex) locations.

Any time that a @RESET pulse is applied (i.e., pin 26 of the Z80 is brought LOW), the CPU automatically goes to the mode 0 condition. This interrupt mode, like the other two maskable interrupt modes, can be set from software by executing the appropriate instruction (in this case, an IM0 instruction).

The mode 1 interrupt is selected by execution of an IM1 instruction. Mode 1 is totally under software control and cannot be accessed by using a hardware action. Once set, the mode 1 interrupt is actuated by bringing the @INT line LOW momentarily. In mode 1, the Z80 will execute a restart to location 00 38 (hex).

The mode 2 interrupt is, perhaps, the most powerful of the Z80 interrupts. It allows an indirect call to any location in memory. The 8080A device (and the Z80 operating in mode 0) permits only eight interrupt lines. But in mode 2, the Z80 can respond to as many as 128 different interrupt lines.

Mode 2 interrupts are said to be vectored, because they can be made to jump to any location in the 65,536 bytes of memory.

INTERRUPT HARDWARE

In this section we will discuss some of the circuitry needed to support the Z80 interrupt capability. Note that the primary emphasis will be on low-cost circuits not necessarily intended originally for use with the Z80.

Interrupt Requests

In the simplest cases, interrupt request lines can be built simply by extending the @INT and/or @NMI lines to the peripheral device. This assumes a very simple arrangement in which only one peripheral is to be

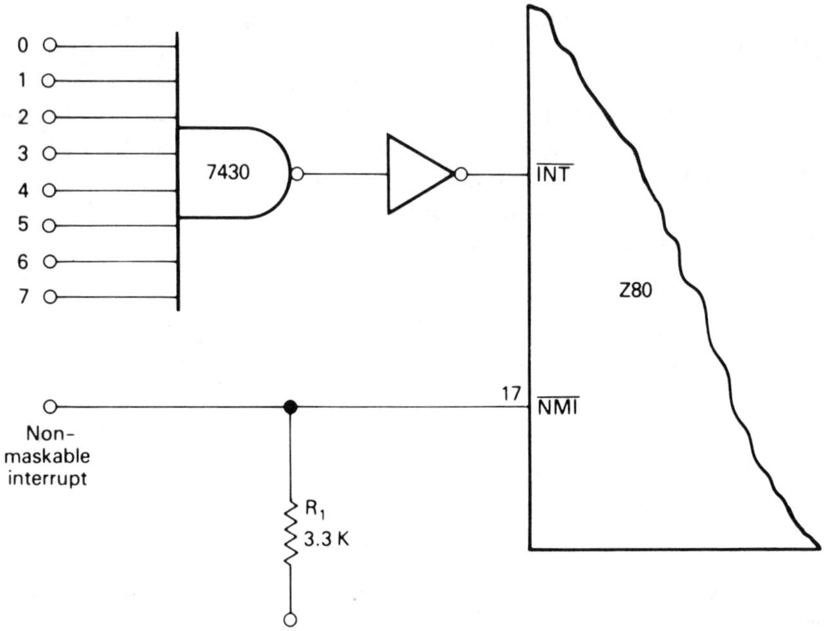

Figure 10-5 Interrupt decoder.

serviced. Figure 10-5 shows how this might be accomplished. The @NMI line (pin 17) is brought out as a nonmaskable interrupt line. The optional pull-up resistor (R1) is used to ensure that pin 17 remains at the HIGH condition and thereby helps reduce noise response.

The @INT line can be treated in exactly the same manner if there is to be but one interrupting peripheral. But in this case, we have demonstrated how the same pin might be used to recognize up to eight interrupts. This arrangement can be used if only mode 0 is anticipated. The peripheral that generates the interrupt then places the correct restart instruction on the data bus. The specific restart instruction received tells the CPU which peripheral initiated the interrupt. The key to this @INT circuit is the eight-input TTL NAND gate (i.e., a 7430 IC). If any one of its inputs, which form @INT request lines, goes LOW, the 7430 output goes HIGH. This forces the output of the inverter LOW, which creates the needed @INT signal at pin 16 of the Z80.

Interrupt Acknowledge

The CPU will always finish executing the current instruction before recognizing an interrupt request. There is, therefore, a slight delay between the initial request and the time when the CPU is ready to process that request. We need some type of signal to tell the peripheral that generated the interrupt request when the CPU is ready to do business. The Z80 samples the @INT line on the rising edge of the last clock pulse of the current instruction. If the @INT line is LOW, the CPU responds by generating an IORQ (input/output request) signal during the next M1 machine cycle. We can, then, accept simultaneous existence of LOW conditions on @IORQ or @M1 to form the interrupt acknowledge signal.

Figure 10-6A shows an interrupt acknowledge scheme that works for a single interrupt line. We assume that one of the interrupt request schemes of Figure 10-5 is also used. The 74125 (IC1/IC2) is a quad, tristate, TTL buffer. Each 74125 contains four noninverting buffer amplifiers that accept TTL inputs and provide TTL outputs. When a control line is HIGH, the associated buffer output will float in the high-impedance tristate mode. But if the control line is brought LOW, the buffer turns on and operates like any other TTL buffer. The control lines for all eight tristate buffers (four from each 74125) are tied together to form a single enable line. The 74125 devices are located inside the peripheral device.

The particular restart instruction designated to service a particular peripheral must be programmed onto the inputs of the 74125s. For example, if we want the peripheral to cause a jump to the RST 10 location (i.e., memory location 00 10), we must place D7 (hex), or 11010111 (binary), on the data bus following the acknowledgment of the interrupt request. We program this value by setting the D0, D1, D2, D4, D6, and D7 inputs of IC1/IC2 to HIGH (binary 1) and the D3 and D5 inputs to LOW (binary 0). This enable line is connected to the inverted output of the NOR gate that detects the interrupt acknowledge condition (i.e., the simultaneous LOW on @IORQ and @M1). The enable line ordinarily remains HIGH, causing the 74125 outputs to float at high impedance. When the brief interrupt acknowledge pulse comes along, this line momentarily drops LOW, thereby transferring the word (D7 hex) at the 74125 inputs to the data bus. The CPU will decode this instruction and perform a restart jump to 00 10 (hex).

Although there is a practical limit to how many tristate outputs one can easily float across the data bus, we find it quite easy to connect all eight allowed in mode 0, and a few more. But how do we differentiate between the peripherals? All will generate the same interrupt request, and these can be handled by using a multi-input NAND gate (see Figure 10-5 again). How do we decode the restart instruction given and then send the interrupt acknowledgment to only the correct peripheral? Chaos would result if we sent the signal to all eight (or more) peripherals at the same time. It is very often used to examine the range of possible binary words that are to be used in any given situation. For the mode 0 interrupt, we are going to use one of eight restart locations, each having its own unique RST op-code. These are listed in Table 10-2. Note that, for all possible states, only three bits change: D3, D4, and D5. The other bits (D0, D1, D2, D6, and D7) remain constant in all cases (in this particular example, they are all HIGH, but the important thing is that they remain at one level in all cases). We can, then, press the 7442 1-of-10 decoder into service once again (see Figure 10-6B). Recall that the 7442 is a 4-bit BCD-to-1-of-10 decoder. The BCD inputs are weighted 1-2-4-8. The 1-2-4 inputs are connected to the D3-D4-D5 lines of the data bus. The 8 line of the 7442 is used as the control line and is connected to the interrupt acknowledge signal.

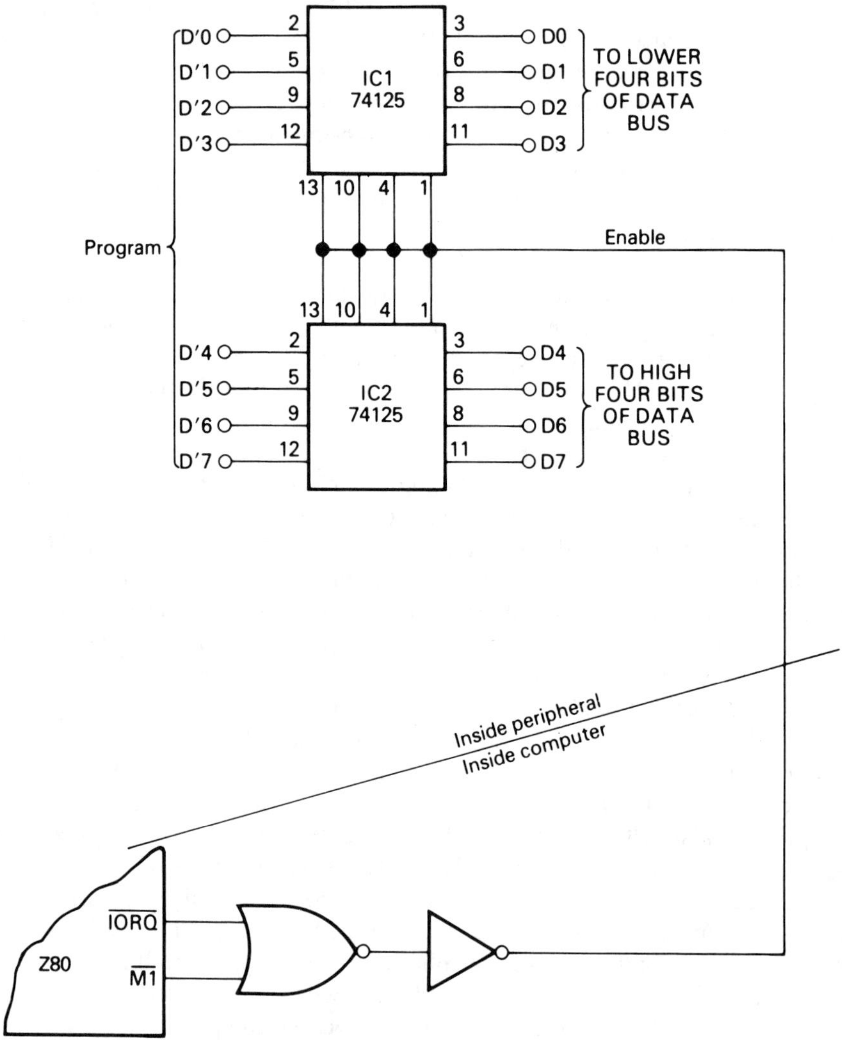

(a)

Figure 10-6 A) Peripheral interface.

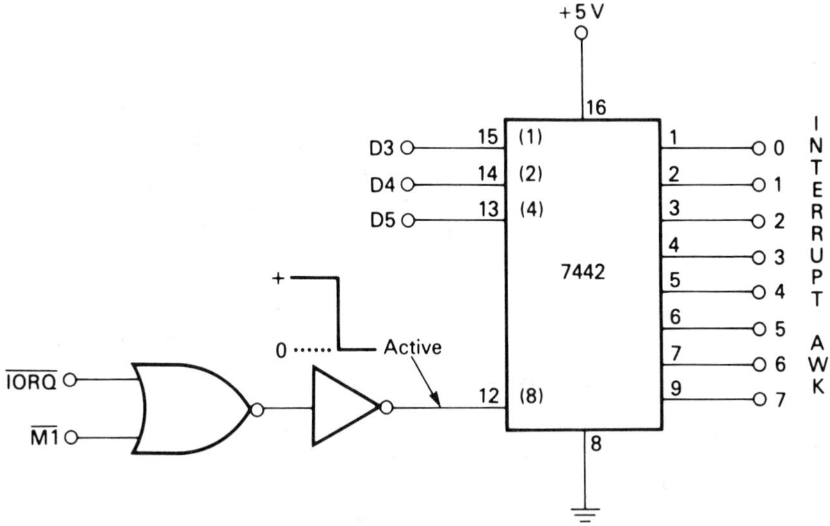

(b)

B) Operation decoder.

In normal, noninterrupt operation, the 8 input of the 7442 is kept HIGH, so the lower eight outputs can never be LOW (when 8 is HIGH, only the 8 and 9 outputs can be active). But when the interrupt acknowledge signal is generated, the 7442 detects the condition of the D3-D5 lines of the data bus and issues the appropriate signal. The only problem that must be considered is the possibility that more than one peripheral will attempt to interrupt at one time. This could cause confusion, to say the least. In a moment we will consider methods for prioritizing the interrupts.

Figure 10-7 shows a decoding scheme that can be used inside the computer and will allow single-line selection for up to eight interrupt lines in mode 0. We are using 74125 quad, tristate buffers in the same manner as in Figure 10-6. But notice in Table 10-2 that the least significant 4 bits of each

TABLE 10-2 RST n CODES FOR INTERRUPTS 0 TO 7

Interrupt	RST n	Hexadecimal	Binary
0	00	C7	11000111
1	08	CF	11001111
2	18	D7	11010111
3	18	DF	11011111
4	20	E7	11100111
5	28	EF	11101111
6	30	F7	11110111
7	38	FF	11111111

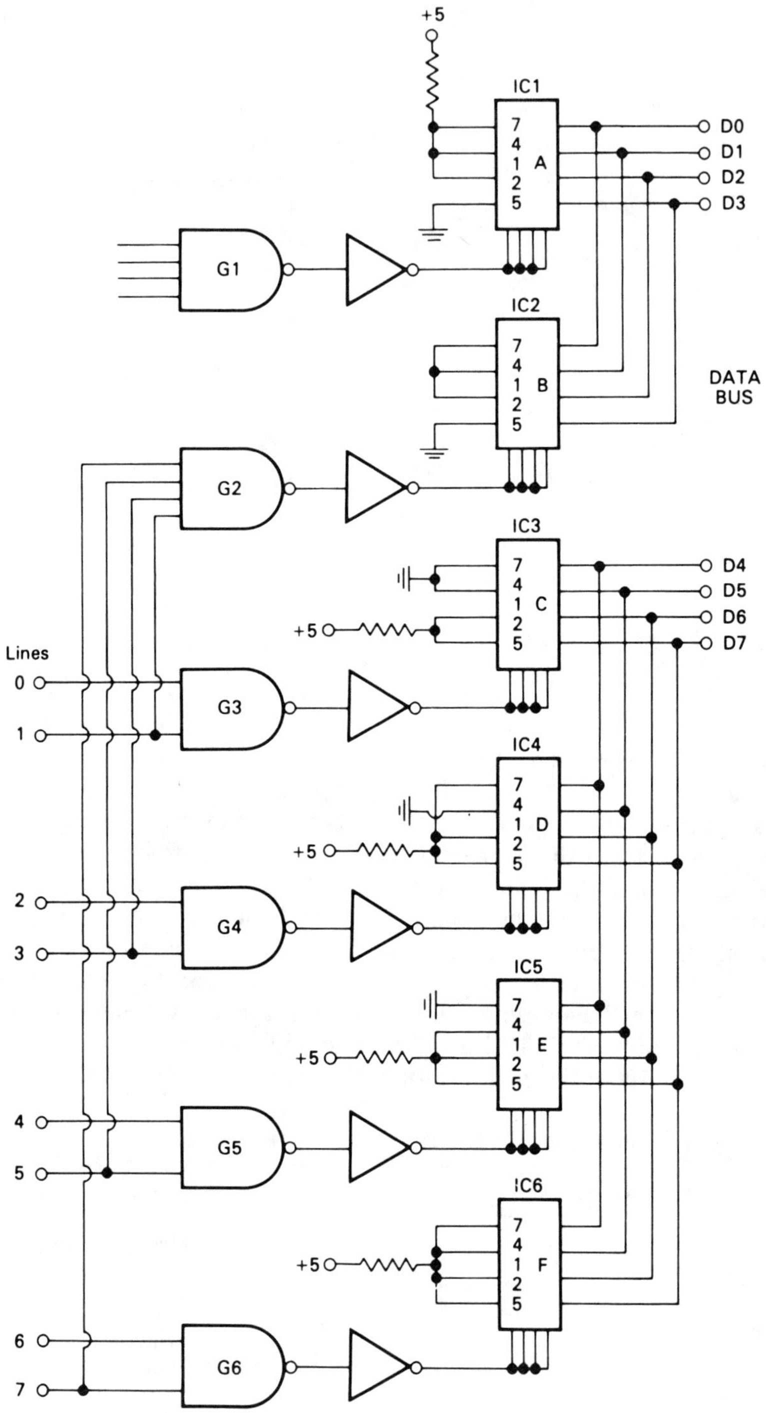

Figure 10-7 Multiple-level decoder.

restart instruction op-code are always either a 7 or F (both hex). Furthermore, the most significant 4 bits will be in one of four possible states, C, D, E, or F. We can, then, create all eight possible op-codes by using only six 74125s and some gates, instead of 16 (as would be required if Figure 10-6 were implemented for all eight). The inputs of the 74125s are programmed as follows:

IC1	7
IC2	F
IC3	C
IC4	D
IC5	E
IC6	F

The key to our decoding scheme is to gate on the enable lines of only the appropriate 74125s. IC1 and IC2 form the code for the least significant half-byte of the op-code. There are four interrupt lines that should turn on IC1, and the other four should turn on IC2. We may use a 7420 four-input NAND gate to select which is turned on. If any input of a NAND gate goes LOW, then its output is HIGH. We connect the respective inputs of gate G1 to those interrupt lines that want a 7 in the least significant spot, that is, 0, 2, 4, and 6 (see Table 10-2). If any of these four interrupt lines goes LOW, then IC1 is turned on and a hex 7 is output to the lower-order half-byte of the data bus. Similarly, gate G2 controls IC2. Its inputs are connected to the 1, 3, 5, and 7 interrupt lines. If any of these lines goes LOW, a hex F is output to the lower-order half-byte of the data bus.

A similar scheme is used to control the higher-order half-byte of the op-code. But in this case, we have four possibilities, each affecting two interrupt lines. IC3 to IC6 form the high-order half-byte of the op-code. Since each of these ICs affects only two interrupt lines, gates G3 to G6 need only two inputs. These are connected as follows:

Gate	Interrupt lines
G3	0, 1
G4	2, 3
G5	4, 5
G6	6, 7

If either interrupt 0 or 1 becomes active, that line will go LOW, causing the output of G3 to go HIGH. This signal is inverted and applied to IC3, which outputs a hex C onto the data bus. Similarly, G4 to G6 will cause the appropriate 74125 to output the correct op-code when their interrupts become active.

The preceding schemes are all relatively simple and involve the use of ordinary TTL-support integrated circuits. But they all also suffer from a common malady. If more than one peripheral device decides to issue an interrupt request, chaos reigns. The logic to prioritize the interrupt response sequence is much more complex than the circuits shown thus far. Fortunately, there are special-purpose integrated circuits, designed for direct interfacing with microprocessor chips, that will allow programming to prioritize and control the interrupts.

You are also permitted to use the Intel 8214 interrupt controller IC (with a little extra logic), even though it was designed for use with the 8080A. Both 8255 and 8257 devices are also useful in Z80 circuits.

Zilog and Mostek, the sources for the Z80, make a Z80-PIO device. This IC is an I/O controller that can handle interrupts.

SERVICING INTERRUPTS

Interrupts are a powerful tool on any programmable digital computer. The designers of the Z80 microprocessor chip, probably well aware of this fact, built into the device four ways to interrupt the CPU: nonmaskable, maskable mode 0, maskable mode 1, and maskable mode 2. Previously, we briefly discussed these interrupts and then concerned ourselves with the hardware aspects of the Z80 interrupt system. In this chapter, we will expand the topic of interrupts by considering the programming aspects of servicing the interrupt request.

Nonmaskable Interrupts

The nonmaskable interrupt is always recognized by the CPU, regardless of the programming being executed. The nonmaskable interrupt goes into effect following the completion of the instruction currently being executed, and it is initiated by bringing the @NMI terminal of the Z80 (i.e., pin 17) LOW. This terminal is sampled by the CPU during the last clock pulse (i.e., T period) of each machine cycle. If @NMI is found to be LOW when this sample is taken, the CPU will automatically begin the interrupt sequence on the next clock pulse.

One principal difference between the nonmaskable interrupt and the maskable interrupts is that the maskables must be enabled by turning on the interrupt flip-flop (IFF1). The nonmaskable does not need to see IFF1 in a SET condition and, in fact, will cause IFF1 to RESET in order to lock out the maskable interrupts (@INT).

The nonmaskable interrupt is very much like a "hardware restart" instruction. In fact, it is an RST 66 instruction (meaning that it will cause a restart instruction to be executed to location 00 66 hex). The nonmaskable interrupt cannot be disabled by software and is always recognized by bringing @NMI LOW. Recall that the restart instructions caused program control to be transferred to one of eight locations in page 0. The principal difference between the eight software restart instructions and the nonmaskable interrupt are (1) @NMI transfers control to a fixed location (address 00 66 hex), and (2) @NMI is *hardware* implemented.

@NMI is used in those situations where it is not prudent to ignore the interrupt. It may be that critical, but transitory, data may be ready to input. Or it may be an alarm condition. A program used to control the environment in a building, for example, probably would want to see no priority higher than the automatic fire alarm. One common application of @NMI when the Z80 is used in a microcomputer is to guard against the problems consequent to a loss of AC mains power. A circuit is built that monitors the AC mains at the primary of the computer's DC power supply. If the AC power drops out for even a few cycles, the circuit generates an @NMI signal to the CPU. The CPU will immediately honor the request and transfer program control to a power loss subroutine. This program is used to transfer all the data in the volatile (i.e., solid-state) memory and the CPU registers/flags into some form of nonvolatile memory (i.e., disk, magnetic tape, etc.). Computers that require this ability must have sufficient backup power stored in batteries, or in the massive filter capacitors of the DC power supply, to execute the power loss subroutine before the energy gives out.

Figure 10-8 shows an example of a typical program sequence for the nonmaskable interrupt. We are executing a program in page 6 (i.e., locations from 60 00 hex). The interrupt service subroutine is stored in locations beginning at 80 00 hex. An interrupt occurs while the instruction at location 60 03 is being executed. The sequence of events is as follows:

1. @NMI occurs while the CPU is executing the instruction located at 60 03.
2. The program counter (PC) is incremented from 60 03 to 60 04, and then its contents are pushed onto the external memory stack.
3. The PC is loaded with 00 66 hex, transferring program control to 00 66 hex.

Before we can service the interrupt, however, we must tend to some housekeeping chores that will allow us to reenter the main program at the point left off, and with no problems. We will want the main program to begin executing at the location that would have been called if the interrupt had not occurred (i.e., 60 04). The *address* of this next location was saved automatically in an external memory stack, but nothing has been done for the flags and other CPU registers. In order to save this environment for use when program control is returned to the main program, we must execute the two exchange instructions (EX and EXX). These are the instructions located at the restart location (00 66 and 00 67 hex). The EX instruction exchanges the

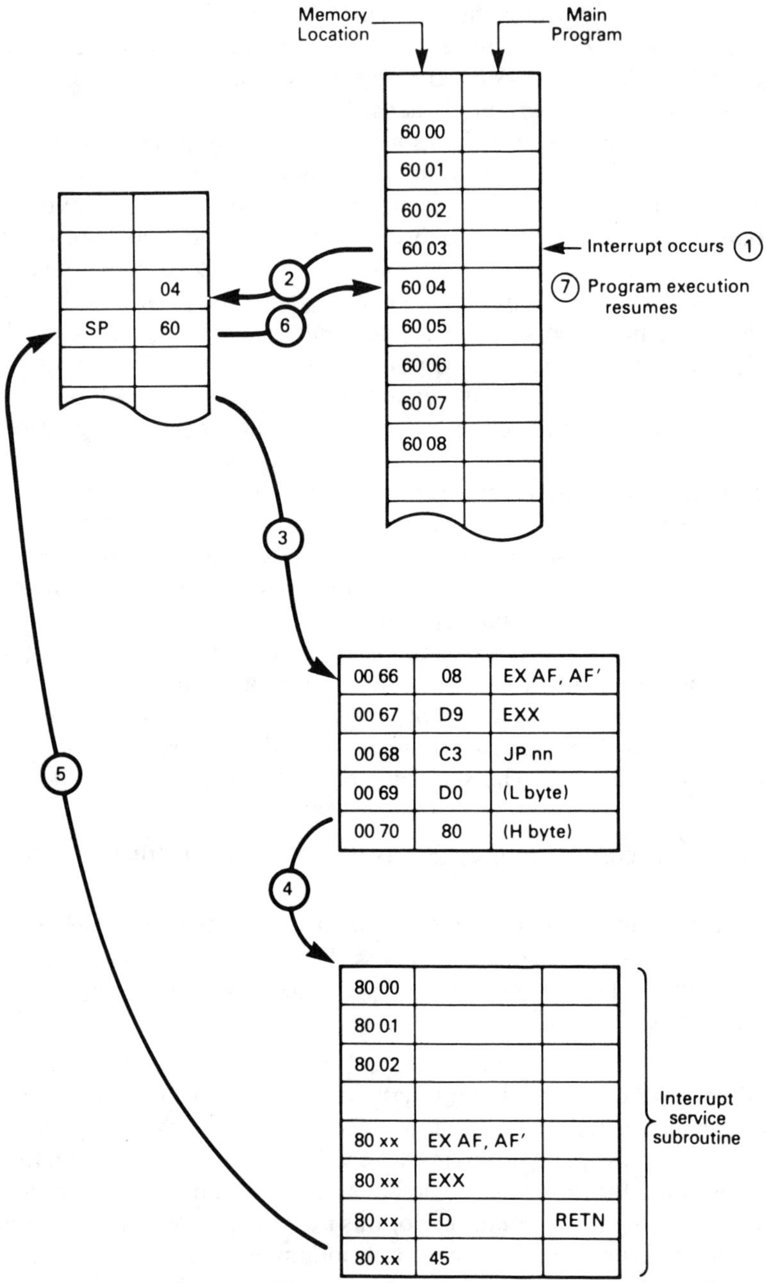

Figure 10-8 Program logic flow.

contents of the AF and A'F' registers, while the EXX instruction causes the other CPU registers to exchange with their alternates (A', F', B', C', D', E', H', and L' are the alternate bank of CPU register in the Z80). The environment (i.e., the contents of the main registers) is now saved in the alternate registers. This will free the main registers for use in the interrupt subroutine and will permit the main program to come back unscratched from the interrupt. Without EX and EXX, trying to figure out where the CPU was would be very difficult.

In some cases, the interrupt service program is short enough that it can be located in the page 0 locations following 00 66 hex. We could, for example, make the first instruction of the service routine at 00 68 hex. But we usually want to save that part of memory for other housekeeping chores (i.e., other restart instructions). In the example shown in Figure 10-8, we execute EX and EXX to save the environment and then jump immediately to location 80 00.

4. The interrupt service program is located higher in memory. In this example, we have located it at 80 00. This program is not shown in detail, because its nature would depend on the type of interrupt being serviced.
5. The last instruction in any nonmaskable interrupt service program *must* be RETN (return from nonmaskable interrupt). This instruction tells the CPU to return control to the main program. RETN returns the contents of the external memory stack to the program counter. Since the PC now contains 60 04 hex, the program resumes at that location. This is the location immediately following the location that was executing when the @NMI signal occurred. Note that, prior to the RETN instruction, we had to reexchange the registers by executing once again the EX and EXX instructions. This will regain the environment lost when the restart-66 occurred.

The nonmaskable interrupt is a hardware function of the Z80 CPU chip. It *cannot* be overridden by the programmer. The maskable interrupt, on the other hand, is designed so that it *can* be overridden by the programmer.

The CPU contains two interrupt flip-flops, labeled IFF1 and IFF2. The first of these, IFF1, is the main interrupt flip-flop, whereas IFF2 is a secondary interrupt flip-flop used to store the condition of IFF1 when a nonmaskable interrupt occurs. We want the CPU restored to its previous state when the nonmaskable interrupt has been serviced. The contents of IFF1 are copied into IFF2 automatically when @NMI is recognized. When RETN is executed, the contents of IFF2 are copied back to IFF1, restoring the condition of IFF1 to that existing when the interrupt occurred. This action completes the restoration of the CPU.

The @NMI will automatically cause the state of IFF1 to be stored in IFF2 and then cause IFF1 to be RESET. This is done to prohibit any additional maskable interrupts during the period that @NMI is being serviced.

Maskable Interrupts

Maskable interrupts can be software-controlled through the use of DI, DO, IM0, IM1, and IM2 instructions. The maskable interrupt is initiated by bringing the @INT terminal on the Z80 (pin 16) LOW momentarily. This action is necessary, but not sufficient, to turn on the interrupt. Recall that IFF1 must be SET before a maskable interrupt is recognized by the CPU. IF IFF1 is RESET, then the @INT command is masked; that is, it is not seen by the CPU, it is ignored. IFF1 is SET by executing IM0, IM1, IM2, or EI instructions. It can RESET by applying a @RESET pulse to pin 16 of the Z80 or by executing a DI (disable interrupt) instruction. There are, then, two ways to turn off the maskable interrupt capability of the CPU.

There are three types of maskable interrupts, designated mode 0, mode 1, and mode 2. Mode 0 is the *default* mode. Unless the programmer demands another mode, by causing the IM1 or IM2 instruction to be executed, mode 0 will be assumed. The CPU is placed in mode 0 as soon as a @RESET signal is received at pin 26 of the Z80. It is usually the practice of designers to automatically apply a power-on @RESET as soon as DC power is applied to the Z80.

Of course, setting any given interrupt mode does not allow the CPU to respond to interrupts. An EI (enable interrupt) instruction must be executed first. Once EI is executed, the interrupt flip-flop (IFF1) is SET, so the CPU will respond to @INT requests (regardless of the mode selected).

Mode 0

Mode 0 is used to make the Z80 think that it is an 8080A microprocessor. This was probably done because one of the objectives of Z80 design was to maintain as much software compatibility between Z80 and the older 8080A as possible. Although there are some differences where timing becomes important, it is a general rule of thumb that 8080A programs will execute on Z80 systems. But the reverse is not true; many Z80 instructions have no 8080A counterparts.

Mode 0 is automatically selected as soon as a @RESET pulse is received. Mode 0 can also be selected through software. The IM0 instruction will cause the CPU to enter mode 0; it is used when the programmer has previously selected one of the other interrupt modes and then wants to return to mode 0 without resetting the CPU.

Like all the maskable interrupts, mode 0 cannot be recognized by the CPU unless the interrupt flip-flop is SET. This flip-flop will be set only if the enable interrupt (EI) instruction is executed. When this is done, the CPU will be ready to respond to maskable interrupt requests.

The mode 0 interrupt requires that the interrupting device place a valid Z80 instruction onto the 8-bit data bus as soon as the interrupt acknowledge signal is generated. In most cases, the instruction used is the 1-byte restart instruction. There are eight unique restart instructions in the Z80 instruction repertoire, and these cause immediate jumps in program control to eight different locations in page 0.

The interrupt service routine should be located at the location in memory where the restart transfers control. For example, if a keyboard causes an interrupt and then jams a restart-10 instruction onto the data bus, the CPU will transfer control to the instruction located at 00 10. If the interrupt service routine is short enough, it might be located in the memory spots immediately following 00 10 (as might well be the case in a simple keyboard input subroutine), or the instruction may be a jump immediate to some location higher in memory. It is very common for programmers to locate these service programs in the top end of the memory available in a particular computer.

Figure 10-9 shows a typical mode 0 response. For the sake of continuity, we are using the same locations as in the nonmaskable interrupt discussion earlier. The program is executing the instruction at location 60 03 when the @INT signal is received by the CPU. The interrupt request is recognized following the completion of the instruction at 60 03, provided that IFF1 is SET. The sequence is as follows:

1. @INT occurs during the execution of the instruction at location 60 03. This is recognized by the CPU during the last clock cycle of that instruction.
2. On the other clock pulse, the CPU acknowledges the interrupt request by causing @IORQ and @M1 to go LOW immediately.
3. When the interrupt acknowledges signal is received, the interrupting device places an RST 10 code on the CPU data bus.
4. The CPU executes the RST 10 by incrementing the PC to 60 04, storing the incremented contents in the external memory stack and then jumping immediately to location 00 10.
5. At location 00 10 the instruction is an immediate jump to location 80 00, where the interrupt service program is found.

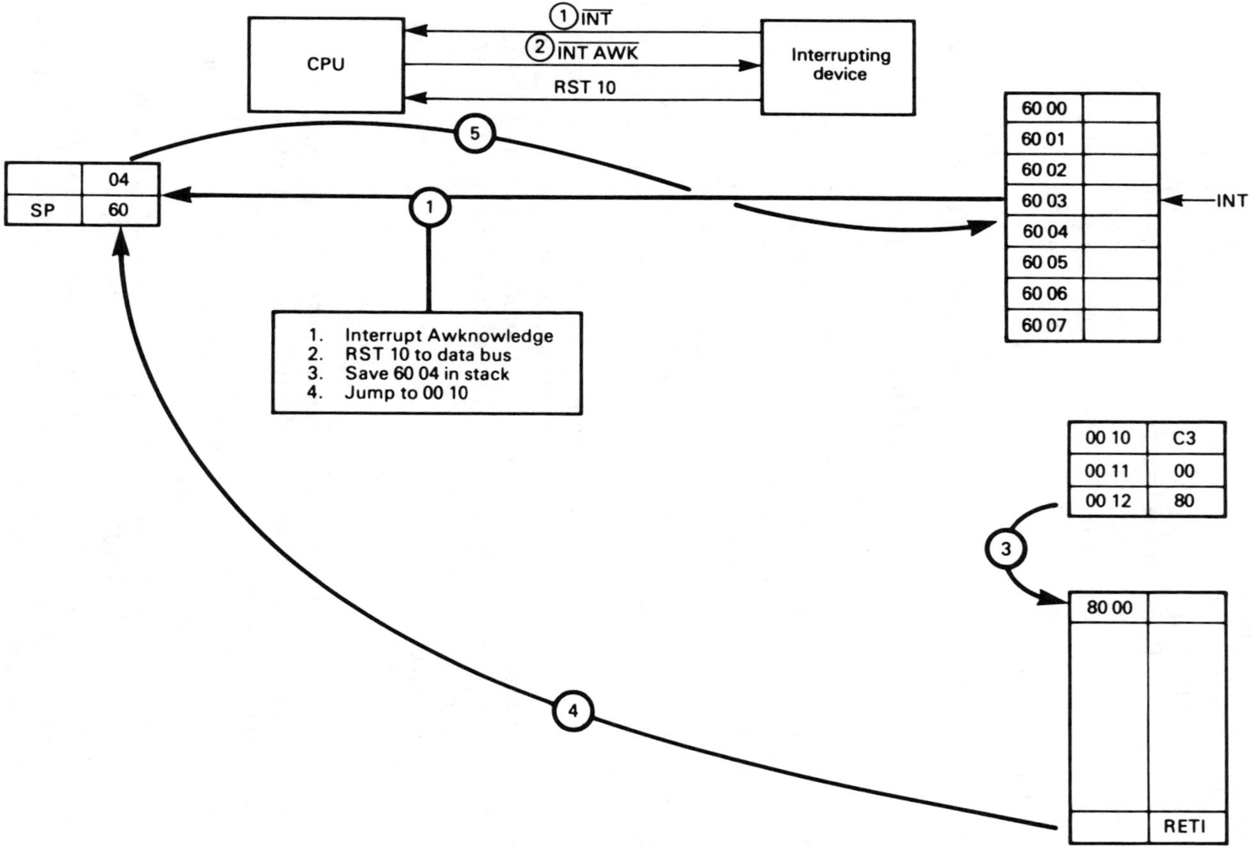

Figure 10-9 Interrupt data flow.

6. Again, the environment must be saved. There are two ways in which this can be done. One is to use the EX and EXX instructions of the previous example. Another is to use the PUSH instructions:

80 00	PUSH AF
80 01	PUSH BC
80 02	PUSH DE
80 03	PUSH HL
80 04	PUSH IY
80 05	PUSH IX

The next instructions would then be instructions of the interrupt service subroutine. When this program is completed, we must execute all of the POP instructions, to bring the contents of the registers back from the memory stack, and an RETI (return from maskable interrupt) instruction, as follows:

80 xx	POP AF
80 xx	POP BC
80 xx	POP DE
80 xx	POP HL
80 xx	POP IY
80 xx	POP IX
80 xx	RETI

7. After the RETN instruction, the CPU will replace the contents of the PC with the data stored in the external stack (60 04). This is the address of the instruction in the main program that would have been executed *next* if the interrupt had not occurred.
8. Program execution resumes at location 60 04.

The mode 0 interrupt preserves some of the compatibility of the Z80 with the Intel 8080A microprocessor. But there is a limitation in this mode. The device will allow only eight interrupt devices, one for each of the eight restart locations.

Interrupt priority encoding is possible by using a priority controller, such as the Intel 8214 (or one of the related devices) or one of the Zilog Z80 peripheral chips.

Mode 1

Mode 1 is not similar to any function of the 8080A device, so it is unique to the Z80, in this respect. It is almost identical to the nonmaskable interrupt, except (1) it is maskable, and (2) it causes a restart jump to location 00 38 instead of 00 66.

The mode 1 interrupt is dependent upon the programmer's setting mode 1 by enabling interrupt flip-flop IFF1 (the EI instruction) and setting mode 1 by executing an IM1 instruction.

The use of mode 1 is similar to the nonmaskable interrupt, except that the priority would be lower than that of a nonmaskable interrupt. It has the advantage that no external logic is needed to cause the restart instruction. It is, then, somewhat faster than the mode 0 operations.

Refer back to the discussion of the nonmaskable interrupt for how this interrupt is serviced. Just be sure to replace in your mind the location 00 66 with 00 38.

Mode 2

The mode 2 interrupt is one of the most powerful microcomputer interrupts. It allows vectored interrupts of up to 128 levels, as opposed to only eight levels in mode 0 and one level in mode 1 and nonmaskable interrupt.

Zilog has conveniently caused the Z80 peripheral control chips (Z80-P1O, Z80-S1O, and Z80-CTC) to allow prioritizing of the interrupts through a daisy-chaining scheme.

The key to the versatility of the mode 2 interrupt is that it is *vectored.* That is, it can use a single 8-bit word to point to any location in memory. The 1-byte address of the interrupt service program is stored in a table of interrupt addresses located somewhere in memory. The location of this table is pointed to by a 2-byte digital word formed from the contents of the interrupt (I) register and the 1-byte word supplied by the interrupting device. The upper 8 bits of this 16-bit pointer are supplied by the I register and must be preloaded by the program. The lower-order 8 bits of the pointer are supplied by the interrupting device.

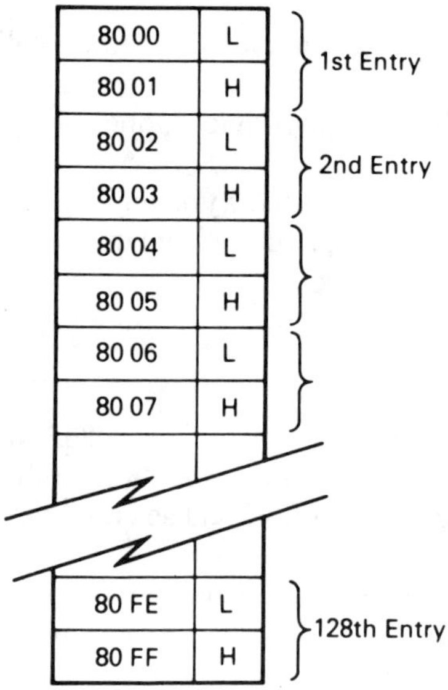

Figure 10-10 Data arrangement.

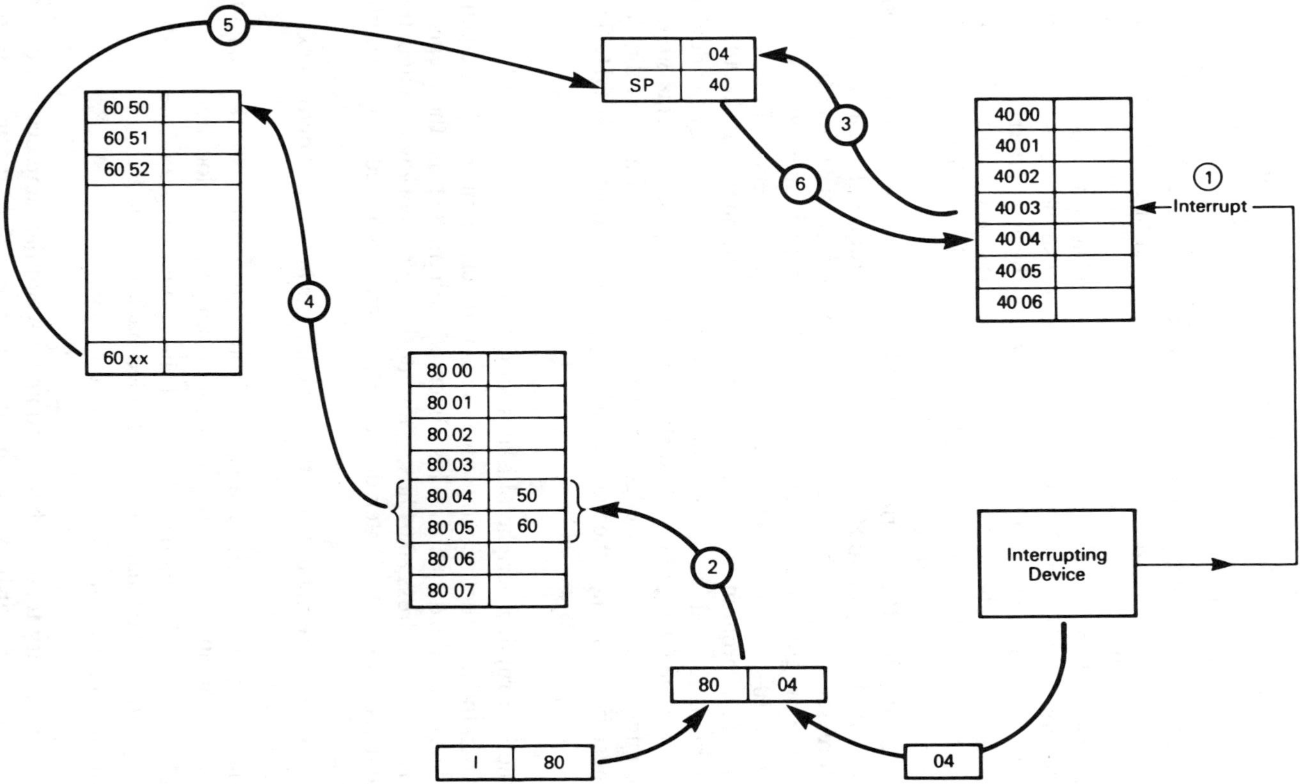

Figure 10-11 Mode-2 interrupt sequence data flow.

There is one restriction on the addresses of the table, which is that they must begin on an even-numbered memory location. All the entries in this table will be 2 bytes in adjacent locations. The first byte of each entry in the table is the low-order byte of the desired address, while the second entry is the high-order byte. One consequence of this constraint is that the least significant bit of the 8-bit word supplied by the interrupting device must be 0.

Figure 10-10 shows an example of such a table. In this case, the programmer elected to locate the table in page 8, and it commences at 80 00 hex. The first entry is found at 80 00 and 80 01. These locations contain the low- and high-order bytes of the address where the first interrupt service program is located. The first part of this address (80 hex) is stored in the I register. The second part is supplied by the interrupting device. Notice that the binary equivalent of 00 ends in a 0.

Similarly, the other entries are found beginning at 80 02, 80 04, 80 06, and so on, all the way up to 80 FE (if 128 levels are required). Each of these table addresses contains the address of a location in memory where the CPU will find the program that serves that particular interrupting device.

Figure 10-11 shows a typical mode 2 interrupt sequence. In this program, the main program is located in page 4 (i.e., beginning at 40 00), the vector table is located in page 8, and the interrupt subroutine for the device shown is in page 6 (beginning at 60 50 hex). The I register contains 80 hex, and the interrupting acknowledge signal is received. The interrupt flip-flop IFF1 must be SET, and the bus request @BUSRQ must be HIGH. The sequence of events is as follows:

1. The interrupting peripheral issues an @INT signal to the CPU.
2. When the interrupt acknowledge signal is received, the peripheral jams 04 hex onto the data bus. This is merged with the 80 from the I register to form the address 80 04 hex. This address in memory will contain the address of the actual interrupt service program required by the peripheral.
3. The PC is incremented, and then its contents are pushed onto an external memory stack.
4. The PC is loaded with the address found at location 80 04. This address is 60 50 hex, so program control jumps to this location.
5. After the last instruction (RETI) of the service program, the PC data saved in the external stack are loaded back into the PC.
6. The main program resumes at location 40 04.

It is necessary to save the environment when the jump occurs, or the CPU will not necessarily be in the same state as before the interrupt occurred. These techniques were discussed earlier in this chapter.

11

BUILDING YOUR OWN IBM PC/XT/AT CLONE

A Viable Alternative for the Low-Budget Laboratory

There is no doubt that the desktop ''personal computer'' has revolutionized both the theoretical and experimental sciences. It puts amazing computing power at the fingertips, as well as providing the ability to perform tremendous chores of data collection and analysis. But the computer can also be expensive for the small laboratory that lacks funding. However, if you already own a machine such as an IBM PC or XT (or clone of these), then it is possible to upgrade to a higher class with minimal investment. In this chapter you will find out some of the approaches—and pitfalls—to making your PC into an XT or an AT.

The IBM PC has 256K of dynamic random-access memory (DRAM) made up of four banks of 4164-150 DRAM chips, with nine chips per bank (Figure 11–1). Almost all modern software appearing on the market today is marked ''. . . requires at least 512K of RPM. . . .'' Therefore, the first upgrade is to add memory. MS-DOS recognizes up to 640K of memory, so I had to add 384K in order to bring my own machine up to date.

The old PC also has a parallel printer port, a monochrome nongraphics (text only) monitor, and a pair of 360K, 5.25-inch, full-height (FH) floppy disk drives (Figure 11-2). No serial (RS-232C) port was available. Current technology offers the serial port and half-height (HH) floppy disk drives, and the 360K FH drives are out. Current machines might use quad-density 5.25-inch HH drives, or 720K 3.5-inch floppy disk drives, or 1.44M 3.5-inch floppy disk drives.

There was no hard disk on old machines unless you owned an XT or XT-clone. Even if you do own one of those machines, the 10 or 20 megabyte

Figure 11-1 IBM PC/XT-compatible add-on memory chips arranged in four banks of nine each.

Figure 11-2 View from right side showing disk drives.

hard disks they used are probably getting a little full . . . so you might want to upgrade.

BASIC UPGRADE

The most basic upgrade that I, and most owners of older PCs, needed was to add more DRAM and a serial port. There are a number of add-on boards that fit into the accessory slots on the motherboard (Figure 11-3A); external connectors for the plug-in boards protrude through the rear panel of the computer (Figure 11-3B). For my upgrade I selected the AST Sixpack because it offers, in one card, up to 384K of additional memory, an asynchronous serial port (RS-232C), an additional parallel port, a game port, and an on-board clock/date function that eliminates the need for typing in the date and time on boot-up.

Shopping around for DRAM chips is well worthwhile. Don't take the prices in mailorder ads too seriously. Ads are prepared several months in advance, but the DRAM market changes on a weekly basis. I found the prices of DRAMs from mailorder sources were generally somewhat cheaper than the ads from the same companies indicated. In one case, the DRAM price was 40 percent less than the advertised price. It is worthwhile to call and ask the price before buying.

When adding a product like the Sixpack you will have to configure the board and the motherboard to recognize the new memory size. Consult the owner's manual for both the plug-in and the computer to find out how to set the switches for the specific memory size that you selected.

I also decided to add a simple graphics card to allow my IBM PC to work with certain modern software. I selected the Hercules-format graphics card because it is minimally acceptable and (primarily) because it would work with the standard green screen TTL monitor that came with the machine many years ago. A new Hercules-format card usually costs less than $80, so it is a reasonable selection for an old computer. Those who want to add a modern monitor, and one that works a whole lot better, can use an EGA or VGA monitor (see below).

ADDING AND CHANGING DISK DRIVES

When many of us bought our PC or XT machines there was only one form of disk drive: the 5.25-inch FH, 360K. But today there are several different forms, including 5.25-inch FH 360K, 5.25-inch HH 360K (Figure 11-4A), 5.25-inch HH 1.2M (quad density), 3.5-inch HH 720K (Figure 11-4B), and 3.5-inch 1.44M HH. While I originally paid $395 for an additional disk drive,

(a)

(b)

Figure 11-3 Expansion slots: A) On motherboard, B) Rear panel.

Figure 11-4 A) 5.25-inch, 360K disk drive mechanism; B) 3.5-inch, 720K disk drive mechanism.

current prices run from around $60 to $80 for 360K drives, to about $120 to $150 for one of the larger-capacity 3.5-inch drives.

One of the things that prompted me to upgrade the old PC was the fact that one of my 5.25-inch 360K FH drives bit the dust. Instead of just replacing the drive with another of the same type I opted to select a different mix. Because many software manufacturers will soon offer their wares in the 3.5-inch, 720K format only, it is wise to select one of these drives. The mix that I selected was one 3.5-inch, 720K HH for Drive B (Figure 11-4B) and one of the original 5.25-inch, 360K drives for Drive A.

There is a problem that confronts owners of older PCs who wish to add a half-height drive. Compare the mounting holes for the drives in the newer machine in Figure 11-5 with the traditional mountings in Figure 11-2. Notice that there are more holes, arranged slightly differently in the newer models. This pattern was designed to fit the HH floppy disk drives, not the full-height. Owners of older machines will have to either fashion an adapter plate of their own or buy a commercially made mounting plate. Jameco Electronics sells their Model TMHD for about $3. Also offered by Jameco are other disk drive mounting hardware that may be needed by some people. In my case, replacing an FH drive with an HH drive left an empty space. The Jameco Model BEZEL, for about $1, covers up the unsightly hole.

Adding a 3.5-inch disk drive to an older machine may not be the straightforward task that it seems. There might be problems with the inter-

Figure 11-5 Disk drive mounts.

Figure 11-6 Half-size expansion card.

nal BIOS (Basic Input Output System) and the version of MS-DOS or PC-DOS that you selected.

The BIOS is a ROM on the motherboard that is the fundamental program of your computer. It allows the computer to read disks (among other tasks) so that the disk operating system (DOS) and applications programs can be input to the computer. Earlier BIOS may not support newer drives, so you will have to buy a later version. Some authorized IBM dealers will sell the latest BIOS for your machine, although you might have to look hard to find one. Alternatively, contact one of those new breed of clone dealers who assemble their own machines. Some of those dealers will sell you a Phoenix, Award, or AMI BIOS chip. I called Phoenix Computer Products, Inc., and they gave me the phone number of a retailer who sold me the Phoenix BIOS Ver. 2.52 (the latest).

Older versions of MS-DOS or PC-DOS will not support the latest floppy disk drives. I used DOS 1.1 and 2.0 for years, but these were not sufficient for the newer 3.5-inch drives. Use DOS 3.2 or higher for 720K drives and 3.3 or higher for 1.44M drives. An anomaly reported on some versions of 3.30 is that the 3.5-inch drive *must* be Drive A; version 3.30A allows any type of drive to be in either A or B slots.

You may also need to obtain a newer floppy disk controller card, and herein lies a pitfall for the unwary. There are several different forms of 3.5-inch drive, and the controller card must be capable of accepting the one that you buy. For that reason I recommend that you buy the disk drive and the controller card from the same source. I bought a Toshiba Model 352KU (720K) drive (Figure 11-4B) and the JE-1043 controller card (Figure 11-6) from Jameco on the basis that their catalog specifically listed these components as compatible with each other. The JE-1043 will control two drives, they can be any of the standard types discussed above, and they can be separately configured on the JE-1043 by moving jumpers around according to the instructions supplied.

When I finished with the basic upgrade of my old IBM PC, it was configured with 640K of memory, a parallel port, a serial port, one of the original 360K FH 5.25-inch floppy drives, a new 3.5-inch 720K drive, and a Hercules graphics card.

HIGHER UPGRADES

The upgrade that I performed on my old IBM PC was sufficient for my needs at the time, and that machine is still in use. But there are other things that could also be done to improve the machine. However, some of these are so extensive that one has to consider whether the task ought to be refurbishing an old machine or building a new machine. Both are viable alternatives for any electronics buff who knows how to use a screwdriver.

Hard Disk Drives

All modern computers should employ either an external or internal (preferable) hard disk drive. Although the original IBM PC/XT hard disk had only 10 megabytes (10M) of storage (which seemed vast in 1985), currently the 40M size is standard and a 20M is considered only minimally acceptable. Several options are available to you for adding hard disk capability.

Owners of older machines, especially the IBM PC (or clones) that did not have a hard disk, may wish to use a "hard card." These products are standard IBM-compatible plug-in cards that fit into one of the slots. It contains a disk drive and a controller card on one plug-in assembly. I added a Shamrock Lepracard Model 330 30M hard card to my old IBM PC.

Various manufacturers offer FH and HH hard disk drives that can boost the performance of your computer considerably. These drives can fit into one of the floppy drive positions on your computer. A common configuration is to use a FH or HH hard disk in the right-hand position formerly occupied by Drive B, and the HH floppies in the left position formerly occupied by the FH Drive A. Many people place a 5.25-inch, 360K (Drive A) and a 3.5-inch 720K drive (Drive B) in the left side position, and a HH or FH hard disk in the right side.

In my own situation, I built an XT-turbo class machine using a Seagate

Figure 11-7 Hard disk drive mechanism.

ST-238R hard disk (Figure 11-7). This same drive (or one of the others) can be used in either a newly constructed machine or an upgraded or refurbished older machine. A hard disk controller will be needed, and these can be bought from the same source as the drive. In fact, it is best policy to buy both the drive and the controller from the same source so that you can make certain that they are compatible. Most mailorder companies will designate the controllers and drives that go together, and many of them offer a special price when the two are purchased together.

More Power

The original IBM PC machine used a 67-watt DC power supply. But if you add a lot of special-purpose cards, or a hard disk drive, then it may become necessary to beef up the DC power supply. Many sources exist for new power supplies (Figure 11-8A) in 135-watt, 150-watt, and even 250-watt capacities; I recommend at least 150 watts. These supplies now cost less than $100, and I've seen some 150-watt models advertised for less than $50. The DC power supply fits into the computer case at the right rear corner (Figure 11-8B) and is attached by two to five screws (depending upon origin) on the rear panel. It can be replaced easily with only a screwdriver or 1/4-inch nut driver.

(a)

(b)

Figure 11-8 A) DC power supply, B) Mounted in cabinet.

New Motherboards

The motherboard is the main printed circuit board in an IBM PC or clone. The motherboard contains the microprocessor chip and most of the main RAM. The motherboard also contains the sockets for the plug-in cards that allow you to custom configure the machine for your own applications. Today there are a number of motherboards to select from. Interestingly enough, replacements for the original IBM PC machine (without hard disk) are hard to find. However, those boards can be replaced with an XT-compatible motherboard. If you have an older IBM PC, then look on the rear panel and count the number of slots for accessories that exist. You will have to make sure that the slots on the new board line up with the older cutouts. Only certain six-slot boards will fit (most XT-clone boards now have eight slots).

If you want to upgrade further than the XT level, then select a board based on the 286 or 386 chips. The so-called "baby AT" boards are compatible with the IBM AT 286 machine but are sized to fit the standard PC/XT cabinet. The standard-size AT board will not fit the older chassis, but new cabinets are relatively easy to obtain.

Figure 11-9 Motherboard.

WARNING: AVOID ESD DAMAGE

Personal computer motherboards, plug-in cards, and the electronics sections of disk drives use components that are sensitive to electrostatic (ESD) potentials that can build up on your body—ESD potentials that you might not even feel can blow out, or shorten the life expectancy of, those electronic components. When working on your PC, always follow good anti-ESD procedures. For example, don't wear wool or those synthetic garments that generate a lot of static electricity. Discharge yourself by touching a grounded point before starting to work. Ideally, wear a grounded ESD wriststrap (available at many electronic parts distributors), and use soldering tools that have grounded tips.

Don't remove boards or components from the packaging until you are ready to use them. Those translucent tubes and plastic bags are ESD treated, so they will prevent damage from static electricity that may be present on your body.

12

SIGNALS: SOME BACKGROUND FACTS

The nature of signals, their relationship to noise and interfering signals, and the effects of digitization (for input to computers) determines appropriate design of the computerized instrument. In this chapter we will take a look at signals and noise and how each affects the design of instrumentation circuits.

Signals can be categorized several ways, but one of the most fundamental is according to their time domain behavior (the other way is by frequency domain). We will assume signals of the form $v = f(t)$ or $i = f(t)$. The classes of signals include: *static, quasistatic*, *periodic*, *repetitive*, *transient, random,* and *chaotic*. Each of these categories has certain properties that can profoundly influence appropriate design decisions.

Static and quasistatic signals. A *static* signal (Figure 12-1A) is, by definition, unchanging over a very long period of time. Such a signal is essentially a DC level, so it must be processed in low-drift DC amplifier circuits. The term *quasistatic* means "nearly unchanging," so a quasistatic signal (Figure 12-1B) refers to a signal that changes so slowly in the time domain that it possesses characteristics more like static signals than dynamic signals.

Periodic signals. A *periodic* signal (Figure 12-1C) is one that repeats itself on a regular—that is, periodic —basis. Examples of periodic signals include sine waves, square waves, sawtooth waves, and triangle waves. The nature of the periodic waveform is such that each waveform is identical

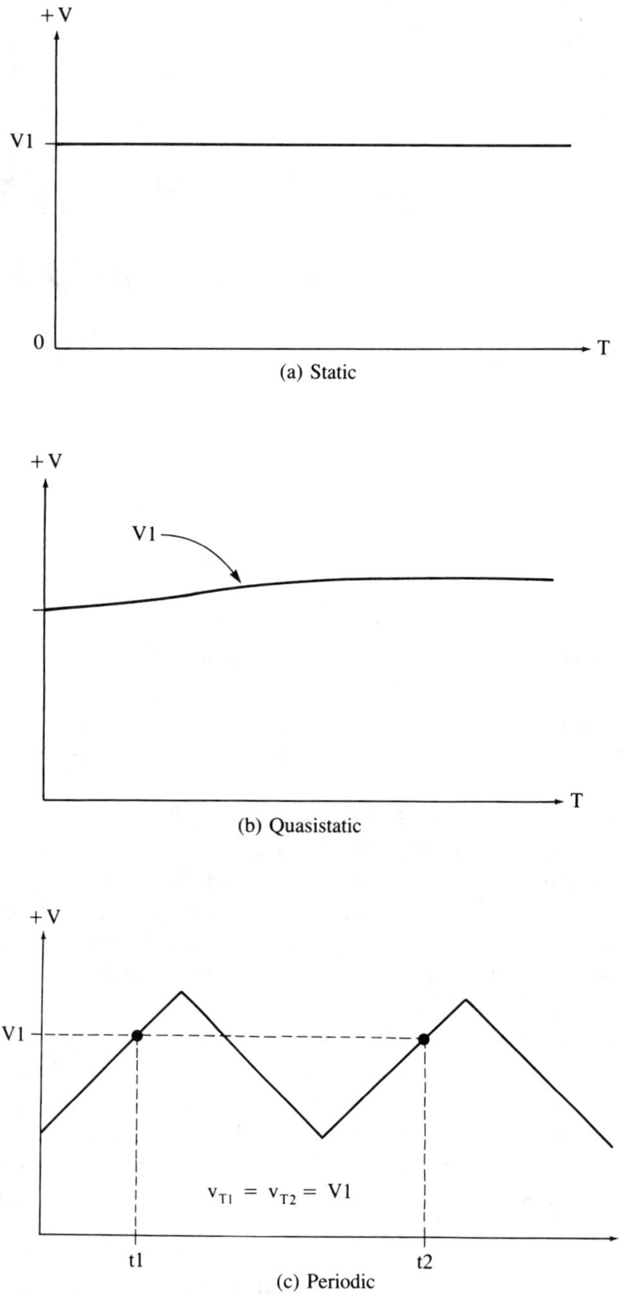

Figure 12-1 Signals: A) Static DC signal, B) Quasistatic (near-DC) signal, C) Periodic signal, D) Repetitive signal, E) Transient signal, F) Quasitransient signal, G) Random or chaotic signal.

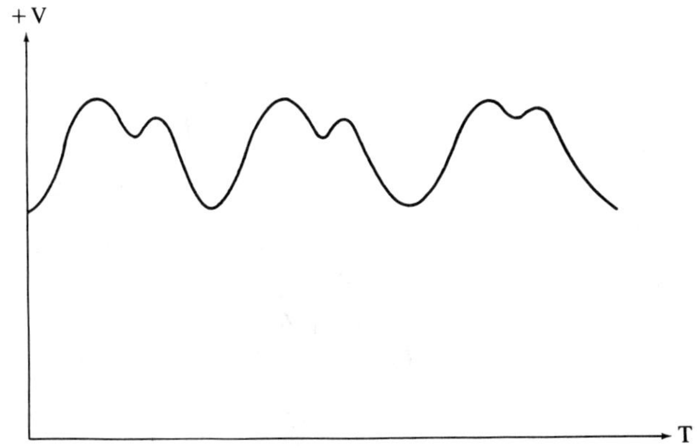

(d) Repetitive

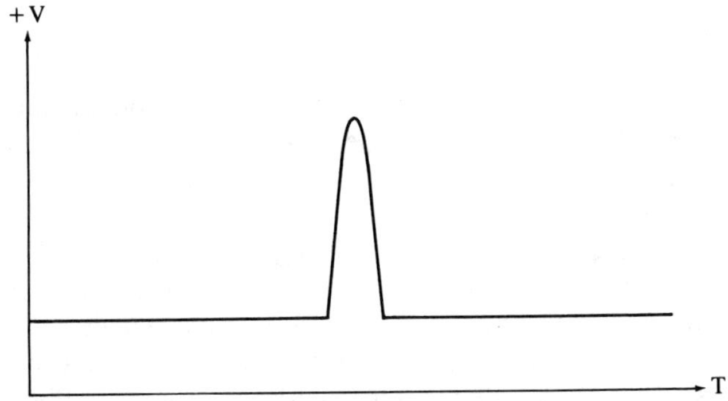

(e) Transient

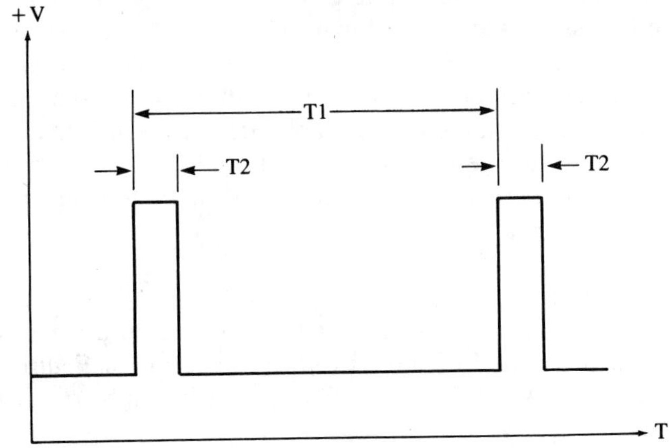

(f) Quasitransient

Figure 12-1 ***(continued)***

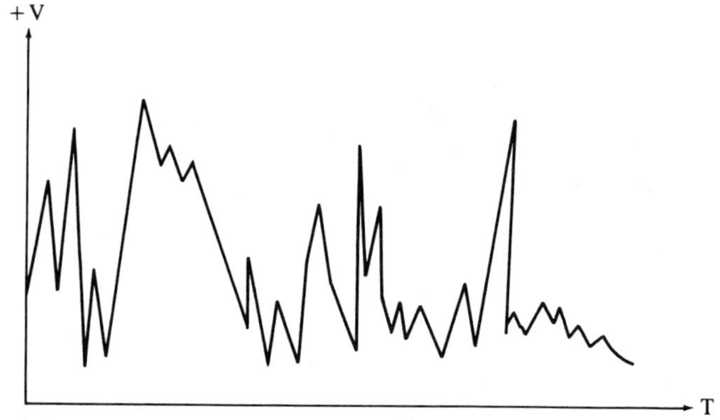

(g) Random or chaotic

Figure 12-1 ***(continued)***

at like points along the time line. In other words, if you advance along the time line by exactly one period (T), then the voltage, polarity, and direction of change of the waveform will be identical.

Repetitive signals. A *repetitive* signal (Figure 12-1D) is quasiperiodic in nature, so it bears some similarity to the periodic waveform. The principal difference between repetitive and periodic signals is seen by comparing the signal at f(t) and f(t + T), where T is the period of the signal. These points might not be identical in repetitive signals, but they are identical in periodic signals. The repetitive signal might contain either transient or stable features that vary from period to period. An example is the human arterial blood pressure waveform (Figure 12-1D). While the waveform tends to vary from a minimum (diastolic) to a maximum (systolic) in a quasiperiodic manner, there are both normal and pathological anomalies from one cycle to another. For example, the amplitudes of the maxima and minima and the repetition rate (i.e., the heart rate) tend to vary quite normally in healthy humans. Repetitive signals may bear characteristics of both transient and periodic signals.

Transient signals. A *transient* signal (Figure 12-1E) is either a one-time event or a periodic event in which the event duration is very short compared with the period of the waveform. It terms of Figure 12-1F, the latter definition means that $t_1 << t_2$. These signals can be treated as if they are transients.

Random and chaotic signals. A *random* signal (Figure 12-1G) is one that is both unpredictable and has either or both of the following properties: a) one or more inputs to the system are unknown, and/or b) the rules by which the system affects the output in response to the inputs are unknown.

A new form of signal recently recognized is the *chaotic* signal. These signals have previously been labelled random, but there is a critical difference between chaotic and random signals. That difference is that with a chaotic signal, even though all inputs and transfer function rules are known, the output signal is nonetheless unpredictable.

The characteristic of these types of signal drive the design of the instrumentation system. An important consideration of all dynamic signals (whether periodic or otherwise) is the frequency response required to faithfully reproduce the signal. For that reason we will turn to the *frequency domain* characteristics of signals.

FOURIER SERIES

All continous periodic signals can be represented by a fundamental sine wave, and a collection of harmonics of that fundamental sine wave, linearly summed together. These frequencies comprise the *Fourier series* of the waveform. The elementary sine wave (Figure 12-2) is described by

$$v = V_m \sin(n\omega t) \tag{12.1}$$

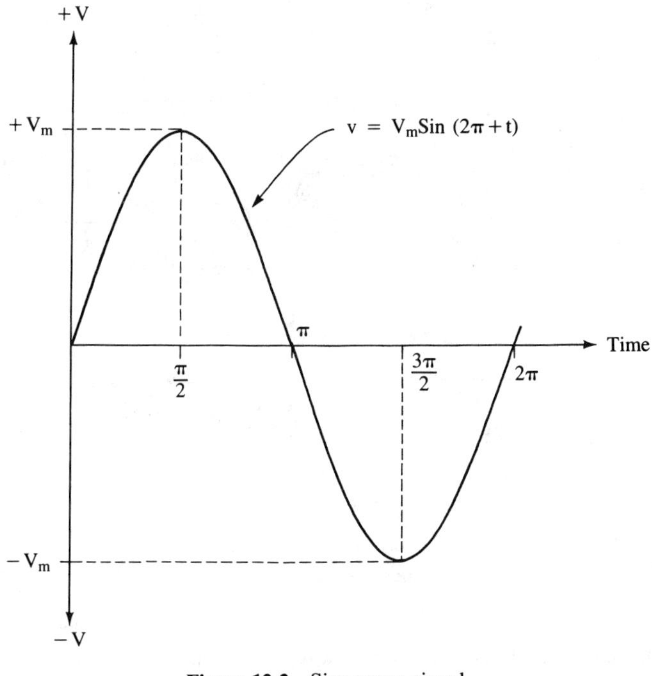

Figure 12-2 Sine-wave signal.

where

v is the instantaneous amplitude of the sine wave

V_m is the peak amplitude of the sine wave

W is the angular frequency ($2\pi F$) of the sine wave

t is the time in seconds

The *period* of the sine wave is the time between reoccurrence of identical events, or $T = 2\pi/\omega = 1/F$ (where F is the frequency in cycles per second).

The Fourier series that makes up a waveform can be found if a given waveform is decomposed into its constituent frequencies either by a bank of frequency-selective filters or by a digital signal-processing algorithm called the *Fast Fourier Transform* (FFT). The Fourier series can also be used to construct a waveform from the ground up. Figure 12-3 shows square wave (Figure 12-3A), sawtooth wave (Figure 12-3B), and peaked wave (Figure

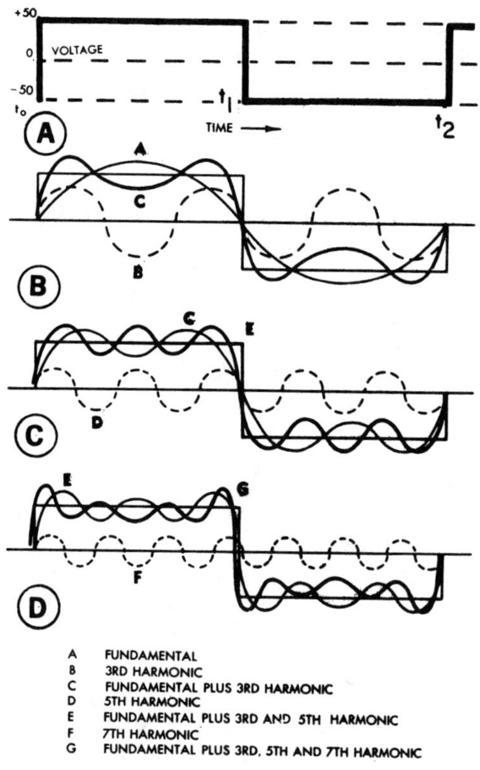

Figure 12-3 Constructing a complex wave from a sine wave and its harmonics: A) Square wave, B) Sawtooth wave, C) Peaked wave.

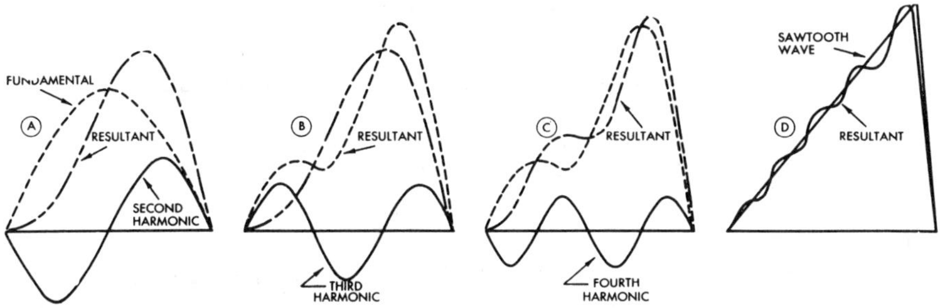

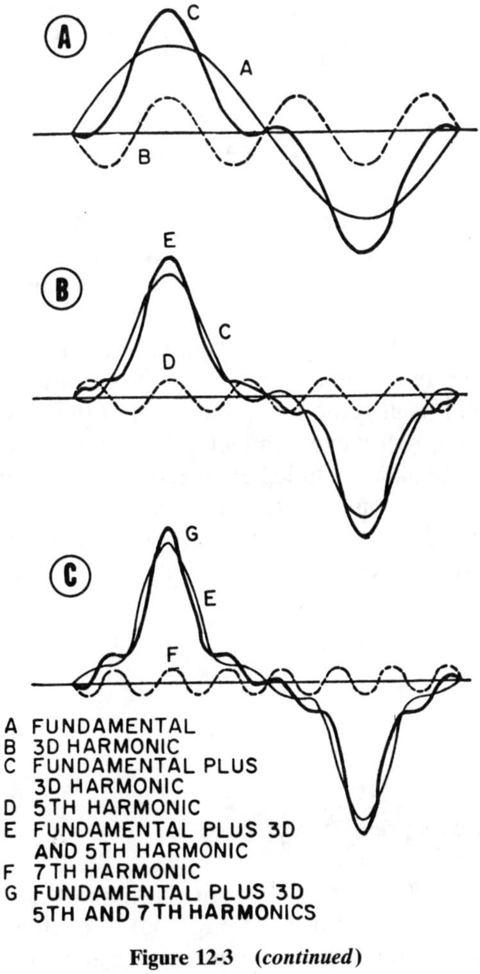

Figure 12-3 ***(continued)***

12-3C) signals constructed from a fundamental sine wave and its harmonic sine and cosine functions.

The Fourier series for any waveform can be expressed in the form

$$F(t) = \frac{a_0}{2} + \sum_{n=1}^{\infty} [a_n \cos (n\omega t) + b_n \sin (n\omega t)]$$

where

a_n and b_n represent the amplitudes of the harmonics (see below)

n is an integer

Other terms are as previously defined

The amplitude coefficients (a_n and b_n) are expressed by

$$a_n = \frac{2}{T}\int_0^T F(t) \cos (n\omega t)\, dt \quad (12.3)$$

and

$$b_n = \frac{2}{T}\int_0^T F(t) \sin (n\omega t)\, dt \quad (12.4)$$

The amplitude terms are nonzero at the specific frequencies determined by the Fourier series. Because only certain frequencies, determined by integer *n*, are allowable, the spectrum of the periodic signal is said to be *discrete*.

The term $a_0/2$ in the Fourier series expression [Equation (12.2)] is the average value of f(t) over one complete cycle (one period) of the waveform. In practical terms, it is also the *DC component* of the waveform. When the waveform possesses half-wave symmetry (i.e., the peak amplitude above zero is equal to the peak amplitude below zero at every point in *t*, or $+V_m = -V_m|$), there is no DC component, so $a_0 = 0$.

An alternative Fourier series expression replaces the $a_n \cos (n\omega t) + b_n \sin (n\omega t)$ with an equivalent expression of another form:

$$F(t) = \frac{2}{T}\sum_{n=1}^{\infty} c_n[n\omega t - \phi_n] \quad (12.5)$$

where

$c_n = [(a_n)^2 + (b_n)^2]^{1/2}$

$\phi_n = \arctan (a_n/b_n)$

All other terms are as previously defined

One can infer certain things about the harmonic content of a waveform by examination of its symmetries. One would conclude from the above equations that the harmonics extend to infinity on all waveforms. Clearly, in practical systems a much less than infinite bandwidth is found, so some of those harmonics must be removed by the normal action of the electronic circuits. Also, it is sometimes found that higher harmonics might not be significant. As n becomes larger, the amplitude coefficients a_n and b_n become smaller. At some point, the amplitude coefficients are reduced sufficiently that their contribution to the shape of the wave is either negligible for the practical purpose at hand or totally unobservable in practical terms. The value of n at which this occurs depends partially on the *rise time* of the waveform. Let's consider a practical example from biomedical instrumentation.

Figure 12-4 shows the human arterial pressure waveform superimposed on the same time line as the electrocardiograph (ECG) waveform.

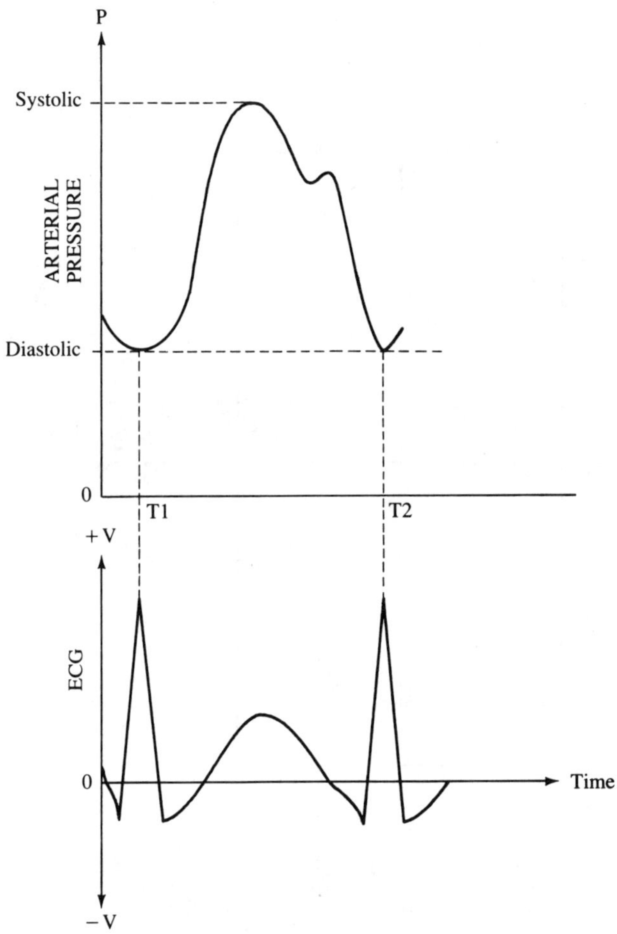

Figure 12-4 ECG and arterial pressure signal.

These waveforms are time-correlated because they both represent different views of the same physical event, i.e., the beating of the human heart. Suppose that the heart rate is 72 beats per minute (BPM), or 1.2 Hz. The pressure waveform has a slower rise time than the ECG waveform, so it contains a smaller number of harmonics. The pressure waveform can be accurately reproduced with about 25 harmonics (e.g.,a 30-Hz bandwidth), while the ECG waveform requires 70 to 80 harmonics for faithful reproduction (e.g., a bandwidth of 100 Hz). In order to adquately process these two waveforms, the instrument must have upper -3 dB frequency responses of 30 Hz and 100 Hz for the pressure and ECG channels, respectively.

Because both pressure and ECG waveforms have significantly rounded features, the lower -3 dB frequency response must be 0.05 Hz.

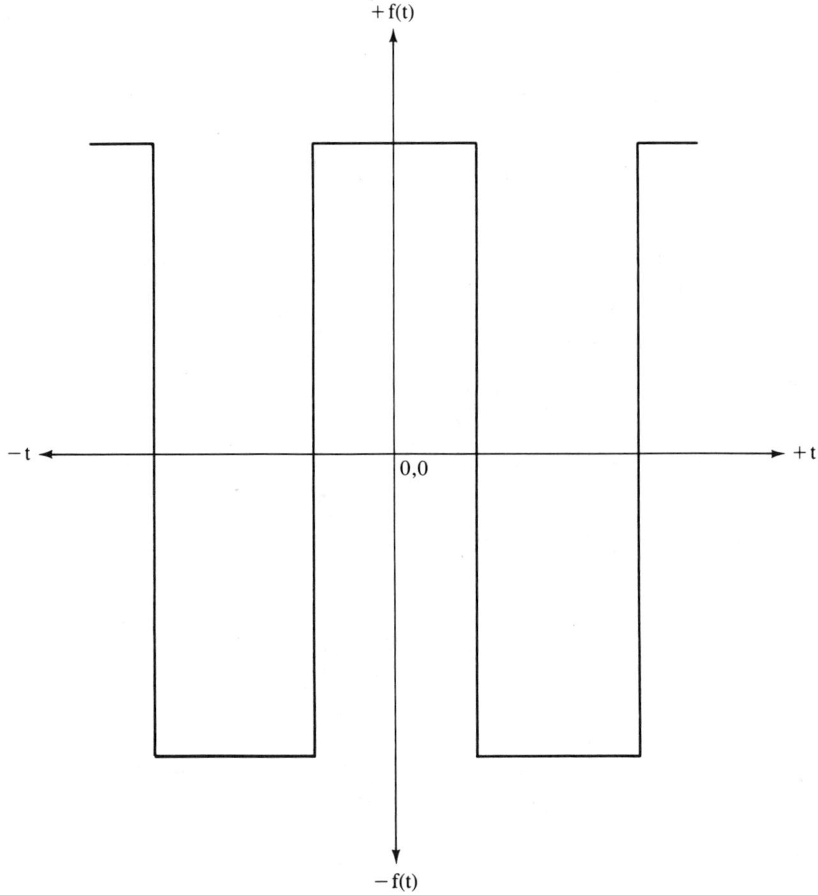

(a)

Figure 12-5 A) Even periodic signal, B) Odd periodic signal.

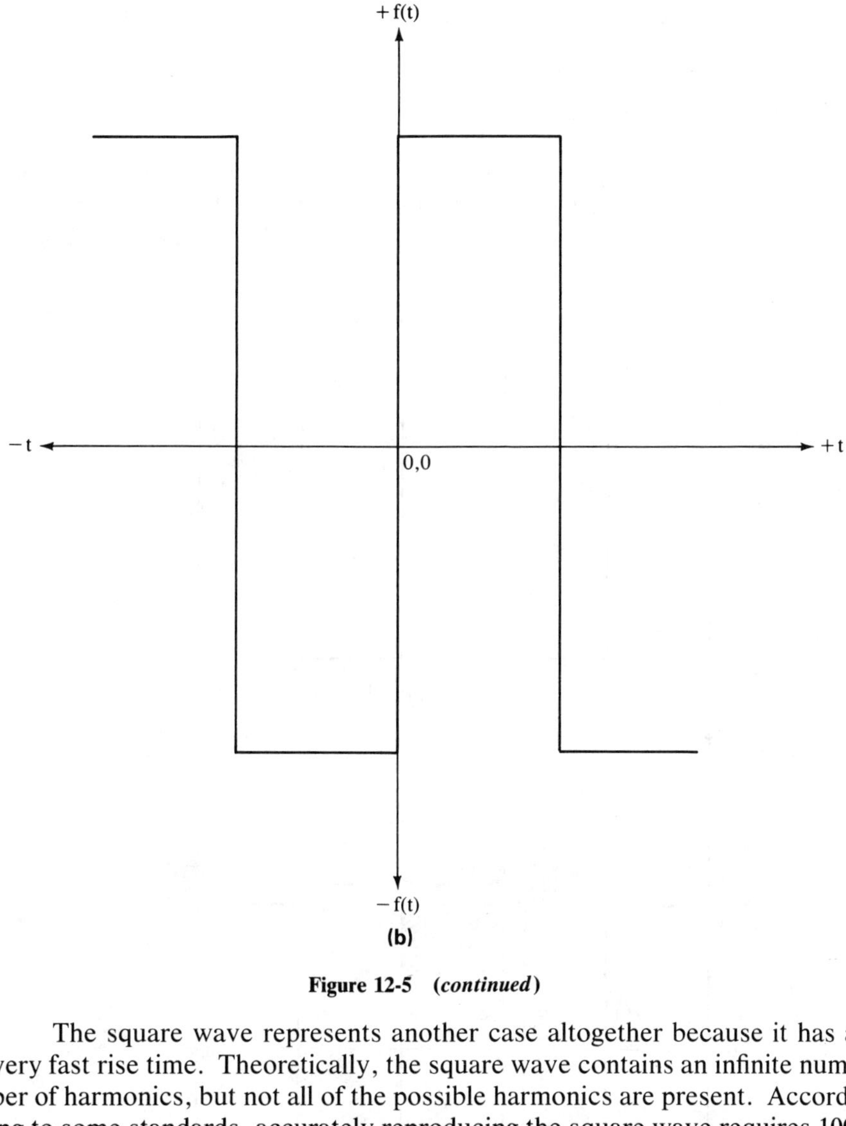

(b)

Figure 12-5 ***(continued)***

The square wave represents another case altogether because it has a very fast rise time. Theoretically, the square wave contains an infinite number of harmonics, but not all of the possible harmonics are present. According to some standards, accurately reproducing the square wave requires 100 harmonics, while others claim that 1000 harmonics are needed. In the case of the square wave, only the odd harmonics are typically found (e.g., 3, 5, 7).

Another factor that determines the profile of the Fourier series of a specific waveform is whether the function is *odd* or *even*. Figure 12-5A shows an odd-function square wave, and Figure 12-5B shows an even-function square wave. The even function is one in which f(t) = f(-t), while for the odd function -f(t) = f(-t). In the even function only cosine harmonics are present, so the sine amplitude coefficient b_n is zero. Similarly, in the odd function only sine harmonics are present, so the cosine amplitude coefficient a_n is zero.

Waveform Symmetry

Symmetry or *asymmetry* can occur in several ways in a waveform (Figure 12-6), and those factors can affect the nature of the Fourier series of the waveform. In Figure 12-6A we see the case of a waveform with a DC

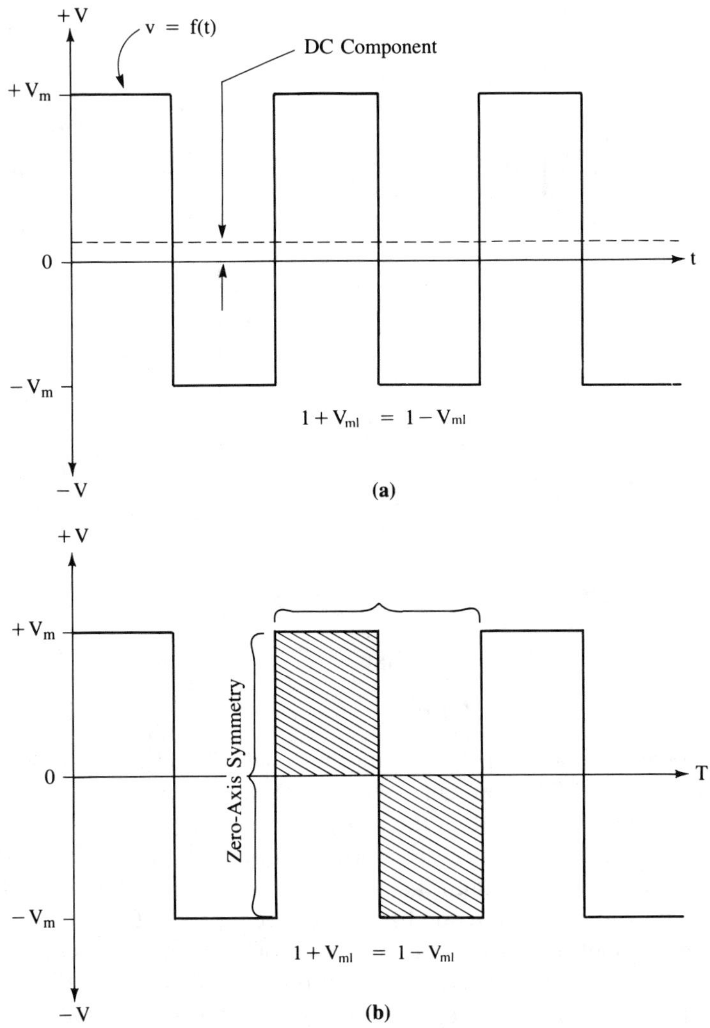

Figure 12-6 A) Nonsymmetrical square wave is not the same on both sides of the zero-volts baseline, B) Zero-axis and half-wave symmetry, C) Baseline mirror symmetry, D) Quarter-wave symmetry.

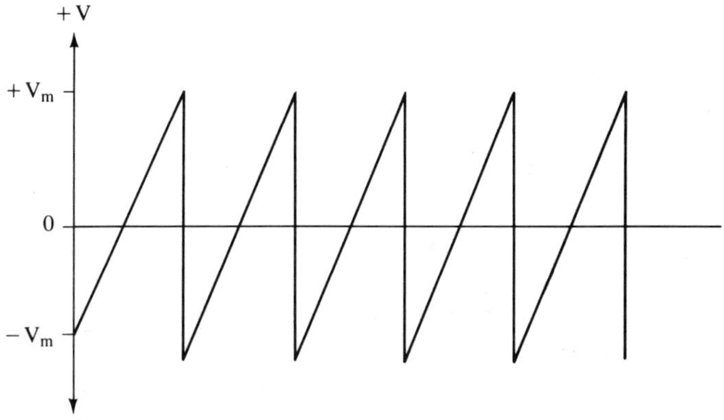

(c)

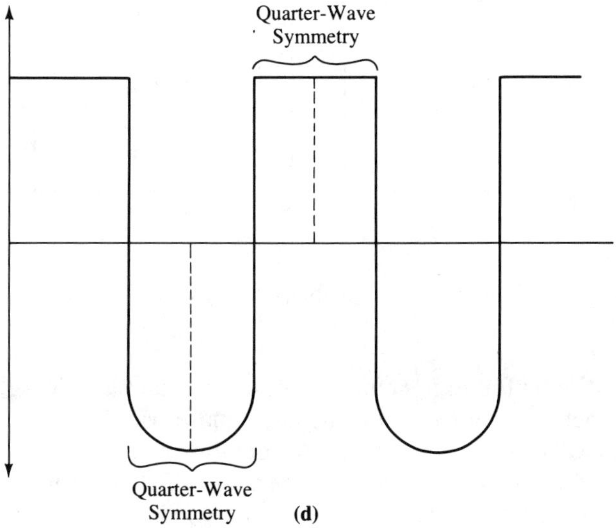

(d)

Figure 12-6 (*continued*)

component. Or, in terms of the Fourier series equation, the term a_0 is nonzero. The DC component represents a case of asymmetry in a signal. This offset can seriously affect instrumentation electronic circuits that are DC-coupled and thereby result in serious artifact.

Two different forms of symmetry are shown in Figure 12-6B. *Zero-axis symmetry* occurs when, on a point-for-point basis, the waveshape and amplitude above the zero baseline is equal to the amplitude below the baseline (or $|+V_m| = |-V_m|$). When a waveform possesses zero-axis symmetry it will usually not contain even harmonics—only odd harmonics are present.

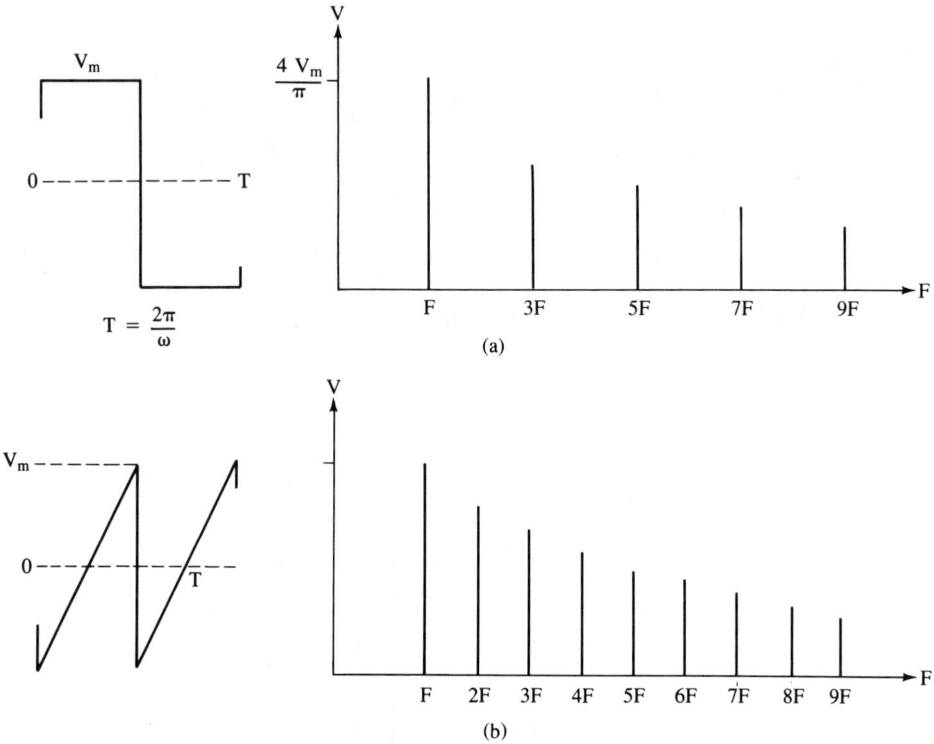

Figure 12-7 A) Square wave and its spectrum, B) Sawtooth wave and its spectrum.

This situation is found in square waves, for example (Figure 12-7A). Zero-axis symmetry is not found in sine and square waves only, however, as the sawtooth waveforms in Figure 12-6C demonstrate.

An exception to the "no even harmonics" general rule is that there will be even harmonics present in the zero-axis symmetrical waveform (Figure 12-7B) *if the even harmonics are inphase with the fundamental sine wave*. This condition will neither produce a DC component nor disturb the zero-axis symmetry.

Also shown in Figue 12-6B is *half-wave symmetry*. In this type of symmetry the *shape* of the wave above the zero baseline is a mirror image of the shape of the waveform below the baseline (see shaded region in Figure 12-6B). Half-wave symmetry also implies a lack of even harmonics.

Quarter-wave symmetry (Figure 12-6D) exists when the left half and right half sides of the waveforms are mirror images of each other on the same side of the zero-axis. Note in Figure 12-6D that above the zero-axis the waveform is like a square wave, and indeed the left- and right-hand sides are mirror images of each other. Similarly, below the zero-axis the rounded waveform has a mirror image relationship between left and right sides. In this case, there is a full set of even harmonics, and any odd harmonics that are present are inphase with the fundamental sine wave.

TRANSIENT SIGNALS

A transient signal is an event that occurs either once only, or occurs randomly over a long period of time, or is periodic but has a very short duration compared with its period (i.e., it is a very short duty cycle event). Many pulse signals fit the latter criterion.

Transient signals are not represented properly by the Fourier series, but they can nonetheless be represented by sine waves in a spectrum. The difference is that the spectrum of the transient signal is *continous* rather than discrete (as in a periodic signal). Consider a transient signal of period *2T*, such as in Figure 12-8A. The *spectral density,* g(W), is:

$$g(\omega) = \int_{-\infty}^{+\infty} F(t)\, e^{-j\omega t}\, dt \tag{12.6}$$

Equation (12.6) represents the spectral density g(W). Given a spectral density, however, the original waveform can be reconstructed from

$$F(t) = \frac{1}{2\pi} \int_{-\infty}^{+\infty} g(\omega)\, e^{j\omega t}\, d\omega \tag{12.7}$$

The shape of the spectral density region is shown in Figure 12-8B. Note that the negative frequencies are a product of the mathematics but do not have physical reality. The shape of Figure 12-8B is expressed by

$$g(\omega) = \frac{\sin \omega t}{\omega t} \tag{12.8}$$

The general form sin X/X is used also for repetitive pulse signals as well as the transient form shown in Figure 12-8B.

SAMPLED SIGNALS

The digital computer is incapable of accepting analog signals but rather requires a digitized representation of that signal. The *analog-to-digital* (A/D) *converter* will convert an input voltage (or current) to a representative binary word. If the A/D converter is either clocked or allowed to run asynchronously according to its own clock, then it will take a continous string of samples of the signal as a function of time. When combined, these signals represent the original analog signal in binary form.

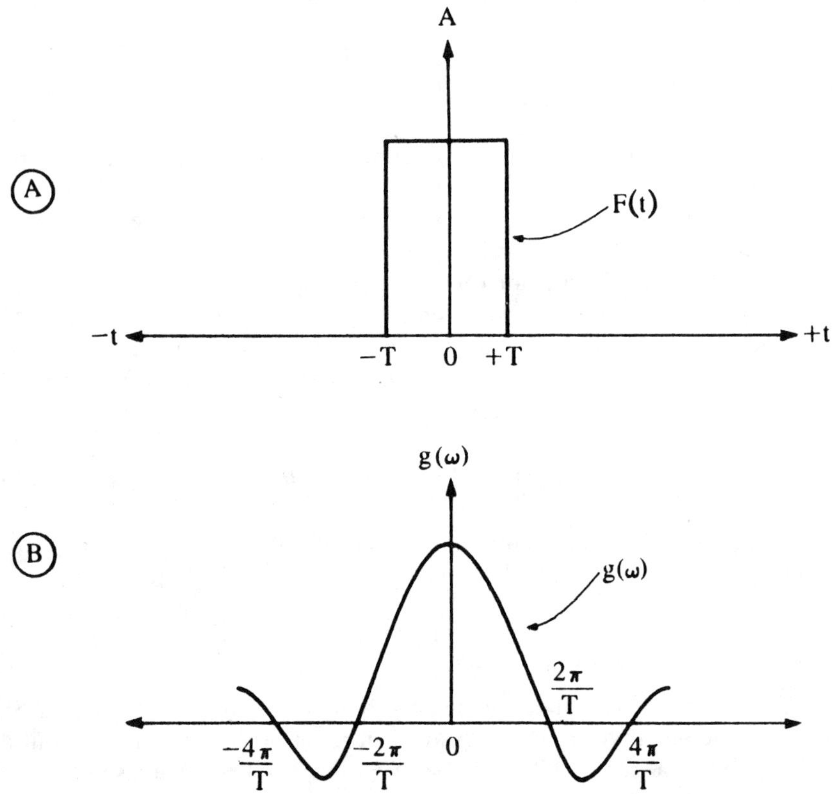

Figure 12-8 Waveform spectrum: A) Transient signal, B) Spectral density.

But the sampled signal is not exactly the same as the original signal, and some effort must be expended to ensure that the representation is as good as possible. Consider Figure 12-9. The waveform in Figure 12-9A is a continous voltage function of time, V(t); in this case a triangle waveform is seen. If the signal is sampled by another signal, p(t), with frequency F_s and sampling period $T = 1/F_s$ (as shown in Figure 12-9B) and then later reconstructed, the waveform may look something like Figure 12-9C. While this may be sufficiently representative of the waveform for many purposes, it would be reconstructed with greater fidelity if the sampling frequency (F_s) were increased.

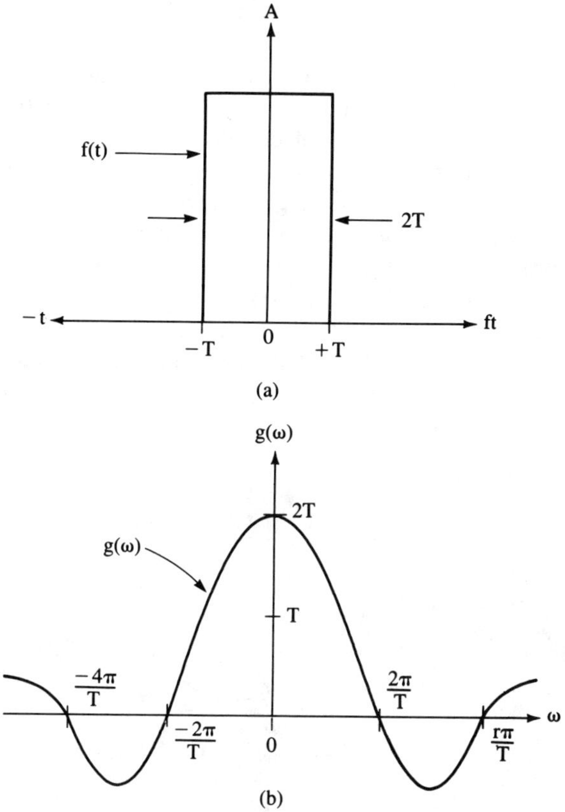

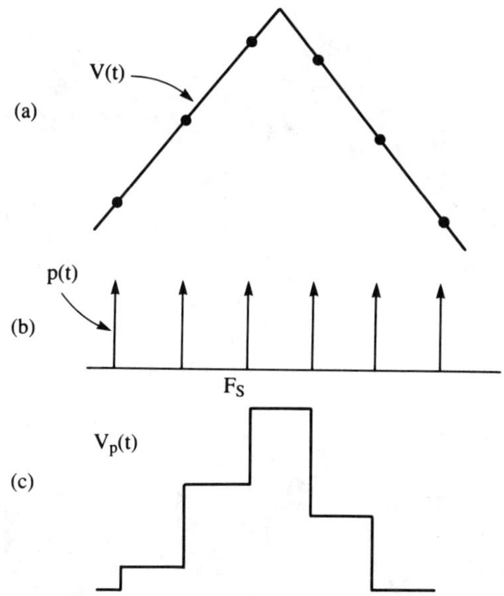

Figure 12-9 Sampling fidelity and error: A) Original waveform, B) Sample pulses, C) Resultant reconstruction.

Figure 12-10 shows another case in which a sine wave, V(t) in Figure 12-10A, is sampled by a pulsed signal, p(t) in Figure 12-10B. The sampling signal, p(t), consists of a train of equally spaced narrow pulses spaced in time by T. The sampling frequency is $F_s = 1/T$. The resultant, shown in Figure 12-10C, is another pulsed signal in which the amplitudes of the pulses represent a sampled version of the original sine wave signal.

The sampling rate, F_s, must by Nyquist's theorem be twice the maximum frequency (F_m) in the Fourier spectrum of the applied analog signal, V(t). In order to reconstruct the original signal after sampling, it is necessary to pass the sampled waveform through a low-pass filter that limits the bandpass to F_s.

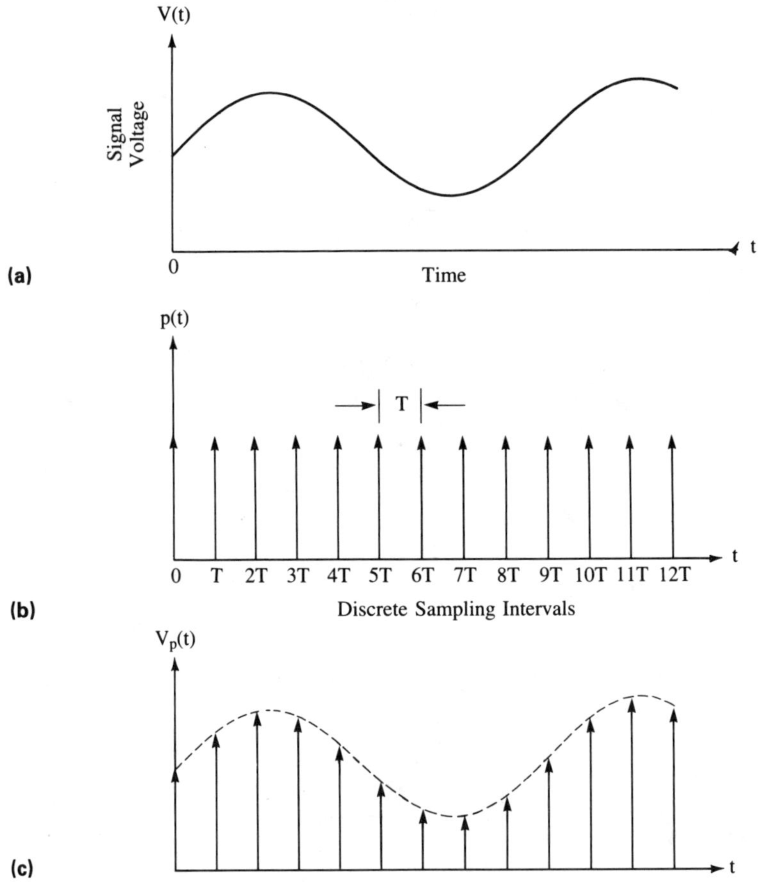

Figure 12-10 Properly sampled waveform: A) Sine wave signal, B) Pulsed sampling signal, C) Resultant signal.

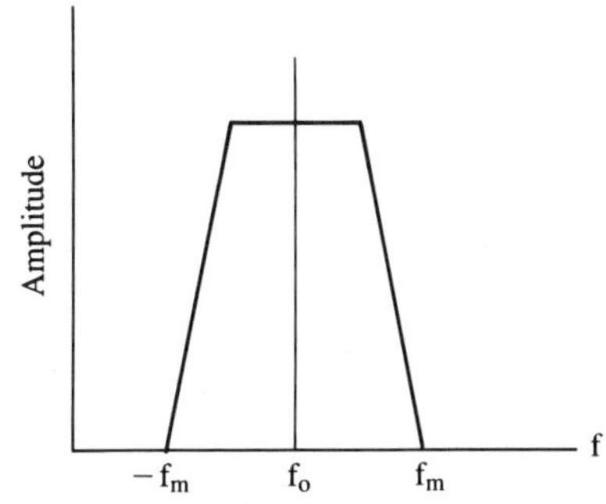

Figure 12-11 Spectrum of sampled signal.

The sampling process is analogous to a form of *amplitude modulation* (AM), in which V(t) is the modulating signal, with spectrum from DC to F_m and p(t) is the carrier frequency. The resultant spectrum is shown partially in Figure 12-11, and it resembles the double sideband with carrier AM spectrum. The spectrum of the modulating signal appears as sidebands around the carrier frequency, shown here as F_0. The actual spectrum is a bit more complex, as shown in Figure 12-12. Like an unfiltered AM radio transmitter, the same spectral information appears not only around the fundamental frequency (F_s) of the carrier (shown at zero in Figure 12-12), but also at the harmonics and subharmonics spaced at intervals of F_s up and down the spectrum.

Provided that the sampling frequency $F_s \geq 2F_m$, the original signal is recoverable from the sampled version by passing it through a low-pass filter with a cutoff frequency F_c, set to pass only the spectrum of the analog signal—but not the sampling frequency. This phenomenon is shown with the dotted line in Figure 12-12.

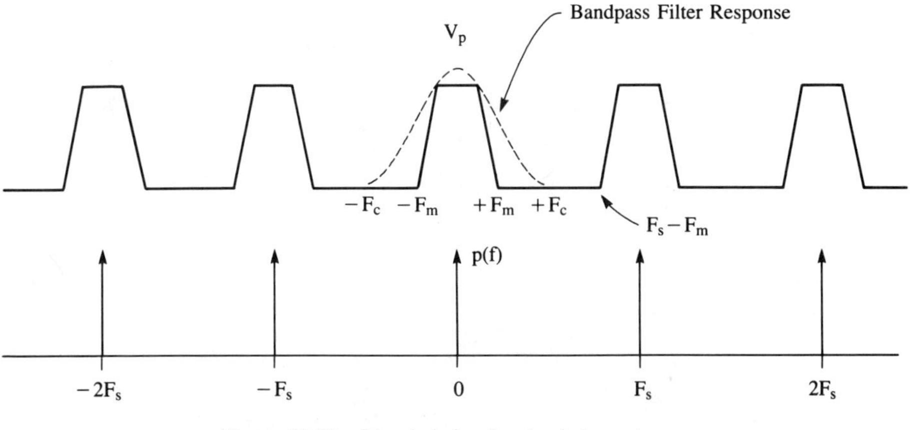

Figure 12-12 Sampled signal extended spectrum.

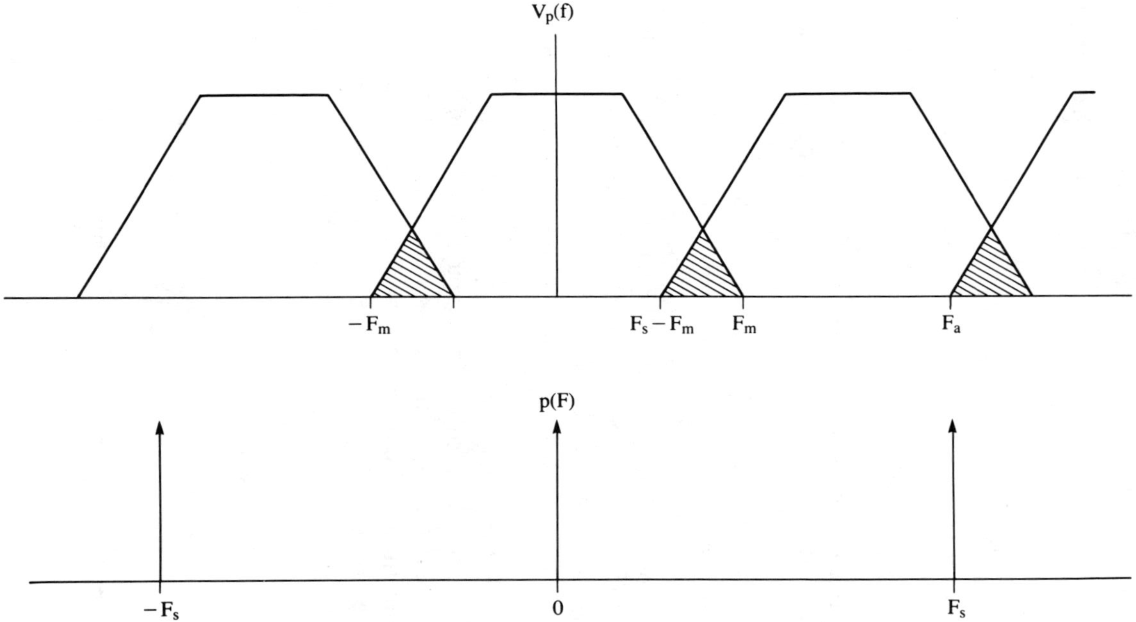

Figure 12-13 Improperly sampled signal.

When the sampling frequency $F_s < 2F_m$, then a problem occurs (see Figure 12-13). The spectrum of the sampled signal looks similar to before, but the regions around each harmonic overlap are such that the value of $-F_m$ for one spectral region is less than $+F_m$ for the next-lower frequency region. This overlap results in a phenomenon called *aliasing*. That is, when the sampled signal is recovered by low-pass filtering, it will produce not the original sine-wave frequency F_0 but a lower frequency equal to $(F_s - F_0)$ . . . and the information carried in the waveform is thus lost or distorted.

The solution, for high-fidelity sampling of the analog waveform for input to a computer, is as follows:

1. Bandwidth limit the signal at the input of the A/D converter with a low-pass filter with a cutoff frequency F_c selected to pass only the maximum frequency in the waveform (F_m) and not the sampling frequency (F_s).
2. Set the sampling frequency F_s *at least* twice the maximum frequency in the applied waveform's Fourier spectrum, i.e., $F_s \geq 2F_m$.

Note: It has been the experience of many designers that some users will not accept a reconstructed sampled waveform if the sample rate is $2F_m$. For example, medical electrocardiograph waveforms, in which $F_m = 100$ Hz, tend to look ''blocky'' (as one nurse remarked) when sampled at 200 Hz and then reconstructed. The user acceptance was much better when the waveform was sampled at 500 Hz, or $5F_m$. While that rate was expensive to accommodate in an 8-bit A/D converter at one time, it is now very low-cost and should be used, Nyquist notwithstanding.

13

TRANSDUCERS AND TRANSDUCTION

Most physical variables that are measured do not lend themselves to direct input into electronic instruments and circuits. Unfortunately, electronic circuits operate only with currents and voltages as inputs. When one is measuring nonelectrical physical quantities it becomes necessary to provide a sensor device that converts physical parameters such as *force*, *displacement,* and *temperature* into proportional voltages or currents. The transducer is such a device.

A *transducer* is a device that converts nonelectrical physical parameters into electrical signals (i.e., currents or voltages) that are proportional to the value of the physical parameter being measured. Transducers take many forms and may be based on a wide variety of physical phenomena. Even when one is measuring the *same* parameter, different instruments may use different types of transducer.

This chapter is not an exhaustive "catalog" treatment covering all transducers—manufacturers' data sheets may be used for that purpose—but we will discuss some of the more common *types* of transducer used in scientific, industrial, medical, and engineering applications.

STRAIN GAGES

All electrical conductors possess electrical resistance. A bar or wire made of such a conductor will have an electrical resistance that is given by

$$R = \frac{\rho L}{A} \tag{13-1}$$

where

R = resistance in ohms

ρ = *resistivity constant,* a property specific to the conductor material, given in units of ohm-centimeters (ohm-cm)

L = length in centimeters (cm)

A = cross-sectional area in square centimeters (cm^2)

Example

A constantan (a 55 percent copper, 45 percent nickel alloy) round wire is 10 cm long and has a radius of 0.01 mm. Find the electrical resistance in ohms. (*Hint*: The resistivity of constantan is 44.2×10^{-6} ohm-cm).

Solution:

$$R = \frac{\rho L}{A}$$

$$= \frac{(44.2 \times 10^{-6} \text{ ohm-cm})(10 \text{ cm})}{0.01 \text{ mm} \times \left|\frac{1 \text{ cm}}{10 \text{ mm}}\right|^2}$$

$$= \frac{4.42 \times 10^{-4} \text{ ohm-cm}^2}{(0.001 \text{ cm})^2}$$

$$= \frac{4.42 \times 10^{-4} \text{ ohm-cm}^2}{\rho\ 10^{-6} \text{ cm}^2} = 141 \text{ ohms}$$

Note that the resistivity factor in Equation (13-1) is a constant, so if length L or area A can be made to vary under the influence of an outside parameter, then the electrical resistance of the wire will change. This phenomenon is called *piezoresistivity*, and it is an example of a transducible property of a material. Piezoresistivity is the change in the electrical resistance of a conductor due to changes in length and cross-sectional area. In piezoresistive materials mechanical deformation of the material produces changes in electrical resistance.

Figure 13-1 shows how an electrical conductor can use the piezoresistivity property to measure *strain*, that is, *force* applied to it in *compression* or *tension*. In Figure 13-1A we have a conductor at rest, in which no forces are acting. The length is given as L_0 and the cross-sectional area as A_0. The resistance of this conductor, from Equation (13-1), is

$$R_0 = \frac{\rho L_0}{A_0} \qquad (13\text{-}2)$$

where

ρ = resistivity as defined previously

R_0 = resistance in ohms when no forces are applied

L_0 = resting (i.e., no force) length (cm)

A_0 = resting cross-sectional area (cm^2)

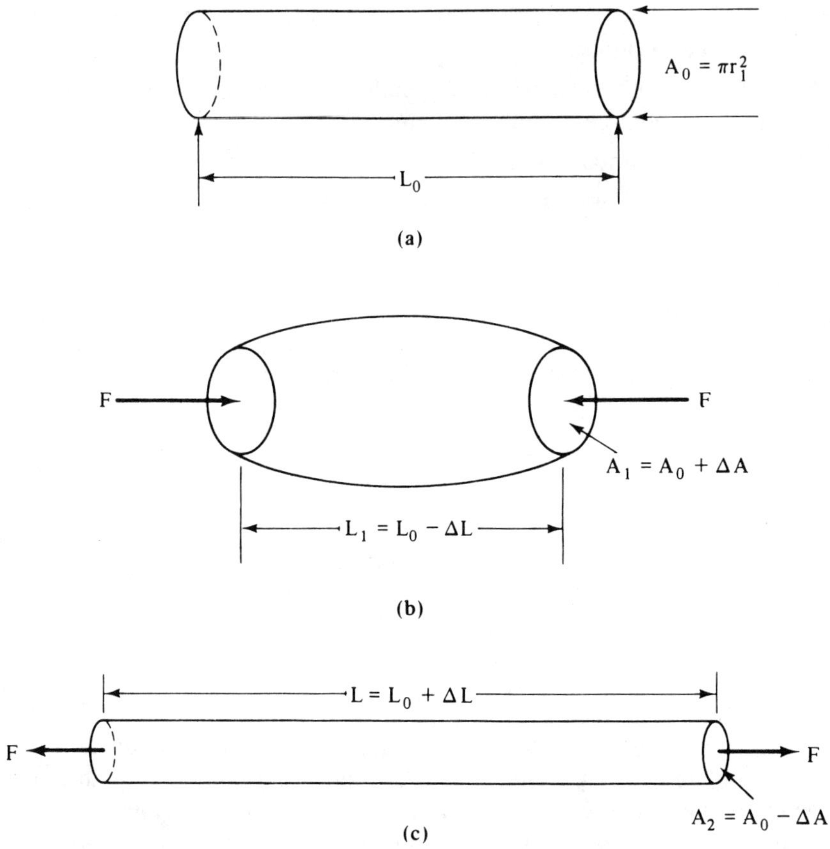

Figure 13-1 Piezoresistive strain gage: A) Conductor at rest, B) In compression, C) In tension.

In Figure 13-1B we see the situation where a compression force of magnitude F is applied along the axis in the inward direction. The conductor will deform, causing the length L_1 to decrease to $(L_0 - \rho L)$ and the cross-sectional area to increase to $(A_0 + \rho A)$. The electrical resistance decreases to $(R_0 - \rho R)$:

$$R_1 = (R_0 - \rho R)\,\rho\,\frac{(L_0 - \rho L)}{A_0 + \rho A} \qquad (13\text{-}3)$$

Similarly, when a tension force of the same magnitude (i.e., F) is applied (i.e., a force that is directed outward along the axis), the length increases to $(L_0 + \rho L)$, and the cross-sectional area decreases to $(A_0 - \rho A)$. See Figure 13-1C. The resistance will increase to

$$R_2 = (R_0 - \rho R)\,\rho\,\frac{L_0 - \rho L}{A_0 + \rho A} \tag{13-4}$$

The *sensitivity* of the strain gage is expressed in terms of unit change of electrical resistance for a unit change in length and is given in the form of a *gage factor S*:

$$S = \frac{\rho R/R}{\rho L/L} \tag{13-5}$$

where

S = gage factor (dimensionless)
R = unstrained resistance of the conductor
ρR = change in resistance due to strain
L = unstrained length of the conductor
ρL = change in length due to strain

We may also express the gage factor in terms of the length and diameter of the conductor. Recall that the diameter is related to the cross-sectional area (i.e., $A = \pi d^2/4 = \pi r^2$), so the relationship between the factor S and these other factors is given by

$$S = 1 + 2\,\frac{\Delta d/d}{\Delta L/L} \tag{13-6}$$

Note that the expression $(\Delta L/L)$ is sometimes denoted by the Greek letter epsilon (e), so Equations (13-5) and (13-6) become

$$S = 1 + 2\,\frac{(\Delta d/d)}{e} \tag{13-7}$$

$$S = \frac{\Delta R/R}{e} \tag{13-8}$$

Gage factors for various metals vary considerably. Constantan, for example, has a gage factor of approximately 2, while certain other common alloys have gage factors between 1 and 2. At least one alloy (92 percent platinum, 8 percent tungsten) has a gage factor of 4. Semiconductor materials such as germanium and silicon can be doped with impurities to provide custom gage factors between 50 and 250. The problem with semiconductor strain gages, however, is that they exhibit a marked sensitivity to temperature changes. Where semiconductor strain gages are used, either a thermally controlled environment or temperature compensating circuitry must be provided.

Bonded and Unbonded Strain Gages

Strain gages can be classified as unbonded or bonded. These categories refer to the method of construction. Figure 13-2 shows both methods of construction.

The unbonded type of strain gage is shown in Figure 13-2A and consists of a wire resistance element stretched taut between two flexible supports. These supports are configured in such a way as to place tension or compression forces on the taut wire when external forces are applied. In the particular example shown, the supports are mounted on a thin metal diaphragm that flexes when a force is applied. Force F1 will cause the flexible supports to spread apart, placing a tension force on the wire and increasing its resistance. Alternatively, when force F2 is applied, the ends of the flexible supports tend to move closer together, effectively placing a compression force on the wire element and thereby reducing its resistance. In actuality, the wire's resting condition is *tautness*, which implies a tension force, so

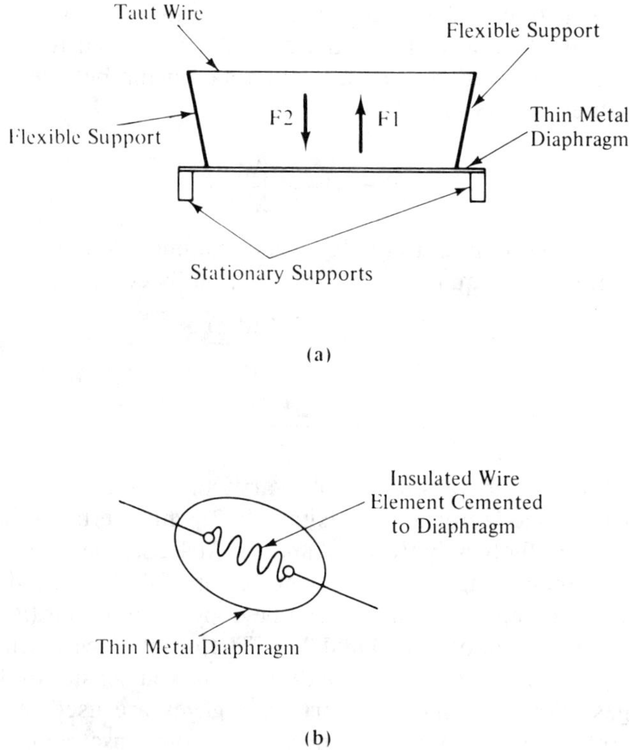

Figure 13-2 A) Unbonded strain gage, B) Bonded strain gage.

F1 increases the tension force from normal, and F2 decreases the normal tension.

The bonded strain gage is shown in Figure 13-2B. In this type of device a wire or semiconductor element is cemented to a thin metal diaphragm. When the diaphragm is flexed, the element deforms to produce a resistance change.

The linearity of both types can be quite good, provided that the elastic limits of the diaphragm and the element are not exceeded. It is also necessary to ensure that the ΔL term is only a very small percentage of L. In the past it has been "standard wisdom" that bonded strain gages are more rugged, but less linear, than unbonded models. Although this may have been true at one time, recent experience has shown that modern manufacturing techniques produce linear, reliable instruments of both types.

STRAIN GAGE CIRCUITRY

Before a strain gage can be useful, it must be connected into a circuit that will convert its resistance changes to a current or voltage output. Most applications are voltage output circuits.

Figure 13-3A shows the *half-bridge* or *voltage-divider* circuit. A strain gage element of resistance R is placed in series with a fixed resistance R1 across a stable, well-regulated voltage source V. The output voltage V_0 is found from the voltage-divider equation

$$V_0 = \frac{V\,R}{R + R1} \tag{13-9}$$

Equation (13-9) describes the output voltage V_0 when the transducer is at rest (i.e., nothing is stimulating the strain gage element). When the element is stimulated, however, its resistance changes a small amount ΔR. To simplify our discussion we will adopt the standard convention used in many texts of letting $h = \Delta R$.

$$V_0 = \frac{V(R + h)}{(R +/- h) + R1} \tag{13-10}$$

Another half-bridge is shown in Figure 13-3B, but in this case the strain gage is in series with a *constant current source* (CCS), which will maintain current I at a constant level regardless of changes in strain gage resistance. The normal output voltage V_0 is

$$V_0 = IR \tag{13-11}$$

$$V_0 = I(R +/- h) \tag{13-12}$$

under stimulated conditions.

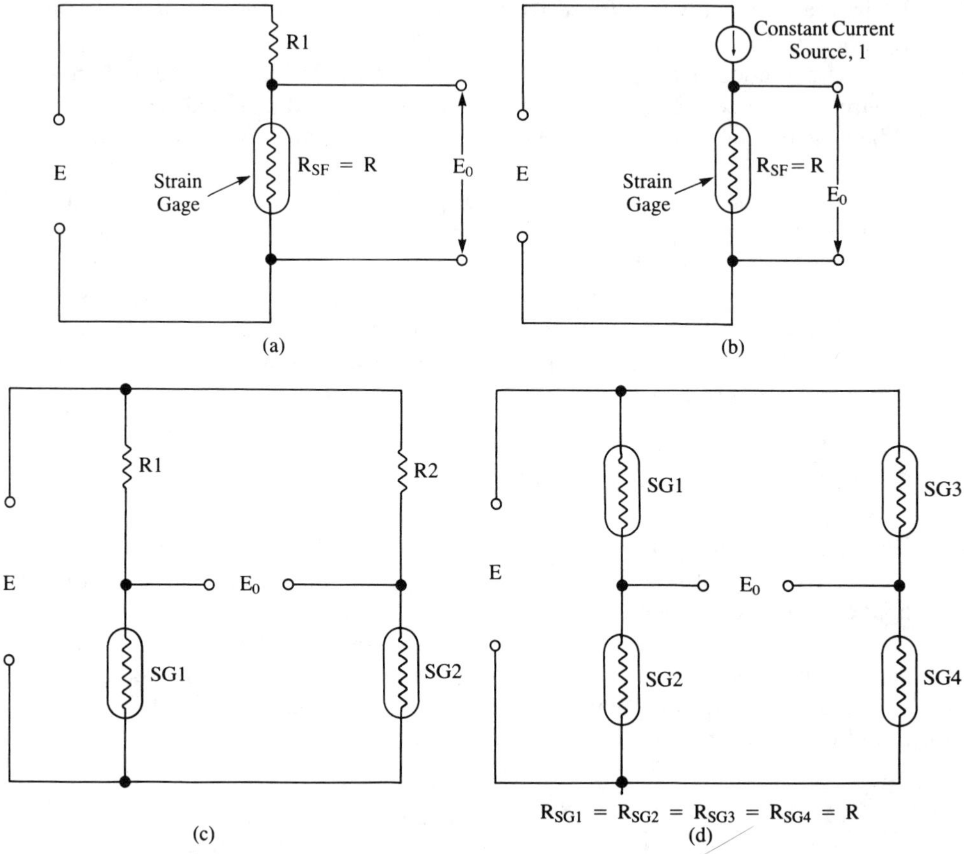

Figure 13-3 A) Voltage-source half-bridge, B) Current-source half-bridge, C) Two-element bridge, D) Four-element bridge.

The Wheatstone Bridge

Many forms of transducer create a variation in an electrical resistance, inductance, or capacitance in response to some physical parameter. These transducers are often in the form of a Wheatstone bridge or one of the related AC bridge circuits. In many cases where the transducer circuit itself is not in the form of a bridge, it may be used in an external bridge circuit that uses other components to form the other arms of the bridge.

The half-bridge circuits suffer from one major defect: output voltage V_0 will always be present regardless of the stimulus. Ideally, in any transducer system, we want V_0 to be zero when the stimulus is also zero and take a value proportional to the stimulus when the stimulus value is nonzero. A

Wheatstone bridge circuit in which one or more strain gage elements form the bridge arms has this property.

Figure 13-3C shows a circuit in which strain gage elements SG1 and SG2 form two bridge arms and fixed resistors R1 and R2 form the other two arms. It is usually the case that SG1 and SG2 will be configured so that their actions oppose each other; that is, under stimulus, SG1 will have a resistance R + h and SG2 will have a resistance R − h, or vice versa.

One of the most linear forms of transducer bridge is the circuit of Figure 13-3D, in which all four bridge arms contain strain gage elements. In most such transducers all four strain gage elements have the same resistance (R), which has a value between 100 and 1000 ohms in most cases.

Recall that the output voltage from a Wheatstone bridge is the difference between the voltages across the two half-bridge dividers. The following equations hold true for bridges in which one, two, or four equal active elements are used. For one active element,

$$V_0 = \frac{V\,h}{4\,R} \tag{13-13}$$

(accurate to +/−5 percent, provided that $h \leq 0.1$). For two active elements,

$$V_0 = \frac{V\,h}{2\,R} \tag{13-14}$$

For four active elements,

$$V_0 = \frac{V\,h}{R} \tag{13-15}$$

where, for all three equations,

V_0 = output potential in volts (V)

V = excitation potential in volts (V)

R = resistance of all bridge arms

h = quantity ΔR, the change in resistance of a bridge arm under stimulus

(These equations apply only for the case where all the bridge arms have equal resistances under zero stimulus conditions.)

Example

A transducer that measures force has a nominal resting resistance of 300 Ω and is excited by +7.5 V DC. When a 980-dyne force is applied, all four equal-resistance bridge elements change resistance by 5.2 ohms. Find the output voltage V_0.

Solution:

$V_0 = Vh/R$

$= (7.5\text{ V})(5.2\text{ ohms}))/(300\text{ ohms})$

$= ((7.5\text{ V})(5.2))/300 = 0.13\text{ V}$

Transducer Sensitivity

When designing electronic instrumentation systems involving strain gage transducers, it is convenient to use the *sensitivity factor* (denoted by the Greek letter psi, ψ), which relates the output voltage in terms of the excitation voltage and the applied stimulus. In most cases, we see a specification giving the number of microvolts or millivolts output per volt of excitation potential per unit of applied stimulus (i.e., $\mu V/V/Q_0$).

$$\psi = V_0/V/Q_0 \tag{13-16}$$

and

$$\psi = \frac{V_0}{V \times Q_0}$$

where
V_0 = output potential
V = one unit of electrical potential (i.e., 1 V)
Q_0 = one unit of stimulus

The sensitivity is often given as a specification by the transducer manufacturer. From it we can predict output voltage for any level of stimulus and excitation potential. The output voltage, then, is found from

$$V_0 = \psi VQ \tag{13-17}$$

where
V_0 = output potential in volts (V)
ψ = sensitivity in $\mu V/V/Q_0$
V = excitation potential in volts (V)
Q = stimulus parameter

Example

A well-known medical arterial blood pressure transducer uses a four-element piezoresistive Wheatstone bridge with a sensitivity of 5 μV per volt of excitation per torr (T) of pressure, that is 5 μV/V/T (*Note*: 1 torr = 1 mm Hg). Find the output voltage if the bridge is excited by 5 V DC and 120 torr of pressure is applied.

Solution:

$$V_0 = \psi VQ$$

$$= \frac{5\psi V}{V - T} \times (5\ V) \times (120\ T)$$

$$= (5 \times 5 \times 120)\ \mu V = 3{,}000\ \mu V$$

Balancing and Calibrating the Bridge

Few, if any, Wheatstone bridge strain gages meet the ideal condition in which all four arms have exactly equal resting resistances. In fact, the bridge resistance specified by the manufacturer is a *nominal* value only. There will inevitably be an *offset voltage* (i.e., $V_0 \leq 0$) when $Q = 0$. Figure 13-4 shows a circuit that will balance the bridge when the stimulus is zero. Potentiometer Rl, which is usually a type that uses 10 or more turns to cover its entire resistance range, is used to inject a balancing current I into the bridge circuit at one of the nodes. R1 is adjusted, with the stimulus at zero, for zero output voltage.

The best calibration method is to apply a precisely known value of stimulus (e.g., apply 100 torr to a pressure transducer) to the transducer and adjust the amplifier following the transducer for the output proper for that level of stimulus. But that practice may prove unreasonably difficult in some cases, so an artificial calibrator is needed to simulate the stimulus. This function is provided by R3 and S1 in Figure 13-4. When S1 is open, the transducer is able to operate normally, but when S1 is closed it *unbalances* the bridge and produces an output voltage V_0 that simulates some standard value of the stimulus. The value of R3 is given by

$$R3 = \frac{R}{4Q\psi} - \frac{R}{2} \tag{13-18}$$

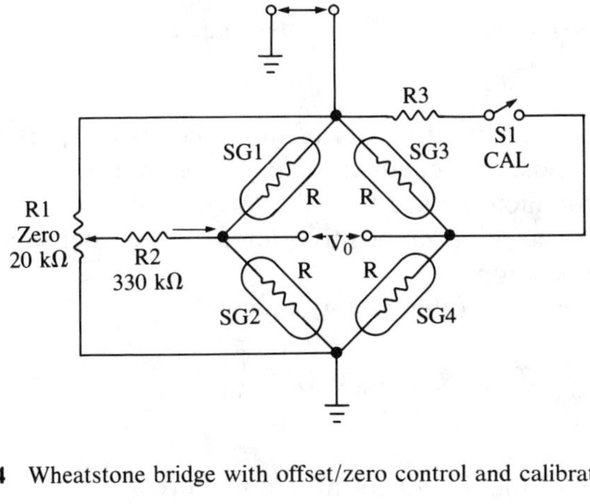

Figure 13-4 Wheatstone bridge with offset/zero control and calibration circuit.

where

R3 = resistance of R3 in ohms

R = nominal resistance of the bridge arms (ohms)

Q = calibrated stimulus parameter

ψ = sensitivity factor (μV/V/Q); note the difference in the units: V instead of μV

TEMPERATURE TRANSDUCERS

A large number of physical phenomena are temperature dependent, so we find quite a variety of electrical temperature transducers on the market. In this discussion, however, we will discuss only three basic types: *thermistor*, *thermocouple*, and *semiconductor pn junctions*.

Thermistors

Metals and most other conductors are temperature sensitive and will change electrical resistance with changes in temperature, as follows:

$$R_t = R_0[1 + \psi(T - T_0)] \qquad (13\text{-}19)$$

where

R_t = resistance in ohms at temperature T

R_0 = resistance in ohms at temperature T_0 (often a standard reference temperature)

T = temperature of the conductor

T_0 = a previous temperature of the conductor at which R_0 was determined

ψ = *temperature coefficient* of the material, a property of the conductor ($^{\circ}C^{-1}$)

The temperature coefficients of most metals are positive, as are the coefficients for most semiconductors (e.g., gold has a value of $+0.004/^{0}C$). Ceramic semiconductors used to make *thermistors* (i.e., thermal resistors) can have either negative or positive temperature coefficients, depending upon their composition.

The resistance of a thermistor is given by

$$R_t = R_0 \text{ EXP } B\left[\frac{1}{T} - \frac{1}{T_0}\right] \qquad (13\text{-}20)$$

where

R_t = resistance of the thermistor at temperature T

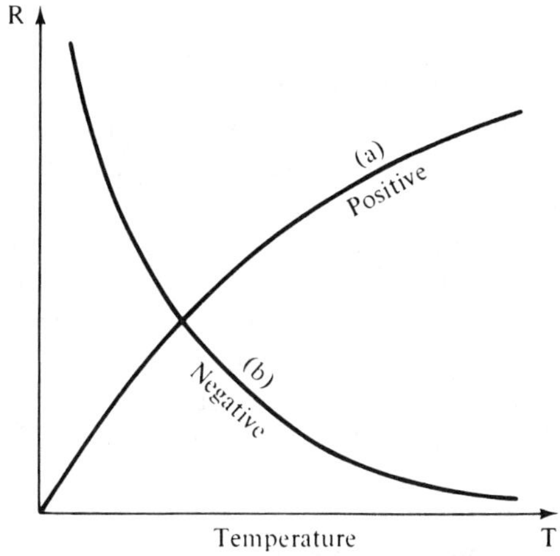

Figure 13-5 PTC and NTC thermistor characteristics.

R_0 = resistance of the thermistor at a reference temperature (usually the ice point, 0° C, or room temperature, 25° C)

e = base of the natural logarithms

T = thermistor temperature in kelvins (K)

T_0 = reference temperature in kelvins (K)

ψ = a property of the material used to make the thermistor

(*Note*: ψ will usually have a value between 1500 and 7000° K.)

Equation (13-20) demonstrates that the response of a thermistor is exponential, as shown in Figure 13-5. Note that both curves are nearly linear over a portion of their ranges, but they become decidedly nonlinear in the remainder of the region. If a wide measurement range is needed, a linearization network will be required.

Thermistor transducers will be used in any of the circuits in Figure 13-3. They will also be found using many packaging arrangements. Figure 13-6 shows a bead thermistor used in medical instruments to continuously monitor a patient's rectal temperature.

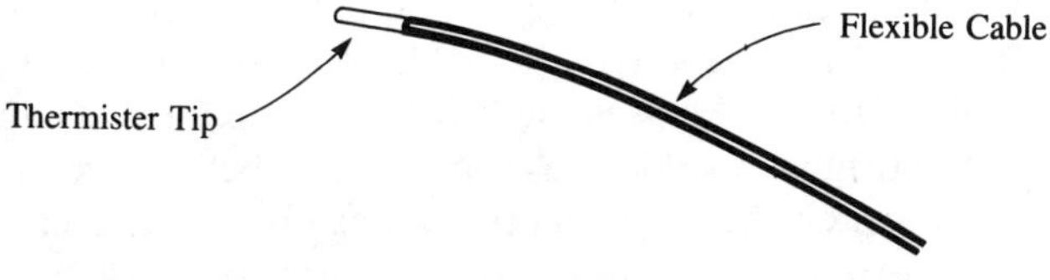

Figure 13-6 Typical thermistor for medical applications.

The equations governing thermistors apply if there is little *self-heating* of the thermistor. Although there are applications where self-heating is used, in straight temperature measurements it is to be avoided. To minimize self-heating it is necessary to control the power dissipation of the thermistor. In most cases, this requirement is met by controlling the applied excitation voltage.

Also of concern in some applications is the *time constant* of the thermistor. The resistance of the thermistor does not jump immediately to the new value when the temperature changes, but rather it requires a small amount of time to stabilize at the new resistance value. This is expressed in terms of the time constant of the thermistor in a manner that is reminiscent of capacitors charging in RC circuits.

Thermocouples

When two dissimilar metals are joined together to form a "vee" as in Figure 13-7A, it is possible to generate an electrical potential merely by heating the junction. This phenomenon, first noted by Seebeck in 1823, is due to different *quantum work functions* for the two metals. Such a junction is called a *thermocouple*. The Seebeck voltage generated by the junction is proportional to the junction temperature and is reasonably linear over wide temperature ranges.

A simple thermocouple is shown in Figure 13-7B; it uses two junctions. One junction is the measurement junction, and it is used as the thermometry probe. The other junction is a reference and is kept at a reference temperature, such as the *ice point* (0° C) or room temperature.

Interestingly, there is an inverse thermocouple phenomenon, called the *Peltier effect*, in which an electrical potential applied across A-B in Figure 13-7B will cause one junction to absorb heat (i.e., get hot) and the other to lose heat (i.e., get cold). Semiconductor thermocouples have been used in small-scale environmental temperature chambers.

Semiconductor Temperature Transducers

Ordinary PN junction diodes exhibit a strong dependence upon temperature. This effect can be easily demonstrated by using an ohmmeter and an ordinary rectifier diode such as the 1N400x-series devices. Connect the ohmmeter so that it forward biases the diode and note the resistance at room temperature. Next, hold a soldering iron or other heat source close to the diode's body and watch the electrical resistance change. In a circuit such as Figure 13-8 the current is held constant, so output voltage V_0 will change with temperature-caused changes in diode resistance.

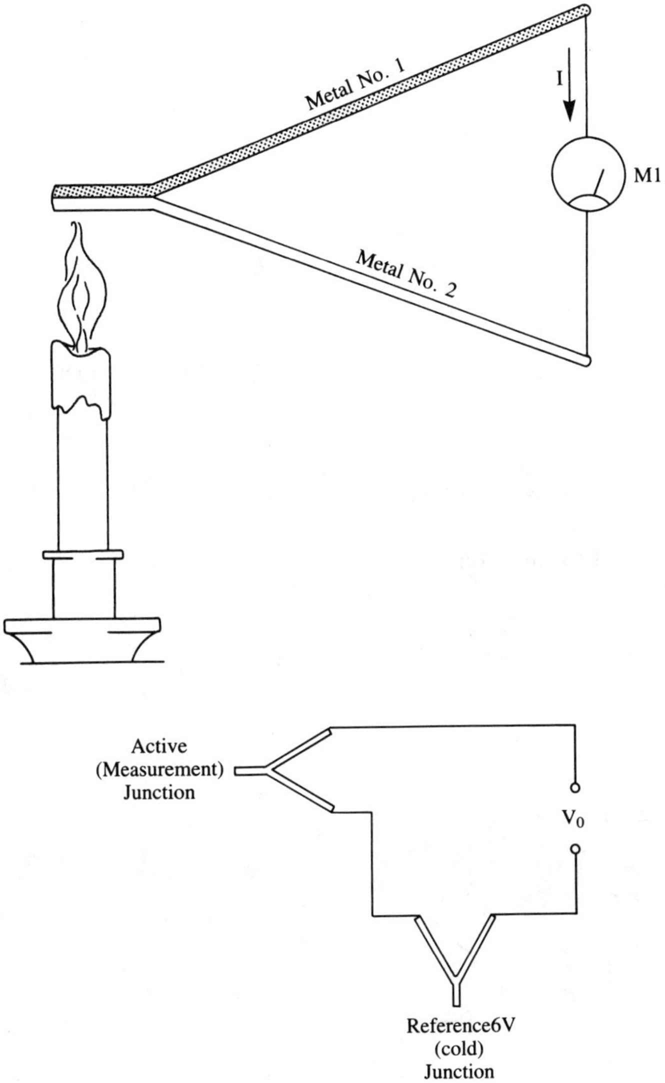

Figure 13-7 A) Basic thermocouple, B) Thermocouple circuit.

Another solid-state temperature transducer is shown in Figure 13-9. In this version, the temperature sensor device is a pair of diode-connected transistors. In any transistor the base-emitter voltage V_{be} is

$$V_{be} = \frac{kT}{q} \log \left[\frac{I_c}{I_s}\right] \tag{13-21}$$

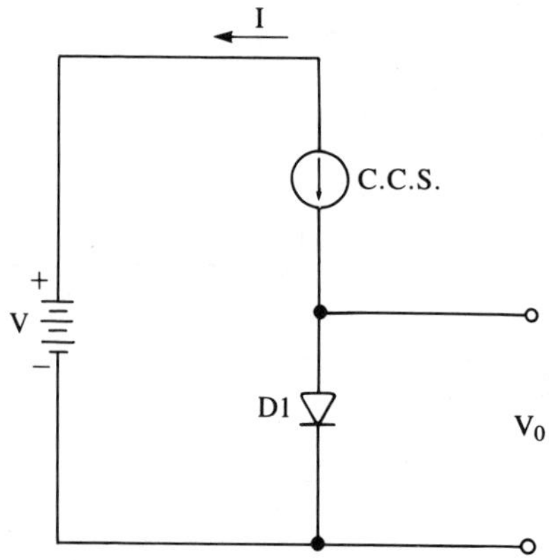

Figure 13-8 Diode temperature sensor with CCS excitation.

where
V_{be} = base-emitter potential in volts (V)
k = Boltzmann's constant (1.38×10^{-23} J/K)
T = temperature in kelvins (K)
q = electronic charge, 1.6×10^{-19} coulomb (C)
I_c = collector current in amperes (A)
I_s = reverse saturation current in amperes (A)

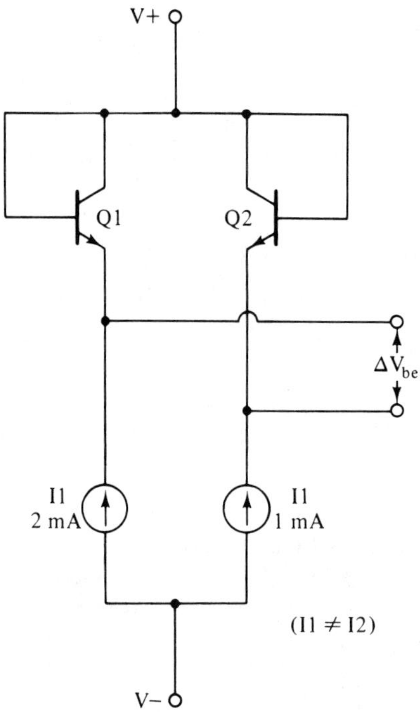

Figure 13-9 PN junction transistors used in temperature sensor circuit.

Note that the *k* and q terms in Equation (13-21) are constants, and both currents can be made to be constant. The only variable, then, is *temperature*.

The circuit of Figure 13-9 uses two transistors connected to provide a differential output voltage (ΔV_{be} that is the difference between $V_{be(Q)}$ and $V_{be(Q2)}$. Combining the expressions for V_{be} for both transistors yields the expression

$$\Delta V_{be} = \frac{kT}{q} \log \left[\frac{I1}{I2}\right] \qquad (13\text{-}22)$$

Note that since Ln 1 = 0, currents Il and I2 *must not be equal.* In general, designers set a ratio of 2:1, for example, I1 = 2 mA and I2 = 1 mA. Since currents I1 and I2 are supplied from constant current sources, the ratio I1/I2 is a constant. Also, it is true that the logarithm of a constant is a constant. Therefore, all terms in Equation (13-22) are constants, except temperature T. Equation (13-22), therefore, may be written in the form

$$\Delta V_{be} = KT \qquad (13\text{-}23)$$

where

$K = (k/q) \log (I1/I2)$

$= ((1.38 \times 10^{-23})/(1.6 \times 10_{-19})) \log (2/1)$

$= 5.98 \times 10^{-5}$ V/K

$= 59.8\ \mu$V/K

We may now rewrite Equation (13-22) in the form

$$\Delta V_{be} = 59.8\ \mu V/^{0}K$$

In most thermometers using the circuit of Figure 13-9, an amplifier increases the output voltage to a level numerically the same as a unit of temperature, so the temperature may be easily read from a digital voltmeter. The most common scale factor is 10 mV/K, so for our transducer the postamplifier requires a gain of

$$A_v = \frac{10\ mV/^{0}K}{59.8\ \mu V \times (1\ mV/10^3\ \mu V)} = 167 \qquad (13.25)$$

INDUCTIVE TRANSDUCERS

Inductance (L) and inductive reactance (X_L) are transducible properties because they can be *varied* by certain mechanical methods. Figure 13-10A shows an example of an inductive Wheatstone bridge. Resistors R1 and R2 form two fixed arms of the bridge, while coils L1 and L2 form variable

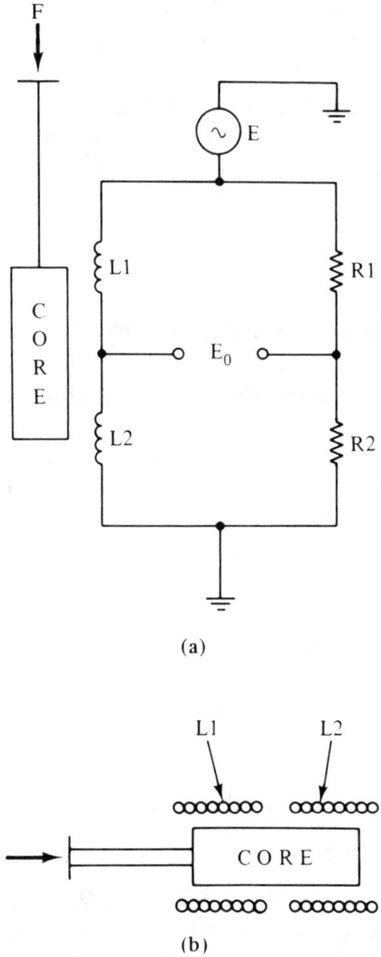

Figure 13-10 Inductive bridge transducer: A) Circuit, B) Structure.

arms. Since inductors are used, the excitation voltage must be AC. In most cases, the AC excitation source will have a frequency between 400 and 5000 Hz and an rms amplitude of 5 to 10 V.

The inductors are constructed coaxially, as shown in Figure 13-10B, with a common core. It is a fundamental property of any inductor that a ferrous core increases its inductance. In the rest condition (i.e., zero stimulus), the core will be positioned equally inside of both coils. If the stimulus moves the core in the direction shown in Figure 13-10B, the core tends to move out of L1 and further into L2. This action reduces the inductive reactance of L1 and increases that of L2, unbalancing the bridge.

LINEAR VARIABLE DIFFERENTIAL TRANSFORMERS

Another form of inductive transformer is the *linear variable differential transformer* (LVDT) shown in Figure 13-11A. The construction of the LVDT (Figure 13-11B) is similar to that of the inductive bridge, except that it also contains a primary winding.

One advantage of the LVDT over the bridge-type transducer is that it provides higher output voltages for small changes in core position. Several commercial models are available that produce 50 to 300 mV/mm. In the

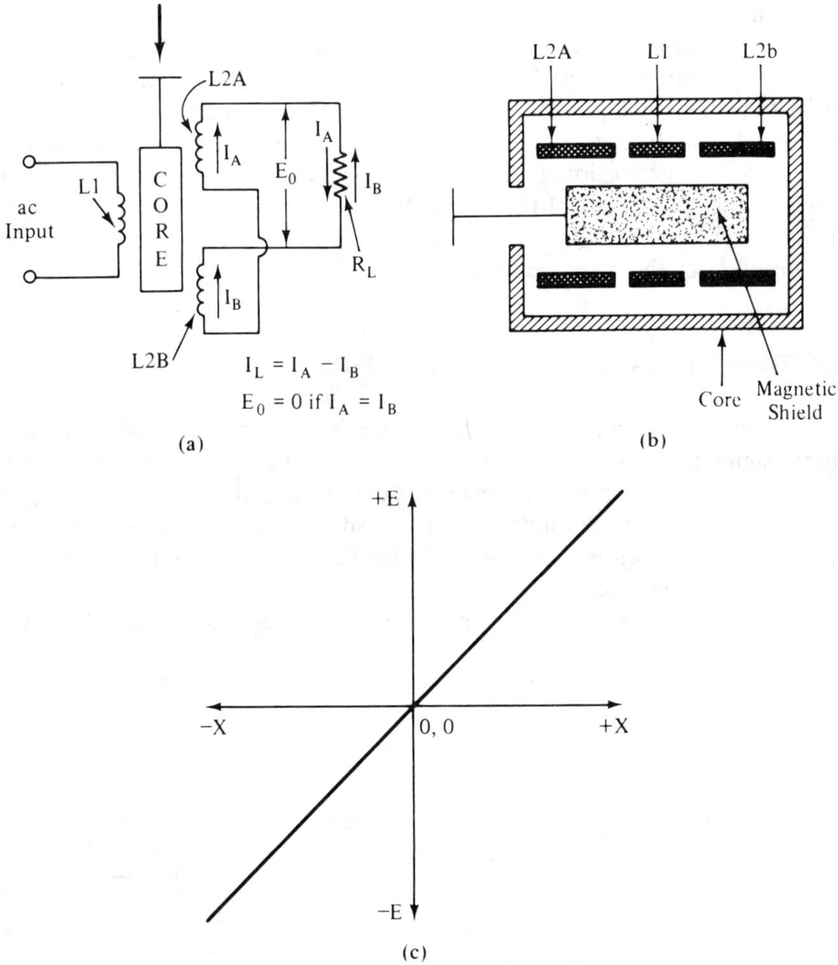

Figure 13-11 LDVT transducer: A) Circuit, B) Structure, C) Transfer function.

latter case, this means that a 1-mm displacement of the core produces a voltage output of 300 mV.

In normal operation, the core is equally inside both secondary coils, L2A and L2B, and an AC carrier is applied to the primary winding. This carrier typically has a frequency between 40 Hz and 20 kHz and an amplitude in the range from 1 to 10 V rms.

Under rest conditions the coupling between the primary and each secondary is equal. The currents flowing in each secondary, then, are equal to each other. Note in Figure 13-11A that the secondary windings are connected in series opposing, so if the secondary winding currents are equal, they will exactly cancel each other in the load. The AC voltage appearing across the load, therefore, is *zero* ($I_A = I_B$).

When the core is moved so that it is more inside L2B and less inside L2A, the coupling between the primary and L2B is greater than the coupling between the primary and L2A. Since this fact makes the two secondary currents no longer equal, the cancellation is not complete. The current in the load I_L is no longer zero. The output voltage appearing across load resistor R_L is proportional to the core displacement, as shown in Figure 13-11C. The *magnitude* of the output voltage is proportional to the *amount* of core displacement, while the *phase* of the output voltage is determined by the *direction* of the displacement.

POSITION-DISPLACEMENT TRANSDUCERS

A position-displacement transducer, or position transducer, will create an output signal that is proportional to the position of some object along a given axis. For very small position ranges we could use a strain gage (i.e., Figure 13-12), but note that the range of such transducers is necessarily very small. Most strain gages either are nonlinear for large displacements or are damaged by large displacements.

The LVDT can be used as a position transducer. Recall that the output

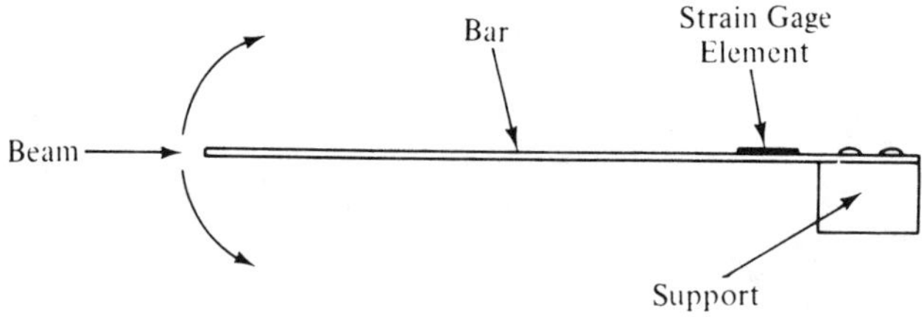

Figure 13-12 Displacement sensor based on strain gage.

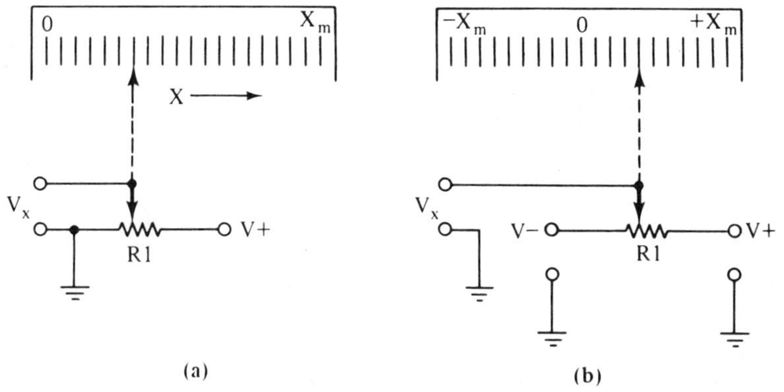

Figure 13-13 Resistor displacement sensors: A) One quadrant, B) Two quadrant.

polarity indicates the direction of movement from a zero-reference position, and the amplitude indicates the magnitude of the displacement. Although the LVDT will accommodate larger displacements than the strain gage, it is still limited in range.

The most common form of position transducer is the potentiometer. For applications that are not too critical, it is often the case that ordinary linear taper potentiometers are sufficient. Rotary models are used for curvilinear motion and slide models for rectilinear motion. In precision applications designers use either regular precision potentiometers or special potentiometers designed specifically as position transducers.

Figure 13-13 shows two possible circuits using potentiometers as position transducers. In Figure 13-13A we see a single-quadrant circuit for use where the zero point (i.e., starting reference) is at one end of the scale. The pointer will always be at some point such that $0 <= X <= X_m$. The potentiometer is connected so that one end is grounded and the other is connected to a precision, regulated voltage source V+. The value of V_x represents X and will be $0 <= V_x <= V+$, such that $V_x = 0$ when $X = 0$, and $V_x = V+$ when $X = X_m$.

A two-quadrant system is shown in Figure 13-13B and is similar to the previous circuit except that, instead of grounding one end of the potentiometer, it is connected to a precision, regulated *negative*-to-ground power source, V−. Figure 13-14 shows the output functions of these two transducers. Figure 13-14A represents the circuit of Figure 13-13A, while Figure 13-14B represents the circuit of Figure 13-13B.

A four-quadrant transducer can be made by placing two circuits such as Figure 13-13B at right angles to each other and arranging linkage so that the output signal varies appropriately.

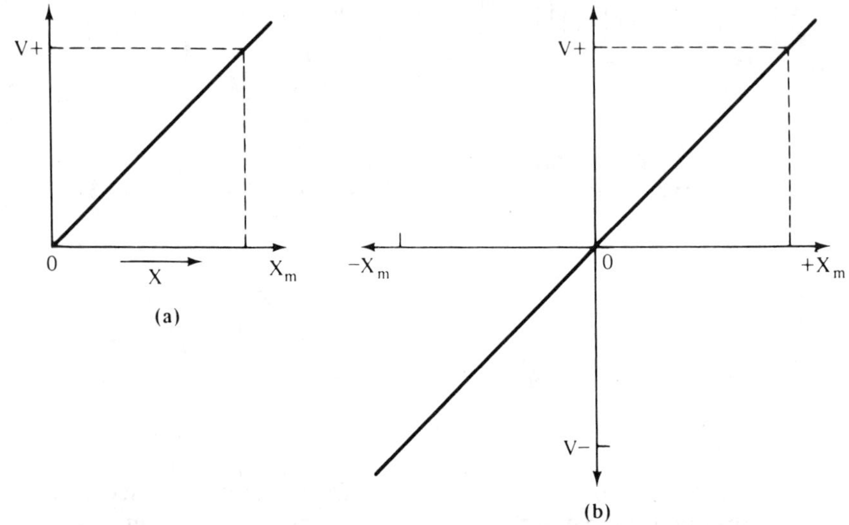

Figure 13-14 A) Output of Figure 13-13A, B) Output of Figure 13-13B.

VELOCITY AND ACCELERATION TRANSDUCERS

Velocity can be defined as displacement per unit of time, and acceleration is the time rate of change of velocity. Since both velocity (v) and acceleration (a) can be related back to position (s), we often find position transducers used to *derive* velocity and acceleration signals. The relationships are as follows:

$$v = \frac{ds}{dt} \tag{13-26}$$

$$a = \frac{dv}{dt} \tag{13-27}$$

$$a = \frac{d^2s}{dt^2} \tag{13-28}$$

Velocity and acceleration are the first and second time derivatives of displacement (i.e., change of position), respectively. We may derive electrical signals proportional to v and a by using an operational amplifier differentiator circuit (see Figure 13-15). The output of the transducer is a time-dependent function of position (i.e., displacement). This signal is differentiated by the stages following to produce the velocity and acceleration signals.

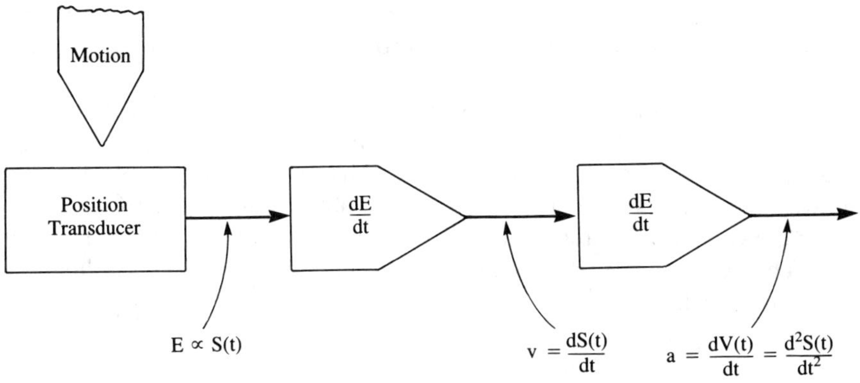

Figure 13-15 Generating accelleration and velocity signals from a displacement transducer.

TACHOMETERS

Alternating- and direct-current generators are also used as velocity transducers (collectively called "tachometers"). In their basic form they will transduce rotary motion (i.e., produce an angular velocity signal), but with appropriate mechanical linkage they will also indicate rectilinear motion. In the case of a DC generator, the output signal is a DC voltage with a magnitude that is proportional to the angular velocity of the armature shaft. The AC generator, or *alternator*, maintains a relatively constant output voltage, but its AC *frequency* is proportional to the angular velocity of the armature shaft.

If a DC output is desired, instead of an AC signal, a circuit similar to Figure 13-16 is used. The AC output of the tachometer is fed to a trigger circuit (either a comparator or Schmitt trigger) so that squared-off pulses are created. These pulses are then differentiated to produce spikelike pulses to

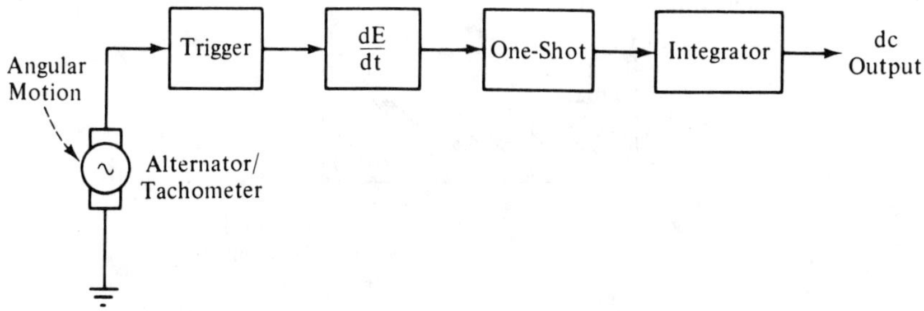

Figure 13-16 Tachometer sensor.

trigger the monostable multivibrator (one-shot). The output of the one-shot is integrated to produce a DC level proportional to the tachometer frequency.

The reason for using the one-shot stage is to produce output pulses that have a *constant amplitude* and *duration*. Only the pulse repetition rate (i.e., number of pulses per unit of time) varies with the input frequency. This fact allows us to integrate the one-shot output to obtain our needed DC signal. If either duration or amplitude varied, the integrator output would be meaningless. This technique is widespread in electronic instruments, so it should be understood well.

FORCE AND PRESSURE TRANSDUCERS

Force transducers can be made by using strain gages or either LVDT or potentiometer displacement transducers. In the case of the displacement transducer (Figure 13-17A) it becomes a force transducer by causing a power spring to either compress or stretch. Recall Hooke's law, which tells us that

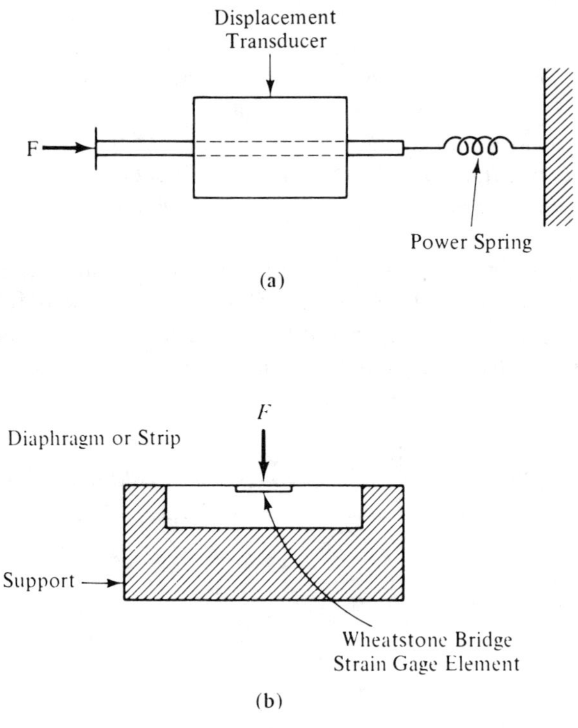

Figure 13-17 Force sensors: A) Displacement, B) Strain gage.

the force required to compress or stretch a spring is proportional to a *constant* and the *displacement* caused by the compression or tension force applied to the spring. So, by using a displacement transducer and a calibrated spring, we are able to measure force.

Strain gages connected to flexible metal bars are also used to measure force, because it requires a certain amount of force to deflect the bar any given amount. Several transducers on the market use this technique; they are advertised as "force-displacement" transducers. Such transducers form the basis of the digital bathroom scales now on the market.

Do not be surprised to see such transducers, especially the smaller types, calibrated in *grams*. We know that the gram is a unit of *mass*, not force, so what this usage refers to is the gravitational force on 1 gram at the earth's surface, roughly 980 dynes. A 1-g weight suspended from the end of the bar in Figure 13-12 will represent a force of 980 dynes.

A side view of a cantilever force transducer is shown in Figure 13-17B. In this device a flexible strip is supported by mounts at either end, and a piezoresistive strain gage is mounted to the underside of the strip. Flexing the strip unbalances the gage's Wheatstone bridge, producing an output voltage.

A related device uses a cup- or barrel-shaped support and a circular diaphragm instead of the strip. Such a device will measure force or pressure, that is, *force per unit of area*.

FLUID PRESSURE TRANSDUCERS

Fluid pressures are measured in a variety of ways, but the most common involve a transducer such as those shown in Figures 13-18 and 13-19.

In the example of Figure 13-18A, a strain gage or LVDT is mounted inside a housing that has a bellows or aneroid assembly exposed to the fluid. More force is applied to the LVDT or gage assembly as the bellows compresses. The compression of the bellows is proportional to the fluid pressure.

An example of the *Bourdon tube* pressure transducer is shown in Figure 13-18B. Such a tube is hollow and curved, but flexible. When a pressure is applied through the inlet port, the tube tends to straighten out. If the end tip is connected to a position/displacement transducer, the transducer output will be proportional to the applied pressure.

Figure 13-19A shows another popular form of fluid transducer. In this version, a diaphragm is mounted on a cylindrical support similar to Figure 13-18. In some cases, a bonded strain gage is attached to the underside of the diaphragm, or flexible supports of an unbonded type are used. In the example shown, the diaphragm is connected to the core drive bar of an inductive transducer or LVDT.

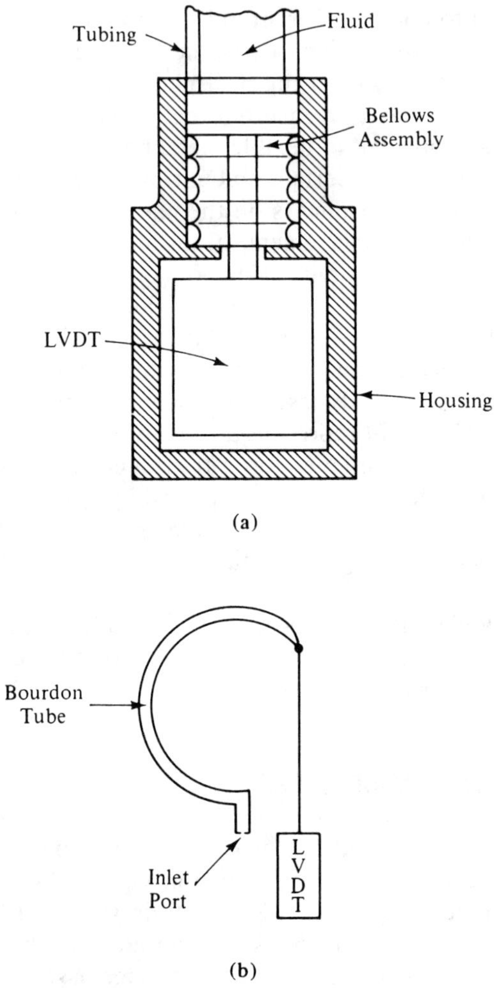

Figure 13-18 A) Bellows pressure sensor, B) Bourdon tube.

Figure 13-19B shows the type of transducer that is typically used in medical electronics to measure human blood pressure. In this device, the hollow fluid-filled dome is fitted with *Luer-lock* fittings, which are standard in medical apparatus.

The fluid transducers shown so far will measure gage pressure (i.e., pressure above atmospheric pressure) because one side of the diaphragm is open to air. A *differential* pressure transducer will measure the difference between pressures applied to the two sides of the diaphragm. Such devices will have two ports marked, such as P1 and P2, or something similar.

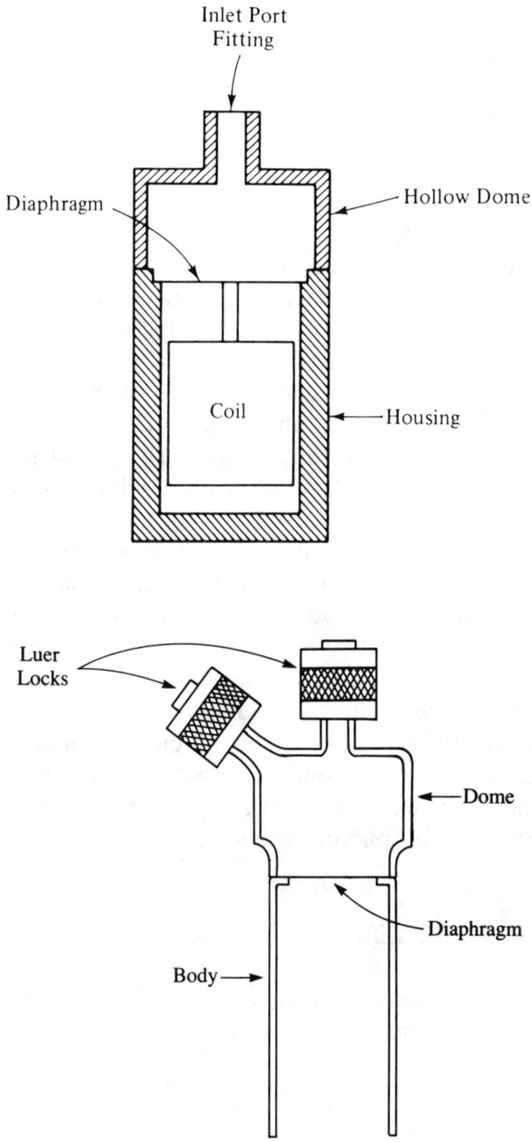

Figure 13-19 A) Inductive pressure sensor, B) Medical pressure sensor.

LIGHT TRANSDUCERS

There are several different phenomena for measuring light, and they create different types of transducer. We will limit the discussion to *photoresistors, photovoltaic cells, photodiodes,* and *phototransistors.*

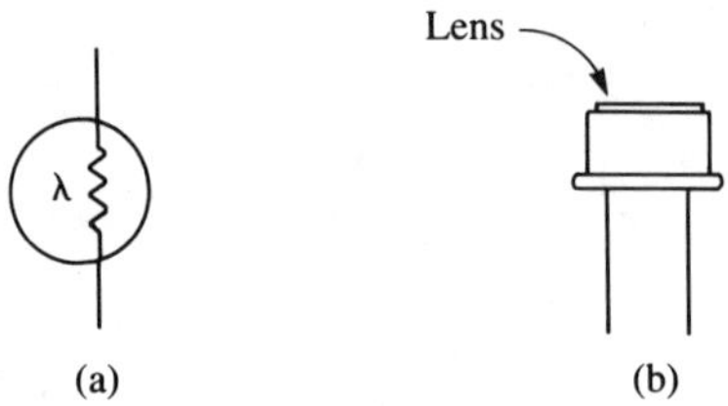

Figure 13-20 Photoresistor cell: A) Symbol, B) Construction.

A photoresistor can be made because certain semiconductor elements show a marked decrease in electrical resistance when exposed to light. Most materials do not change linearly with increased light intensity, but certain combinations such as cadmium sulfide (CdS) and cadmium selenide (CdSe) are effective. These cells operate over a spectrum from "near-infrared" through most of the visible light range and can be made to operate at light levels of 10^{-3} to 10^{+3} footcandles (i.e., 10^{-3} to 70 mW/cm^2). Figure 13-20A shows the photoresistor circuit symbol, while Figure 13-20B shows an example of a photoresistor.

A photovoltaic cell, or "solar cell" as it is sometimes called, will produce an electrical current when connected to a load. Both silicon (Si) and selenium (Se) types are known. The Si type covers the visible and near-infrared spectrum, at intensities between 10^{-3} and 10^{+3} mW/cm^2. The Se cell, on the other hand, operates at intensities of 10^{-1} to 10^2 mW/cm^2 but accepts a spectrum of near-infrared to ultraviolet light.

Semiconductor PN junctions under sufficient illumination will respond to light. They tend to be photoconductive when heavily reverse-biased and photovoltaic when forward-biased. These phenomena have led to a whole family of photodiodes and phototransistors.

CAPACITIVE TRANSDUCERS

A parallel plate capacitor can be made by positioning two conductive planes parallel to each other. The capacitance is given by

$$C = \frac{kKa}{d} \qquad (13\text{-}29)$$

where

C = capacitance in farads (F) or a subunit (μF, pF)

k = a units constant

K = dielectric constant of the material used in the space between the plates (K for air is 1)

A = area of the plates "shading" each other

d = the distance between the plates

Figure 13-21 shows several forms of capacitance transducer. In Figure 13-21A we see a rotary plate capacitor that is not unlike the variable capacitors used to tune radio transmitters and receivers. The capacitance of this unit is proportional to the amount of area on the fixed plate that is covered (i.e., "shaded") by the moving plate. This type of transducer will give signals proportional to curvilinear displacement or angular velocity.

A rectilinear capacitance transducer is shown in Figure 13-21B; it consists of a fixed cylinder and a moving cylinder. These pieces are configured so that the moving piece fits inside the fixed piece but is insulated from it.

The two types of capacitive transducer discussed so far vary capacitance by changing the shaded area of two conductive surfaces. Figure 13-21C, on the other hand, shows a transducer that varies the *spacing* between surfaces, that is, the *d* term in Equation (13-29). In this device, the metal surfaces are a fixed plate and a thin diaphragm. The dielectric is either air or a vacuum. Such devices are often used as capacitance microphones.

Capacitance transducers can be used in several ways. One method is to use the varying capacitance to frequency modulate an RF oscillator. This

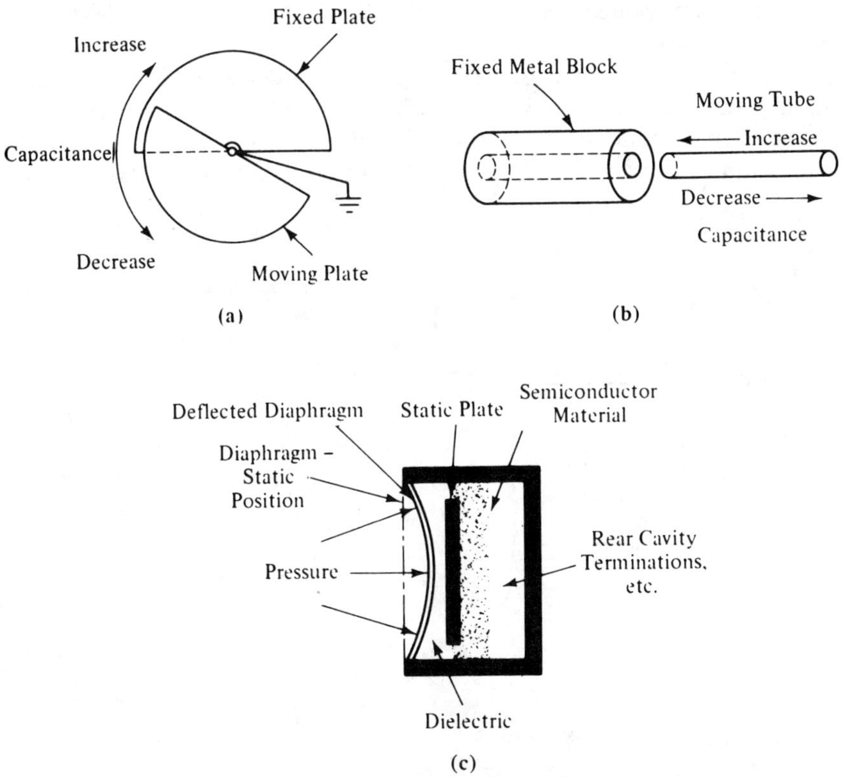

Figure 13-21 Capacitor transducers: A) Butterfly, B) Piston, C) Solid-state.

is the method employed with capacitance microphones (those built like Figure 13-21C, not the newer electrotet type). Another method is to use the capacitance transducer in an AC bridge circuit.

REFERENCE SOURCES

The voltage reference in most electronics circuits, including transducer excitation, is the zener diode, as shown in Figure 13-22A. The zener diode uses a controlled avalanche point to maintain a constant voltage.

The *I versus V curve* for a zener diode is shown in Figure 13-22B. At applied voltages from V_z to V_g, the current through the zener diode is merely a tiny reverse leakage current I_L. At voltages above V_g, which is usually 0.6 to 0.7 V in silicon diodes, the zener operates in exactly the same manner as any other silicon rectifier diode. The current increases rapidly with applied forward voltage.

It is in the area of V_z, however, that the zener diode is unique: the voltage across the diode (which is a reverse bias) remains constant despite increases in the applied voltage. The result is a reasonably reliable reference voltage.

Unfortunately, zener diodes are not too precise (V_z is merely nominal) and may tend to vary with temperature. Some manufacturers, such as Ferranti Semiconductor and National Semiconductor, offer band-gap zeners that are an improvement. National also offers the LM-199 device, which

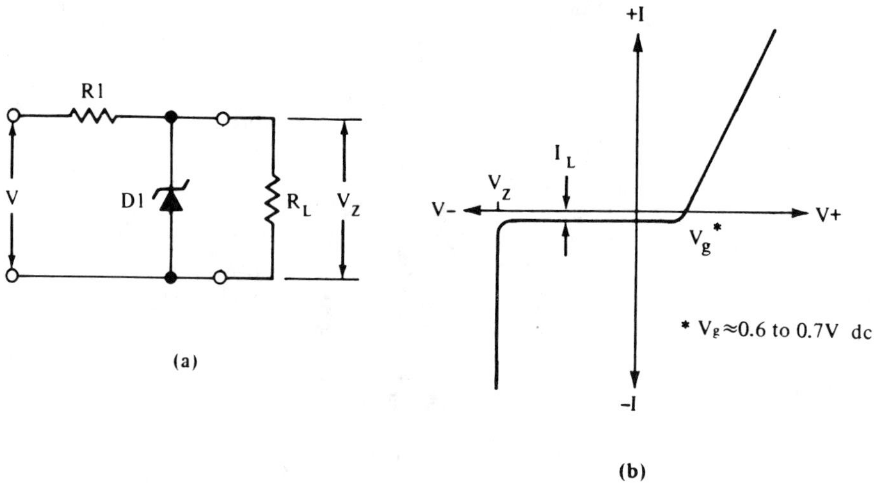

Figure 13-22 A) Zener diode circuit, B) I-vs.-V (current versus voltage) characteristic.

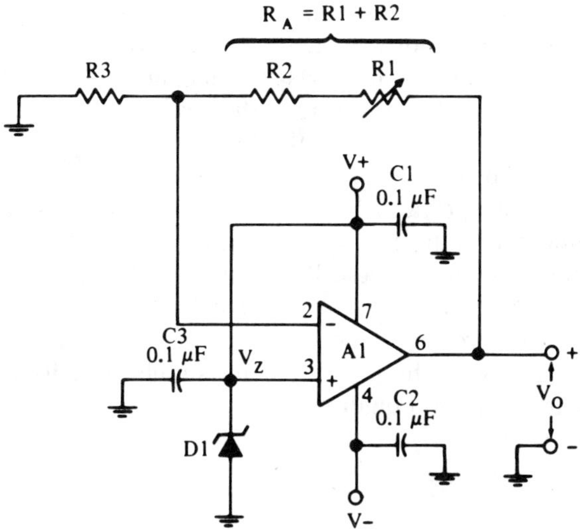

Figure 13-23 Zener-based reference source.

contains an on-chip heater that keeps the embedded zener element at a constant temperature, thereby providing superior temperature performance.

Figure 13-23 shows an operational amplifier circuit that can be used to buffer the zener diode and provide precision control over the output voltage. It also allows us to provide any output voltage (within the output range of the operational amplifier) from any zener diode. This fact allows us to use the 6.2-V LM-199 device, or the 1.26- and 2.45-V band-gap devices to produce 2.56-, 5.00-, or 10.00-V reference supplies. In essence, the circuit is a noninverting follower with gain with the zener potential (V_z) as the input voltage. The output potential is given by

$$V_O = V_z\left[\frac{R1 + R2}{R3} + 1\right] \tag{13-30}$$

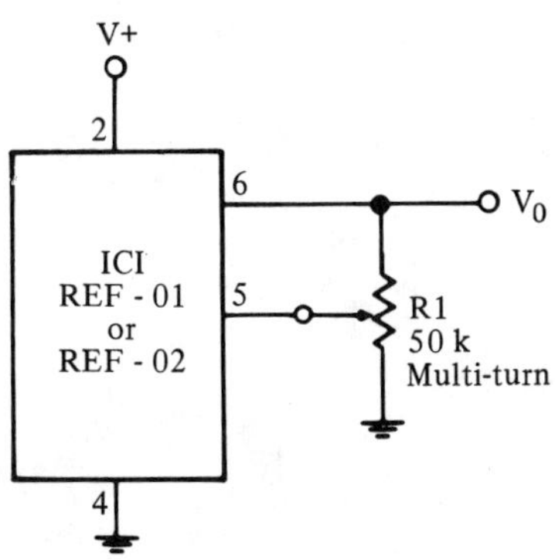

Figure 13-24 REF-01/02 reference source.

The potentiometer (Rl) should be a quality 10- to 20-turn trimmer type in order to adjust the output voltage to a precise value. It should be of high quality so that temperature variations and mechanical shock do not cause its value to change. The other resistors in the circuit should be precision wire-wound resistors in order to obtain the low temperature coefficient of such resistors. We may use either an ordinary zener diode, a band-gap zener, a reference-grade zener, or an LM-199-type device. The last alternative is to use an IC reference voltage source, of which there are many now on the market. Perhaps the oldest is the Precision Monolithics, Inc., type REF-01 (10 V) or REF-02 (5 V), as shown in Figure 13-24. These devices can be adjusted to a very precise value close to their rated voltages. As in the circuit previously given, the potentiometer should be a 10- or 20-turn trimmer potentiometer of good quality.

14

AMPLIFIERS FOR SIGNALS PROCESSING

The Analog Subsystem Is Important!

Although many modern instruments are based on microcomputer technologies, many of the instrumentation applications for microcomputers (and microprocessors) involve analog signals. Many of the sensors or natural signals that represent various physical parameters are either available only in an analog version, or the analog version is a lot less costly than a digital version. Medical, scientific, engineering, and industrial instrumentation, for example, use many analog transducers. Human blood pressure is usually measured by special transducers that are either piezoresistive or inductive Wheatstone bridges or piezoelectric quartz sensors. Temperature is usually measured by thermistors, thermocouples, or PN junction sensors, which output an analog signal. Indeed, a wide variety of signal transducers are available that output only an analog signal. Therefore, due consideration needs to be given to the attributes of the analog processing stages in an instrument.

Amplifiers are used for signal enhancement in the analog subsystem. These amplifiers may be used only to build up the low-level transducer output signal so that it is compatible with the input range of the analog-to-digital converter used in the microcomputer. In other cases, we might actually do some signal processing in the analog subsystem. This statement might be pure heresy to the digitally oriented person, but it is often the case that analog processing is the better engineering choice.

One must remember that the engineer's job is not necessarily to use the most modern technology, but to get the job done in the most economic and efficient manner possible. There are times when that requirement means using a small amount of analog preprocessing. It is especially necessary when the computer memory size is limited by some other uncontrollable factor, or when the computer cycle time is too slow for the particular digital algorithm to be executed on the signal. In other cases, especially in 4- or 8-bit microcomputers, the word length limits resolution too much, yet such resolution is available in an analog subassembly. An engineer from Burr-Brown Research Corporation, a leading maker of both analog devices and data-acquisition products for computers, has stated that one of their analog signal-processing hybrid modules has the resolution to compete with a 20-bit computer. An 8-bit microcomputer would require three successive operations to obtain the triple precision needed to beat that specification.

On the other hand, the analog devoteee must realize that the digital method is usually the best for any signal-processing application. For the most part, we will limit analog processing to the amplification (and possibly filtering) needed to make the signal compatible with the A/D converter at the input of the computer. One must, however, keep in mind that other applications may require analog processing.

Some may criticize the inclusion of operational amplifier and transducer information in a book on microcomputer interfacing. However, if the computer is to be interfaced to the real world of scientific, medical, engineering, and industrial signals (which is a major market for interface engineering), it is highly pertinent to include in our deliberations material on the transducers and signal processors suitable to that world.

OPERATIONAL AMPLIFIERS: AN INTRODUCTION

The operational amplifier has been in existence for several decades, but only in the last 15 or 20 years has it come into its own as an almost universal electronic building block. The term operational is derived from the fact that these devices were originally designed for use in analog computers to solve *mathematical operations*. The range of circuit applications today has increased immensely, so the operational amplifier has survived and prospered, even though analog computers, in which they were once a principal constituent, are now in almost total eclipse.

Keep in mind, however, that even though the programmable analog computer is no longer used extensively, many instruments are little more than a nonprogrammable, dedicated-to-one-chore, analog computer with a numeric readout of some sort.

In this chapter we will examine the gross, or large-scale, properties of the basic operational amplifier, and we will learn to derive the transfer equations for most common operational amplifier circuits using only Ohm's law, Kirchhoff's law, and the properties common to all true operational amplifiers.

One of the profound beauties of the modern, integrated circuit operational amplifier is its simplicity when viewed from the outside world. Of course, the inner workings are complex, but they are of little interest in our discussion of the operational amplifier's gross properties. We will limit our discussion somewhat by considering the operational amplifier as a *black box*, and that approach allows for a very simple analysis in which we relate the performance to the universal transfer function for all electronic circuits: V_{out}/V_{in}.

PROPERTIES OF THE IDEAL OPERATIONAL AMPLIFIER

An ideal operational amplifier is a gain block, or black box if you prefer, that has the following general properties:

1. Infinite open-loop (i.e., no feedback) gain ($A_{vol} = \infty$).
2. Infinite input impedance ($Z_{in} = \infty$).
3. Zero output impedance ($Z_0 = 0$).
4. Infinite bandwidth ($f_0 = \infty$).
5. Zero noise generation.

Of course, it is not possible to obtain a real IC operational amplifier that meets these properties—they are ideal—but if we read ''infinite'' as ''very, very high'' and ''zero'' as ''very, very low,'' then the approximations of the ideal situation are accurate. Real IC operational amplifiers, for example, can have an open-loop voltage gain from 50,000 to over 1,000,000, so it can be classed as infinite for practical purposes, and the equations work in most cases.

DIFFERENTIAL INPUTS

Figure 14-1 shows the basic symbol for the common operational amplifier, including power terminals. In many schematics of operational amplifier circuits, the V_{CC} and V_{EE} power terminals are deleted, so the drawing will be less ''busy.''

Note that there are two input terminals, which are labeled (−) and (+). The terminal labeled (−) is the inverting input. The output signal will be out of phase with signals applied to this input terminal (i.e., there will be a 180-degree phase shift). The terminal labeled (+) is the noninverting input, so output signals will be in phase with signals applied to this input. It is important to remember that these input look into equal open-loop gains, so they will have equal but opposite effects on the output voltage.

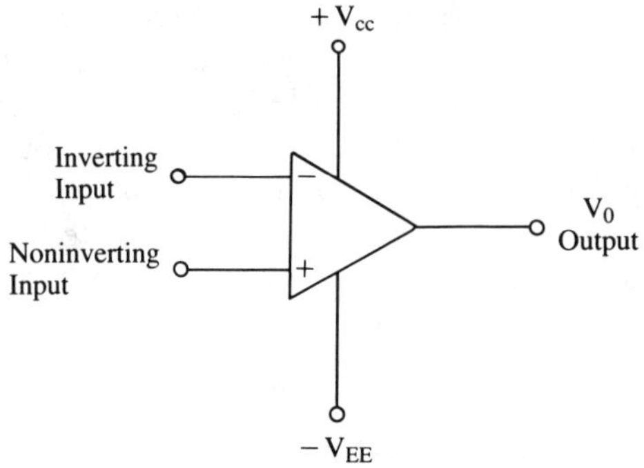

Figure 14-1 Standard operational amplifier symbol.

At this point let us add one further property to our list of ideal properties:

6. Differential inputs follow each other.

This property implies that the two inputs will behave as if they were at the same electrical potential, especially under static conditions. In Figure 14-2 we see an inverting follower circuit in which the noninverting (+) input is grounded. The sixth property allows us (in fact requires us) to treat the inverting (−) input as if it were also grounded. Many textbooks and magazine articles like to call this phenomenon a "virtual" ground, but such a term serves only to confuse the reader. It is better to accept as a basic axiom of operational amplifier circuitry that, for purposes of calculation and voltage measurement, the (−) input will act as if grounded whenever the (+) input is actually grounded.

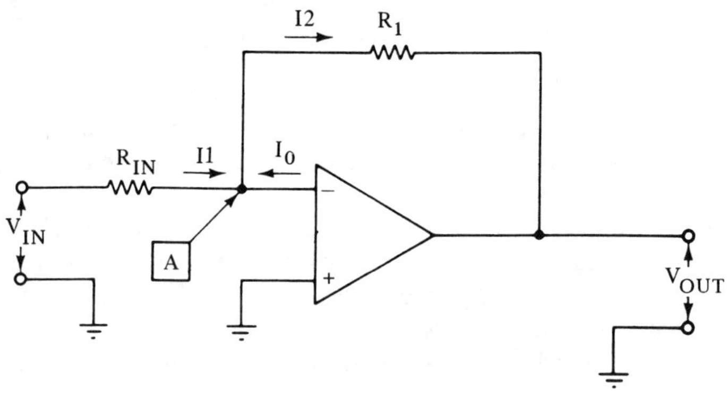

Figure 14-2 Inverting follower circuit.

ANALYSIS USING KIRCHHOFF'S AND OHM'S LAWS

We know from Kirchhoff's current law that the algebraic sum of all currents entering and leaving a point in a circuit must be zero. The total current flow into and out of point A in Figure 14-2, then, must be zero. Three possible currents exist at this point: input current I2 and any currents flowing into or out of the (−) input terminal of the operational amplifier, I_0. But according to ideal property number 2, the input impedance of this type of device is infinite. Ohm's law tells us that

$$I_0 = \frac{V}{Z_{in}} \qquad (14\text{-}1)$$

Current I_0 is zero, because V/Z_{in} is zero. So if current I_0 is equal to zero, we conclude that $I1 + I2 = 0$ (Kirchhoff's current law). Since this is true, then

$$I2 = -I1 \qquad (14\text{-}2)$$

We also know that

$$I1 = \frac{V_{in}}{R_{in}} \qquad (14\text{-}3)$$

and

$$I2 = \frac{V_{out}}{R_F} \qquad (14\text{-}4)$$

By substituting Equation (14-3) and (14-4) into Equation (14-2), we obtain the result

$$\frac{V_{out}}{R_F} = \frac{-V_{in}}{R_{in}} \qquad (14\text{-}5)$$

Solving for V_{out} gives us the following transfer function, normally given in operational amplifier literature for an inverting amplifier:

$$V_{out} = -V_{in} \times \left[\frac{R_F}{R_{in}}\right] \qquad (14\text{-}6)$$

The term R_F/R_{in} is the voltage gain factor and is usually designated by the symbol A_V, which is written as

$$A_V = \frac{-R_F}{R_{in}} \qquad (14\text{-}7)$$

We sometimes encounter Equation (14-6) written using the left-hand side of Equation (14-7):

$$V_{out} = -A_V V_{in} \qquad (14\text{-}8)$$

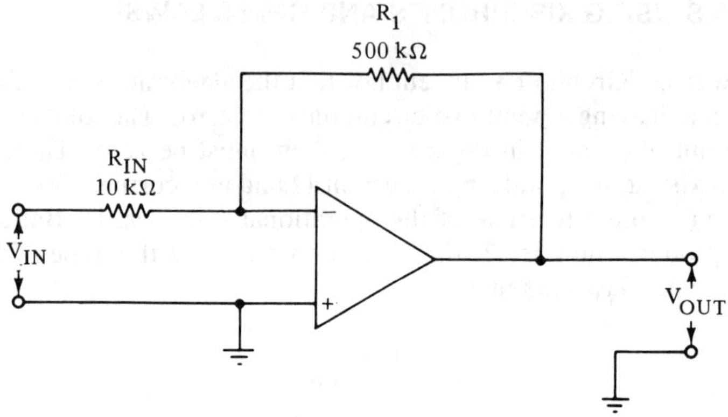

Figure 14-3 Gain-of-50 inverting follower.

When designing simple inverting followers using operational amplifiers, use Equations (14-7) and (14-8). Let us look at a specific example. Suppose that we have a requirement for an amplifier with a gain of 50. We want to drive this amplifier from a source that has an output impedance of 1000 Ω. A standard rule of thumb for designers to follow is to make the input impedance not less than ten times the source impedance, so in this case the amplifier must have a source impedance that is equal to or greater than 10,000 Ω (10 kΩ). This requirement sets the value of the input resistor at 10 kΩ or higher, but in this example we select a 10-kΩ value for R_{in}.

$$A_V = \frac{R_F}{R_{in}} \tag{14-9}$$

$$50 = \frac{R_F}{10{,}000\ \Omega} \tag{14-10}$$

$$R_F = 500{,}000\ \Omega$$

Our gain-of-50 amplifier will be as shown in Figure 14-3.

NONINVERTING FOLLOWERS

The inverting follower circuits of Figures 14-2 and 14-3 suffer badly from low input impedance, especially at higher gains, because the input impedance is the value of R_{in}. This problem becomes especially acute when we attempt to obtain even moderately high gain figures from low-cost devices. Although some types of operational amplifier allow the use of 500-kΩ to 2-MΩ input resistors, they are costly to purchase and often uneconomical to use. The noninverting follower of Figure 14-4 solves this problem by using the input impedance problem very nicely, because the input impedance of the op-amp is typically very high (ideal property number 2).

We may once again resort to Kirchhoff's current law to derive the transfer equation from our basic ideal properties. By property number 6 we know that the inputs follow each other, so the inverting input can be treated as if it were at the same potential as the noninverting input, which is V_{in}, the input signal voltage. We know that

$$I1 = I2 \tag{14-11}$$

$$I1 = \frac{V_{in}}{R_{in}} \tag{14-12}$$

$$I2 = \frac{V_{out} - V_{in}}{R_F} \tag{14-13}$$

By substituting Equations (14-12) and (14-13) into Equation (14-11), we obtain

$$I1 = I2 \tag{14-14}$$

$$\frac{V_{in}}{R_{in}} = \frac{V_{out} - V_{in}}{R_F} \tag{14-15}$$

Solving Equation (14-15) for V_{out} results in the transfer equation for the noninverting follower amplifier circuit. Multiply both sides by R_F:

$$\frac{R_F V_{in}}{R_{in}} = V_{out} - V_{in} \tag{14-16}$$

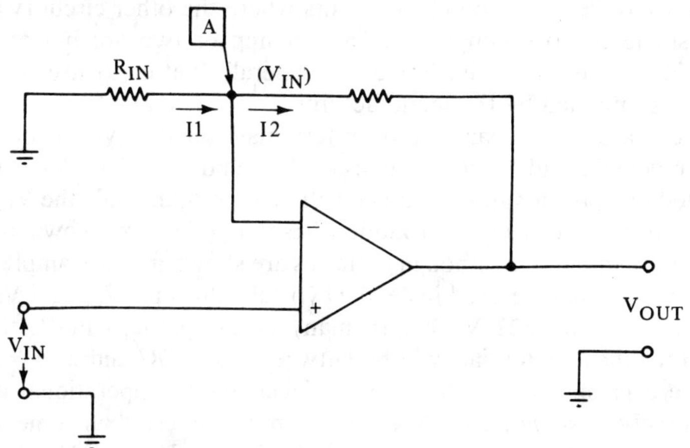

Figure 14-4 Noninverting follower circuit.

Add V_{in} to both sides:

$$\frac{R_F V_{in}}{R_{in}} + V_{in} = V_{out} \tag{14-17}$$

Factor out V_{in}:

$$V_{in} \times \left[\frac{R_F}{R_{in}} + 1\right] = V_{out} \tag{14-18}$$

In this discussion we have arrived at both of the transfer functions commonly used in operational amplifier design by using only the basic properties, Ohm's law, and Kirchhoff's current law. We may safely assume that the operational amplifier is merely a feedback device that generates a current that exactly cancels the input current. Figure 14-5 gives a synopsis of the characteristics of the most popular operational amplifier configurations. The unity gain noninverting follower of Figure 14-5C is a special case of the circuit in Figure 14-5B, in which $R_F/R_{in} = 0$. In this case, the transfer equation becomes

$$\begin{aligned} V_{out} &= V_{in}(0 + 1) \\ &= V_{in}(1) \\ &= V_{in} \end{aligned} \tag{14-19}$$

OPERATIONAL AMPLIFIER POWER SOURCES

Although almost every circuit using operational amplifiers uses a dual polarity power supply, it is possible to operate the device with a single polarity supply. An example of single-supply operation might be in equipment designed for mobile operation or in circuits where the other circuitry requires only a single polarity supply, and an op-amp or two are but a minority feature in the design. It is, however, generally better to use the bipolar supplies as intended by the manufacturer.

There are two separate power terminals on the typical operational amplifier device, and these are marked V_{CC} and V_{EE}. The V_{CC} supply is connected to a power supply that is positive to ground, while the V_{EE} supply is negative with respect to ground. These supplies are shown in Figure 14-6. Keep in mind that although batteries are shown in the example, regular power supplies may be used instead. Typical values for V_{CC} and V_{EE} range from ±3 V DC to ±22 V DC. In many cases, perhaps most, the value selected for these potentials will be between ±9 V DC and ±15 V DC.

There is one further constraint placed on the operational amplifier power supply *in some cases:* $V_{CC} - V_{EE}$ must be less than some specified voltage, usually 30 V. So if V_{CC} is +18 V DC, then V_{EE} must be not greater than (30 − 18), or 12 V DC.

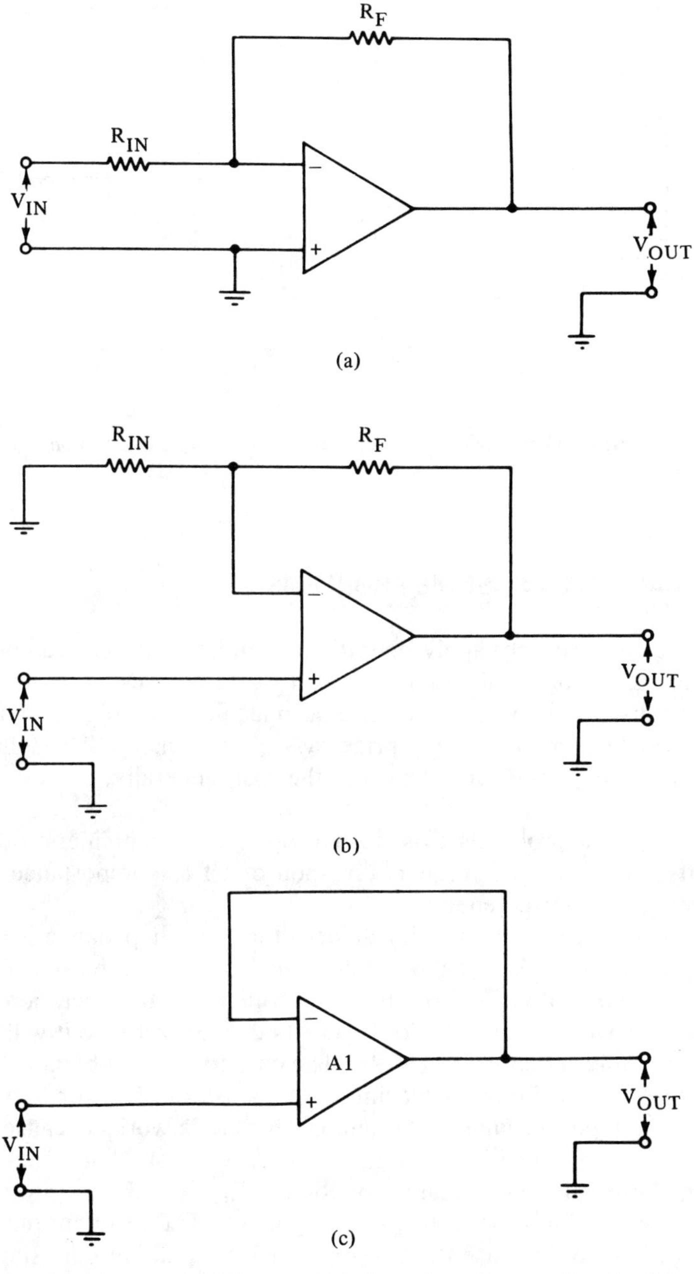

Figure 14-5 A) Inverting follower, B) Noninverting follower with gain, C) Unity gain noninverting follower.

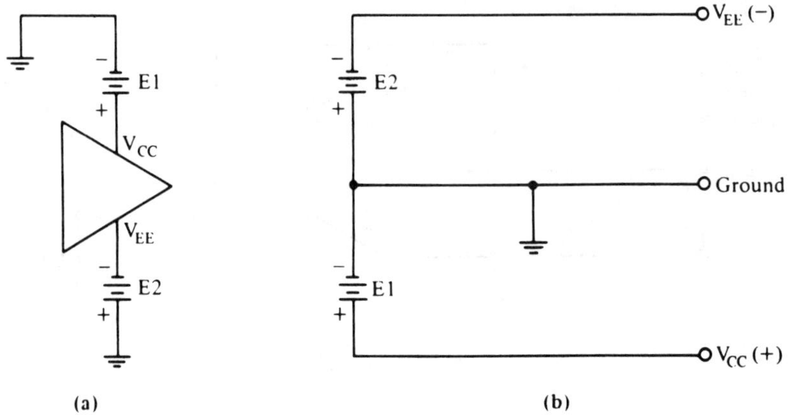

Figure 14-6 Operational amplifier power supply configuration.

PRACTICAL DEVICES: SOME PROBLEMS

Before we can properly apply operational amplifiers in real equipment we must learn some of the limitations of real-world devices. The devices that we have considered up until now have been ideal, so they do not exist. Real IC operational amplifiers carry price tags of less than half a dollar up to several dozen dollars each. The lower the cost, generally, the less ideal the device.

Three main problems exist in real operational amplifiers: offset current, offset voltage, and frequency response. Of less importance in many cases is internal noise generation.

In real operational amplifier devices the input impedance is less than infinite, and this implies that a small input bias current exists. The input current may flow into or out of the input terminals of the operational amplifier. In other words, current I_0 of Figure 14-2 is not zero, so it will produce an output voltage equal to $-I_0 \times R_F$. The cure for this problem is shown in Figure 14-7, and it involves placing a compensation resistor between the noninverting input terminal and ground. This tactic works because the currents in the respective inputs are approximately equal. Since resistor R_C is equal to the parallel combination of R_F and R_{in}, it will generate the same voltage drop that appears at the inverting input. The resultant output voltage, then, is zero, because the two inputs have equal but opposite polarity effect on the output.

Output offset voltage is the value of V_{out} that will exist if the input end of the R_{in} is grounded (i.e., $V_{in} = 0$). In the ideal device, V_{out} would be zero under this condition, but in real devices there may be some offset potential

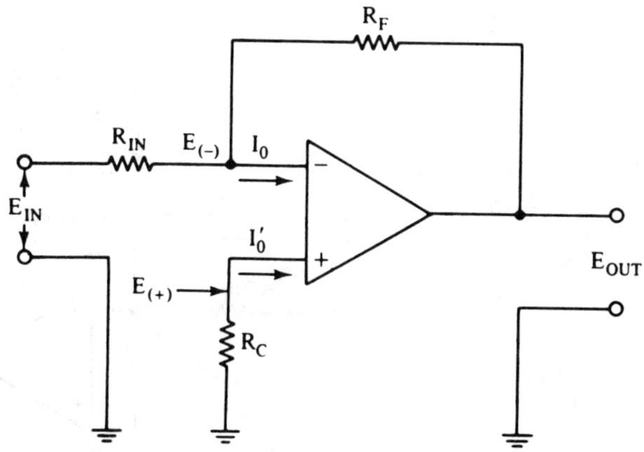

Figure 14-7 Inverting follower with offset compensation resistor.

present. This output potential can be forced to zero by any of the circuits in Figure 14-8.

The circuit in Figure 14-8A uses a pair of offset null terminals found on many, but not all, operational amplifiers. Although many IC operational amplifiers use this technique, some do not. Alternatively, the offset range may be insufficient in some cases. In either event, we may use the circuit of Figure 14-8B to solve the problem.

The offset null circuit of Figure 14-8B creates a current flowing in resistor R1 to the summing junction of the operational amplifier. Since the offset current may flow either into or out of the input terminal, the null control circuit must be able to supply currents of both polarities. Because of this requirement, the ends of the potentiometer (R1) are connected to V_{CC} and V_{EE}.

In many cases, it is found that the offset is small compared with normally expected values of input signal voltage. This is especially true in low-gain applications, in which case the nominal offset current will create such a low output error that no action need be taken. In still other cases, the offset of each stage in a cascade chain of amplifiers may be small, but their cumulative effect may be a large offset error. In this type of situation, it is usually sufficient to null only one of the stages late in the chain (i.e., close to the output stage).

In those circuits where the offset is small, but critical, it may be useful to replace R1 and R2 of Figure 14-8B with one of the resistor networks of Figures 14-8C through 14-8E. These perform essentially the same function but have superior resolution. That is, there is a smaller change in output voltage for a single turn of the potentiometer. This type of circuit will have a

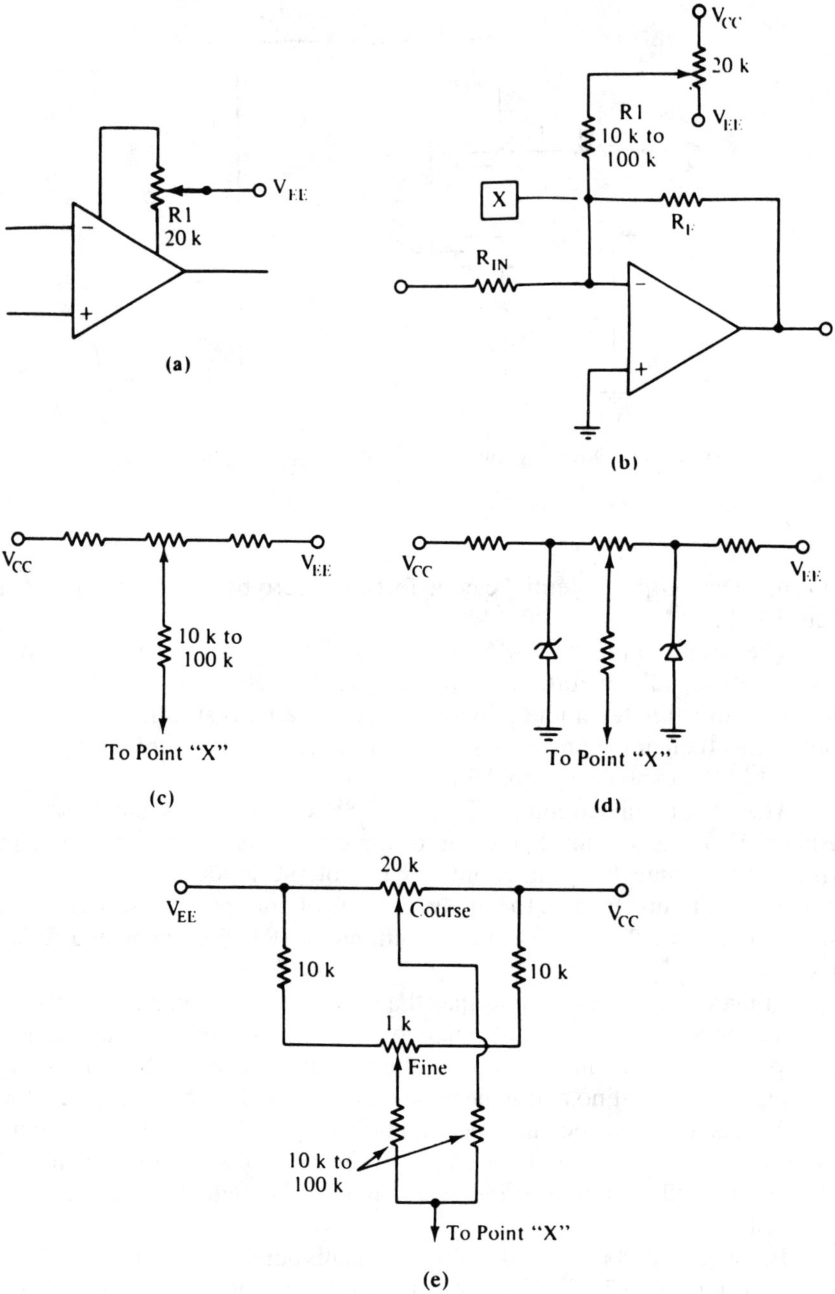

Figure 14-8 A) Using offset null terminals on op-amp, B) Universal offset null circuit, C)–E) Finer-resolution versions of B).

superior resolution in any event, but even further improvement is possible if a 10-turn (or more) potentiometer is used.

DC DIFFERENTIAL AMPLIFIERS

The fact that an IC operational amplifier has two complementary inputs, inverting and noninverting, makes it natural to use it for application as a differential amplifier. These circuits produce an output voltage that is proportional to the difference between two ground-referenced input voltages. Recall from our previous discussion that the two inputs of an operational amplifier have equal but opposite effect on the output voltage. If the same voltage or two equal voltages are applied to the two inputs (i.e., a common-mode voltage, E3 in Figure 14-1), the output voltage will be zero. The transfer equation for a differential amplifier is

$$V_{out} = A_V(V1 - V2) \tag{14-20}$$

So if V1 = V2, then $V_{out} = 0$.

The circuit of Figure 14-9 shows a simple differential amplifier using a single IC operational amplifier. The voltage gain of this circuit is given by

$$A_v = \frac{R3}{R1} \tag{14-21}$$

provided that R1 = R2 and R3 = R4.

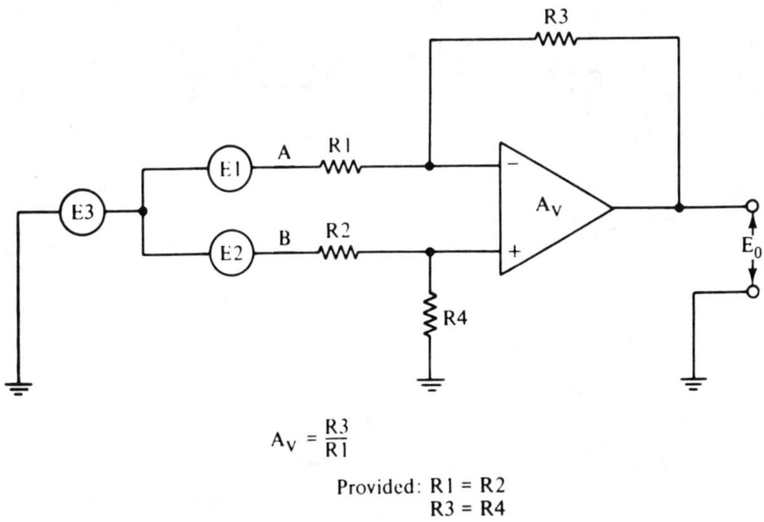

Figure 14-9 DC differential amplifier.

The main appeal of this circuit is that it is economical, as it requires but one IC operational amplifier. It will reject common-mode voltages reasonably well if the equal value resistors are well matched. A serious problem exists for some applications, however, because the circuit has a relatively low input impedance. Additionally, with the problems existing in real operational amplifiers, this circuit may be a little difficult to tame in high-gain applications. As a result, designers frequently use an alternate circuit in these cases.

In recent years, the instrumentation amplifier (IA) of Figure 14-10 has become popular because it alleviates most of the problems associated with the circuit of Figure 14-9. The input stages are noninverting followers, so they will have a characteristically high input impedance. Typical values run to as much as 1000 MΩ.

The instrumentation amplifier is relatively tolerant of different resistor ratios used to create voltage gain. In the simplest case, the differential voltage gain is given by

$$A_V = \frac{2R3}{R1} \tag{14-22}$$

provided that R3 = R2 and R4 = R5 = R6 = R7.

It is interesting to note that the common-mode rejection ratio is not seriously degraded by mismatch of resistors R2 and R3; only the gain is

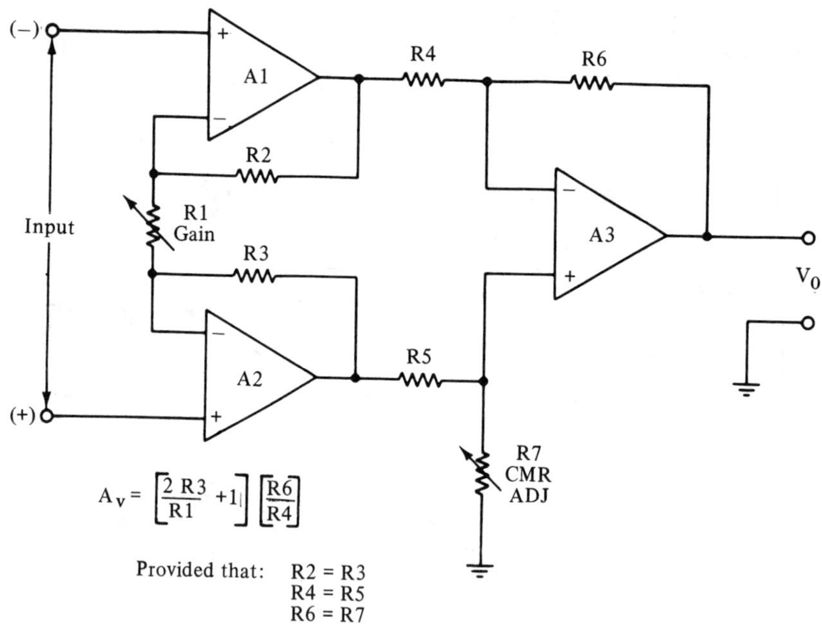

Figure 14-10 Instrumentation differential amplifier.

affected. If these resistors are mismatched, a differential voltage gain error will be introduced.

The situation created by Equation (14-22) results in having the gain of A3 equal to unity (i.e., 1), which is a waste. If gain in A3 is desired, Equation (14-22) must be rewritten into the form

$$A_V = \left[\frac{2R3}{R1} + 1\right]\left[\frac{R7}{R5}\right] \tag{14-23}$$

One further equation that may be of interest is the general expression from which the other instrumentation amplifier transfer equations are derived:

$$A_V = \frac{R7(R1 + R2 + R3)}{R1R6} \tag{14-24}$$

which remains valid provided that the ratio R7/R6 = R5/R4.

Equation (14-24) is especially useful, since you need not be concerned with matched pairs of precision resistors, but only that their ratios be equal.

PRACTICAL CIRCUIT

In this section we will consider a practical design example using the instrumentation amplifier circuit. The particular problem requires a frequency response to 100 kHz and that the input lines be shielded. But the latter requirement would also deteriorate the signal at high frequencies because of the shunt capacitance of the input cables. To overcome this problem, a high-frequency compensation control is built into the amplifier. Voltage gain is approximately 10.

The circuit to the preamplifier is shown in Figure 14-11. It is the instrumentation amplifier of Figure 14-10 with some modifications. When the frequency response is less than 10 kHz or so, we may use any of the 741-family devices (i.e., 741, 747, 1456, and 1458), but premium performance demands a better operational amplifier. In this case, one of the most economical is the RCA CA3140, although an L156 would also suffice.

Common-mode rejection can be adjusted to compensate for any mismatch in the resistors or IC devices by adjusting R10. This potentiometer is adjusted for zero output when the same signal is applied simultaneously to both inputs.

The frequency response characteristics of this preamplifier are shown in Figures 14-12 through 14-15. The input in each case is a 1000-Hz square wave from a function generator. The waveform in Figure 14-12 shows the output signal when resistor R9 is set with its wiper close to the ground end. Notice that it is essentially square and shows only a small amount of roll-off of high frequencies. The waveform in Figure 14-13 is the same signal when

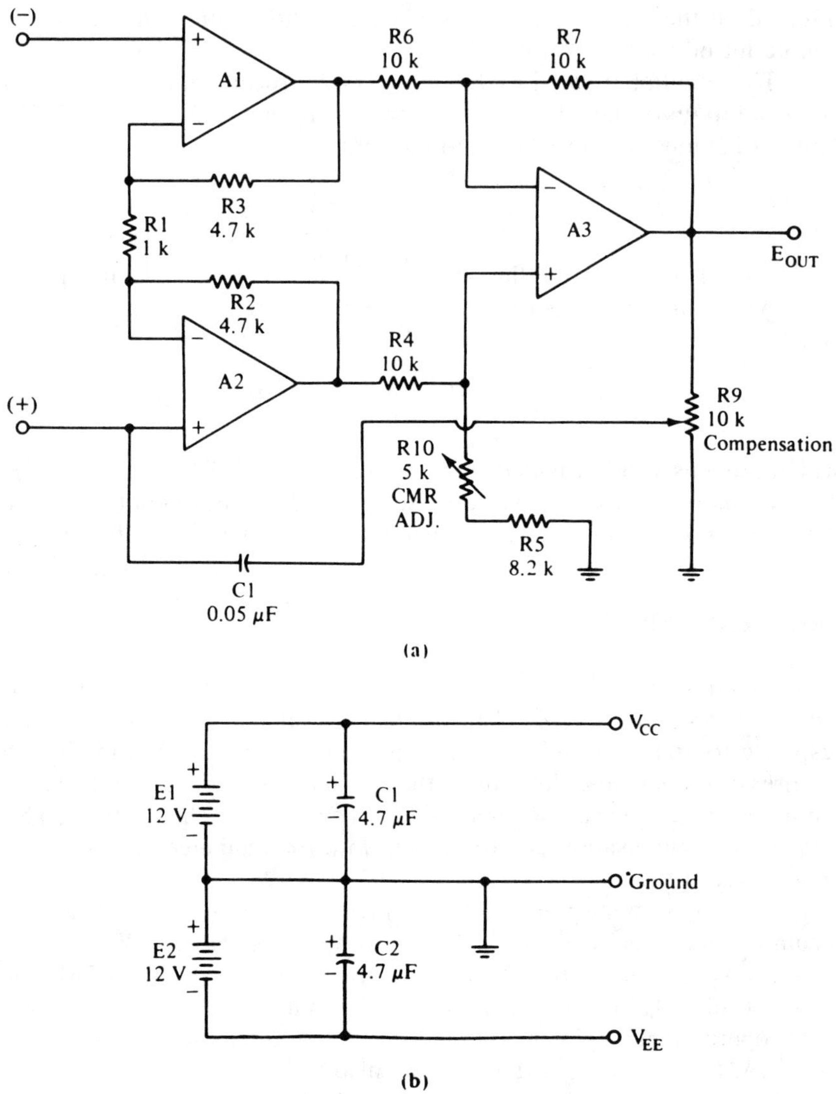

Figure 14-11 A) Instrumentation amplifier with CMRR control, B) DC power supply.

R9 is at maximum resistance. This creates a small amount of regenerative (i.e., positive) feedback: although it is not sufficient to start oscillation, it will enhance amplification of high frequencies.

The problem of oscillation can be quite serious (Figure 14-15), however, if certain precautions are not taken, most of which involve limiting the

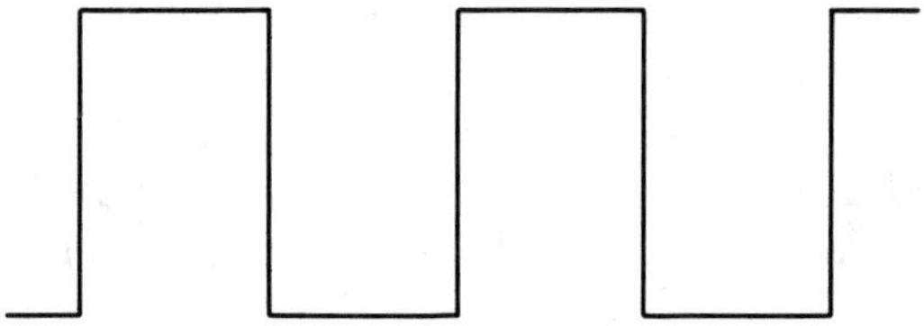

Figure 14-12 Square-wave signal.

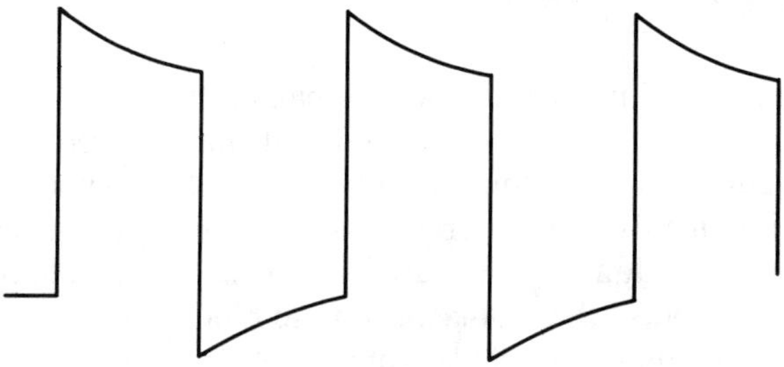

Figure 14-13 Peaked square-wave signal.

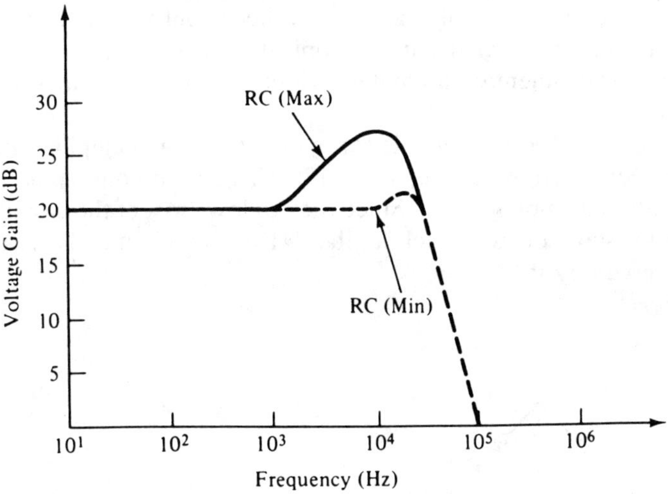

Figure 14-14 Amplifier frequency response.

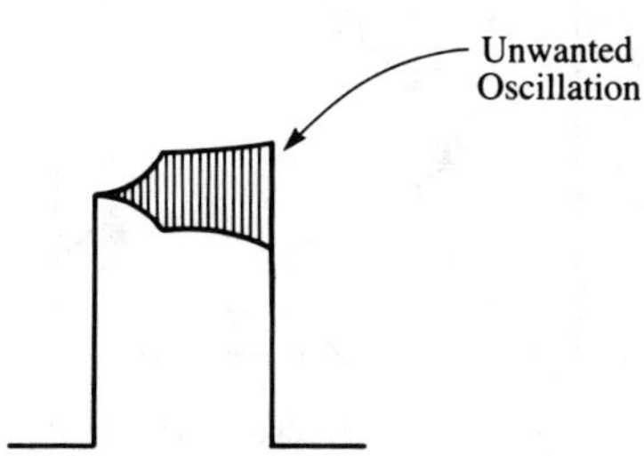

Figure 14-15 Unwanted oscillation on square-wave peak.

amplitude of the feedback signal. This goal is realized by using a 2200-Ω resistor in series with the potentiometer.

Another source of oscillation is the value of C1. When a 0,001-μF capacitor is used at C1, an 80-kHz oscillation is created. The frequency response is shown in Figure 14-14. To obtain any particular response curve, modify the values of C1 and R9.

DIFFERENTIAL AMPLIFIER APPLICATIONS

Differential amplifiers find application in many different instrumentation situations. Of course, it should be realized that they are required wherever a differential signal voltage is found. Less obvious, perhaps, is that they are used to acquire signals or to operate in control systems in the presence of large noise signals. Many medical applications, for example, use the differential amplifier because they look for minute bipotentials in the presence of strong common-mode 60-Hz fields from the AC power mains.

Another class of applications is the amplification of the output signal from a Wheatstone bridge; this is shown in Figure 14-16. If one side of the bridge's excitation potential is grounded, the output voltage is a differential signal voltage. This signal can be applied to the inputs of a differential amplifier or instrumentation amplifier to create an amplified, single-ended, output voltage.

A "rear end" stage suitable for many operational amplifier instrumentation projects is shown in Figure 14-17. This circuit consists of three low-cost operational amplifier ICs. Since they follow most of the circuit gain, we may use low-cost devices such as the 741 in this circuit. The gain of this circuit is given by $R2/10^4$.

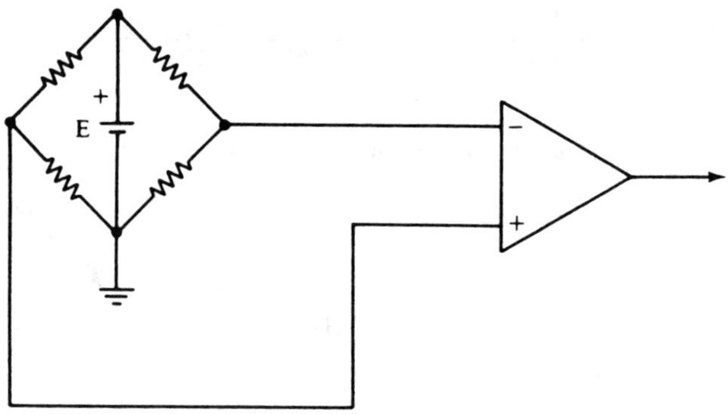

Figure 14-16 Bridge amplifier circuit.

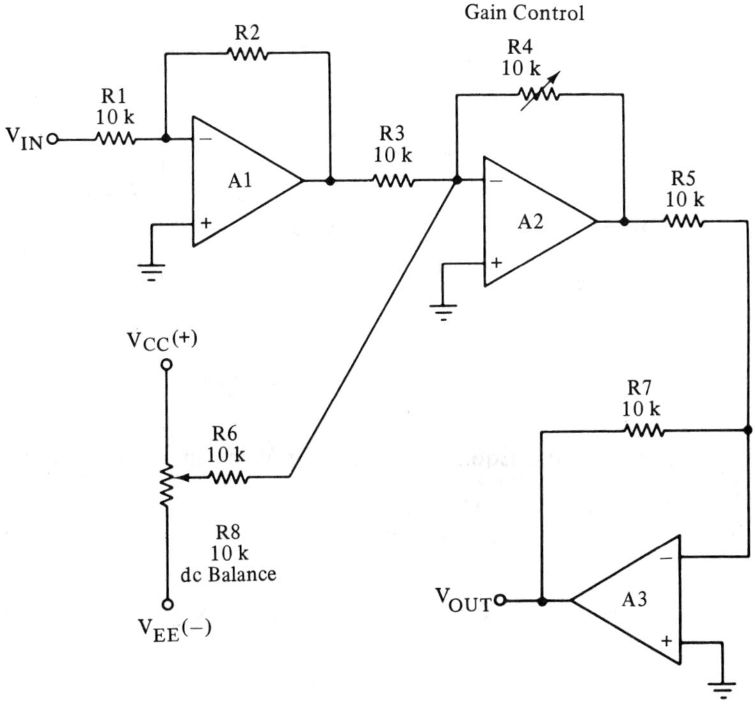

Figure 14-17 Post amplifier.

INTEGRATORS

Figure 14-18 shows the basic operational amplifier integrator circuit. The transfer equation for this circuit may be derived in the same manner as before, with due consideration for C1.

$$I2 = -I1 \tag{14-25}$$

but

$$I1 = \frac{V_{in}}{R1} \tag{14-26}$$

and

$$I2 = C1 \frac{dV_0}{dt} \tag{14-27}$$

Substituting Equations (14-26) and (14-27) into Equation (14-25) results in

$$\frac{C1dV_0}{dt} = \frac{-V_{in}}{R1} \tag{14-28}$$

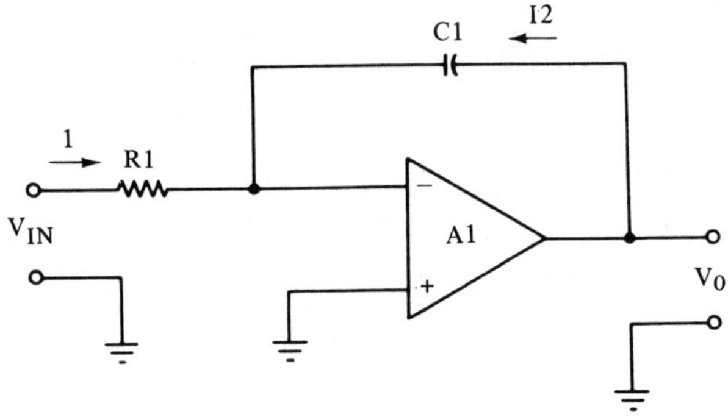

Figure 14-18 Op-amp integrator circuit.

We may now solve Equation (14-28) for V_0 by integrating both sides.

$$\int \frac{C1dV_0}{dt}\,dt = -\int \frac{V_{in}}{R1} \tag{14-29}$$

$$C1V_0 = \frac{-1}{R1}\int V_{in}\,dt \tag{14-30}$$

$$V_0 = \frac{-1}{R1C1}\int V_{in}\,dt \tag{14-31}$$

Equation (14-31), then, is the transfer equation for the operational amplifier integrator circuit.

Example

A constant potential of 2 V is applied to the input of the integrator in Figure 14-18 for 3 s. Find the output potential if R1 = 1 MΩ and C1 = 0.5 μF.

Solution:

$$V_0 = \frac{-1}{R1C1}\int V_{in}\,dt$$

$$V_0 = \frac{-V_{in}}{R1C1}\int dt$$

$$V_0 = \frac{(-2\ V)(t)}{(10^6\ \Omega)(5 \times 10^{27}\ F)}\int_0^3 \tag{14-32}$$

$$V_0 = \frac{(-2\ V)(3\ s)}{(5 \times 10^{22}\ s)} - 0 = -12\ V$$

Note that the gain of the integrator is given by the term 1/R1C1. If small values of R1 and C1 are used, the gain can be very large. For example, if R1 = 100 kΩ and C1 = 0.001 μF, the gain is 10,000. A very small input voltage in that case will saturate the output very quickly. In general, the time constant R1C1 should be longer than the period of the input waveform.

DIFFERENTIATORS

An operational amplifier differentiator is formed by reversing the roles of R1 and C1 in the integrator, as shown in Figure 14-19. We know that

$$I2 = -I1 \tag{14-32}$$

$$I1 = \frac{C1dV_{in}}{dt} \tag{14-33}$$

and

$$I2 = \frac{V_0}{R1} \tag{14-34}$$

Substituting Equations (14-33) and (14-34) into Equation (14-32) results in

$$\frac{V_0}{R1} = \frac{-C1dV_{in}}{dt} \tag{14-35}$$

Solving Equation (14-35) for V_0 gives us the transfer equation for an operational amplifier differentiator circuit:

$$V_0 = -R1C1\,\frac{dV_{in}}{dt} \tag{14-36}$$

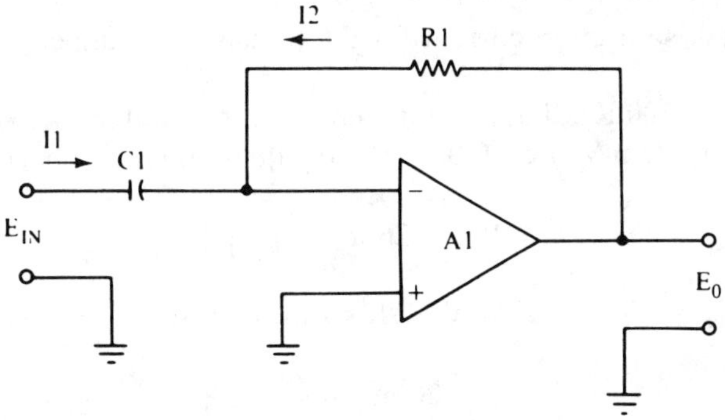

Figure 14-19 Op-amp differentiator circuit.

Example

A 12-V/s ramp function voltage is applied to the input of an operational amplifier differentiator, in which Rl = 1 MΩ and C1 = 0.2 μF. What is the output voltage?

Solution:

$$V_0 = -R1C1 \frac{dV_{in}}{dt}$$
$$= -(10^6\ \Omega)(2 \times 10^{-7}\ F)(12\ V/s) \qquad (14\text{-}36)$$
$$= -(2 \times 10^{-1}\ s)(12\ V/s) = -2.4\ V$$

The differentiator time constant R1C1 should be set very short relative to the period of the waveform being differentiated, or in the case of square waves, triangle waves, and certain other signals, the time constant should be short compared with the rise time of the leading edge.

LOGARITHMIC AND ANTILOG AMPLIFIERS

Figure 14-20A shows an elementary logarithmic amplifier circuit using a bipolar transistor in the feedback loop. We know that the collector current bears a logarithmic relationship to the base-emitter potential V_{be}:

$$V_{be} = \frac{kT}{q} \log \frac{I_c}{I_s} \qquad (14\text{-}37)$$

where

V_{be} = base-emitter potential in volts (V)

k = Boltzmann's constant, 1.38×10^{-23} joules/kelvin (J/K)

T = temperature in kelvins (K)

q = electronic charge, 1.6×10^{-19} coulombs (C)

I_c = collector current in amperes (A)

I_s = reverse saturation current for the transistor in amperes (A)

At 27° C (300 K), that is, room temperature, the term kT/q evaluates to approximately 26 mV (i.e., 0.026 V), so Equation (14-38) becomes

$$V_{be} = 26\ \text{mV} \log \left[\frac{I_c}{I_s}\right] \qquad (14\text{-}38)$$

But $V_{be} = V_0$, and $I_c = V_{in}/R1$, so we may safely say that

$$V_0 = 26\ \text{mV} \log \left[\frac{V_{in}}{I_s\ R1}\right] \qquad (14\text{-}39)$$

But I_s is a constant if the temperature is also a constant, and Rl is constant under all conditions, so, by the rule that the logarithm of a constant

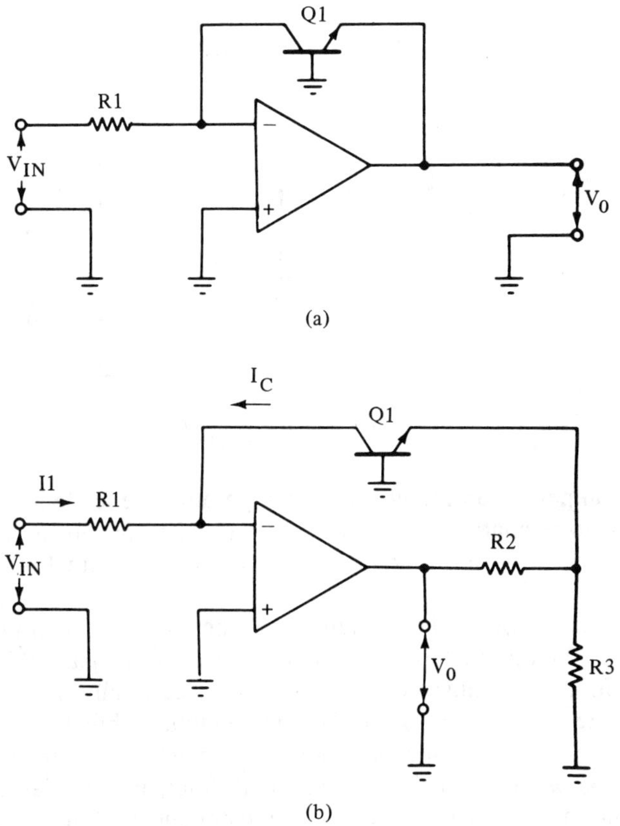

Figure 14-20 A) Logarithmic amplifier, B) Compensated log-amp.

is also a constant, we may state that Equation (14-39) is the transfer function for the natural logarithmic amplifier. For base-10 logarithms,

$$V_0 = 60 \text{ mV} \log_{10} \left[\frac{V_{in}}{I_s R1} \right] \tag{14-40}$$

The relationship of Equations (14-39) and (14-40) allows us to construct amplifiers with logarithmic properties. If Q1 is in the feedback loop of an operational amplifier, the output voltage V_0 will be proportional to the logarithm of input voltage V_{in}. If, on the other hand, the transistor is connected in series with the input of the operational amplifier (see Figure 14-21), the circuit becomes an antilog amplifier.

Both of these circuits exhibit a strong dependence on temperature, as evidenced by the *T* term in Equation (14-37). In actual practice, then, some form of temperature correction must be used. Two forms of temperature correction are commonly used: compensation and stabilization.

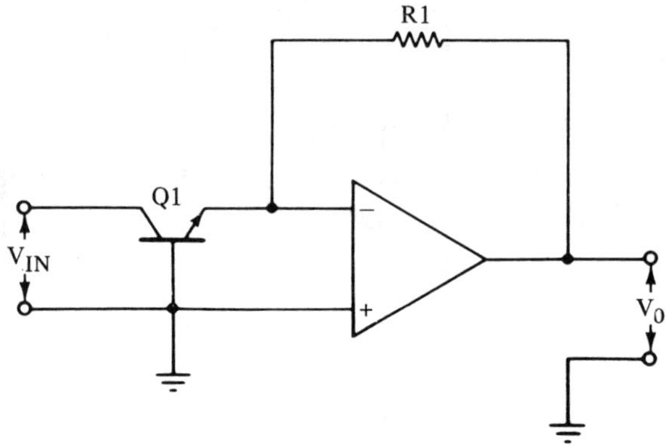

Figure 14-21 Antilog amplifier.

The compensation method uses temperature-dependent resistors, or, thermistors, to regulate the gain of the circuit with changes in temperature. For example, it is common practice to make R3 in Figure 14-20B a thermistor.

The stabilization method requires that the temperature of Q1, and preferably that of the op-amp also, be held constant. In the past, this has meant that the components must be kept inside an electrically heated oven, but today other techniques are used. One manufacturer builds a temperature-controlled hybrid logarithmic amplifier by nesting the op-amp and transistor on the same substrate as a class-A amplifier. Such an amplifier, under zero-signal conditions, dissipates very nearly constant heat. After the chip comes to equilibrium, the temperature will remain constant.

In the case of the antilog amplifier,

$$I_c = \frac{V_0}{R1} \tag{14-41}$$

and

$$V_{in} = V_{be} \tag{14-42}$$

$$V_{in} = 26 \text{ mV} \log\left[\frac{V_0}{R1 I_s}\right] \tag{14-43}$$

CURRENT-TO-VOLTAGE CONVERTERS

Most analog recording devices, such as oscilloscopes or graphic recorders, are voltage-input devices. That is, they require a *voltage* for an input signal. When measuring or recording a current, however, they require some sort of *voltage-to-current converter* circuit.

Two examples of operational amplifier versions are shown in Figure 14-22. In the first example, Figure 14- 22A, a small-value resistor R is placed in series with the current I, which produces a voltage equal to IR. This potential is seen by the operational amplifier as a valid input voltage. The output voltage is

$$V_0 = \frac{-IRR_F}{R_{in}}$$

provided that

$$R << R_{in}$$

$$R << R_F$$

The circuit shown in Figure 14-22B is used for small currents. The output voltage in that circuit is given by

$$V_0 = -I_{in}R_F \qquad (14\text{-}44)$$

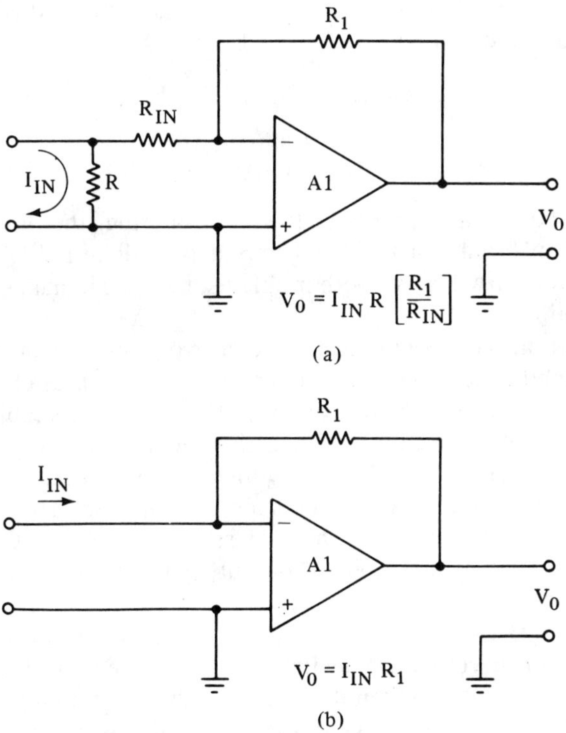

Figure 14-22 A) Small resistor produces IR drop that is proportional to the current flow, B) Active operational amplifier I-to-V converter.

CHOPPER AMPLIFIERS

DC amplifiers have a certain inherent drift and tend to be noisy. These factors are not too important in low- and medium-gain applications (i.e., gains less than 1000), but they loom very large indeed at high gain. For example, a 50-μV/°C drift figure in a $\times 100$ amplifier produces an output voltage of

$$(50\ \mu\text{V/°C}) \times 100 = 5\ \text{mV/°C} \tag{14-45}$$

which is tolerable in many cases. But in a $\times 100{,}0000$ amplifier, the output voltage would be

$$(50\ \mu\text{V/°C}) \times 10^5 = 5\ \text{V/°C}$$

and that amount of drift will probably obscure any real signals in a very short time.

Similarly, noise can be a problem in high-gain applications, where it had been negligible in most low- to medium-gain applications. Operational amplifier noise is usually specified in terms of *nanovolts per square root hertz* (i.e., $\text{noise}_{rms} = \text{nV(Hz)}^{1/2}$]. A typical low-cost operational amplifier has a noise specification of 100 $\text{nV(Hz)}^{1/2}$ so that at a bandwidth of 10 kHz the noise amplitude will be

$$\begin{aligned} \text{Noise}_{rms} &= 100\ \text{nV}\ (10^4\ \text{Hz})^{1/2} \\ &= 100\ \text{nV}\ (10^2) \\ &= 10^4\ \text{nV} = 10^{-5}\ \text{V} \end{aligned}$$

In a $\times 100$ amplifier, without low-pass filtering, the output amplitude will be only 1 mV, but in a $\times 100{,}000$ amplifier it will be 1 V. A circuit called a *chopper amplifier* will solve both problems, because it makes use of an AC-coupled amplifier.

The drift problem is cured because of two properties of AC amplifiers; one is the inability to pass low-frequency (i.e., near-DC) changes such as those caused by drift, and the other is the ability to regulate the stage through the use of heavy doses of negative feedback.

Many low analog signals are very low frequency (i.e., in the DC to 30-Hz range) and will not pass through such an amplifier. The answer to this problem is to *chop* the signal so that it passes through the AC amplifier and then to demodulate the amplifier output signal to recover the original waveshape, but at a higher amplitude.

Figure 14-23A shows the basic chopper circuit. The traditional chopper is a vibrator-driven SPDT switch (S1) connected so that it alternately grounds first the input and then the output of the AC amplifier. An example of a chopped waveform is shown in Figure 14-23C. A low-pass filter following the amplifier filters out any residual chopper hash and any miscellaneous noise signals that may be present.

Most of these mechanical choppers use a chop rate of 400 Hz, although 60-, 100-, 200-, and 500-Hz choppers are also known. The main criterion for the chop rate is that it be twice the highest component frequency that is present in the input waveform. In other words, it must obey Nyquist's criterion.

A differential chopper amplifier is shown in Figure 14-23B. In this circuit an input transformer with a center-tapped primary is used. One input terminal is connected to the transformer center tap, while the other input terminal is switched back and forth between the two ends of the primary winding.

A synchronous demodulator following the AC amplifier detects the signal and restores the original, but now amplified, waveshape. Again, a low-pass filter smooths out the signal.

The modern chopper amplifier may not use mechanical vibrator switches as the chopper. A pair of CMOS or junction field effect transistor (JFET) electronic switches driven out of phase with each other will perform the same job. Some monolithic or hybrid function module chopper amplifiers use a varactor switching bridge for the chopper.

The chopper amplifier limits noise because of the low-pass filter and because the amplifier can have a narrow bandpass centered on the chopper frequency.

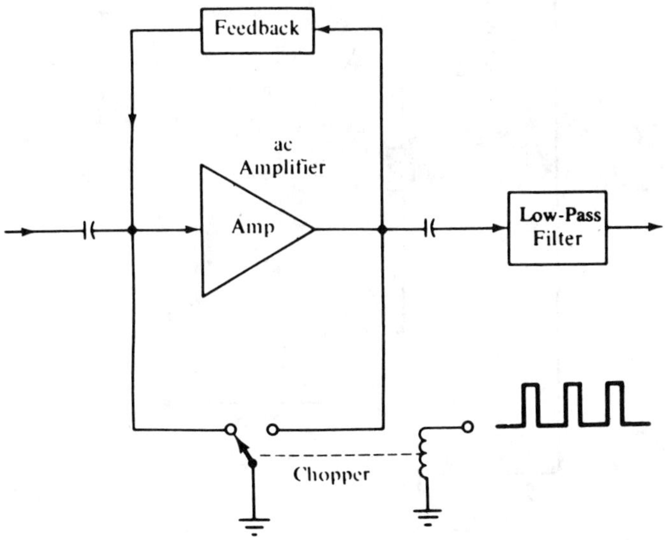

(a)

Figure 14-23 A) Single-ended chopper amplifier.

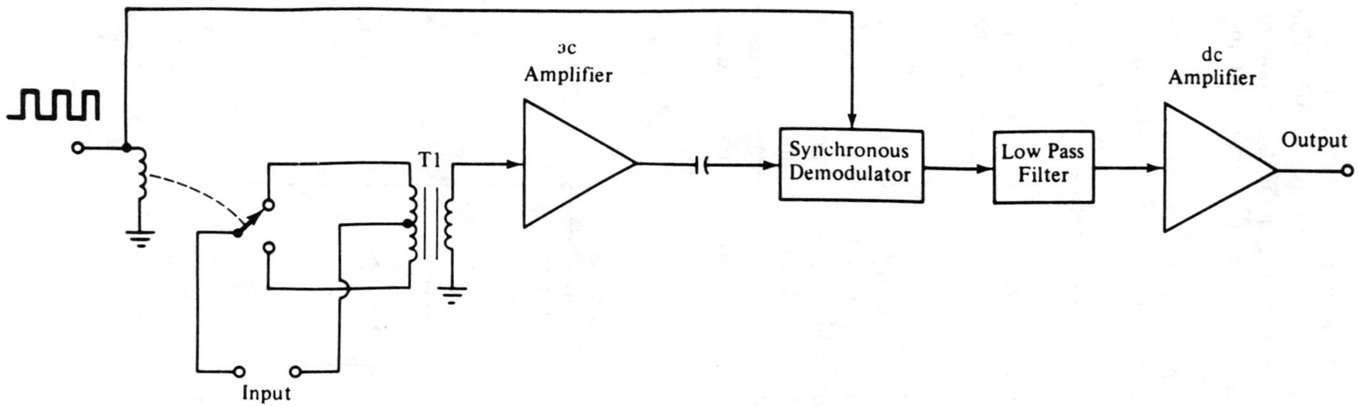

(b)

Figure 14-23 B) Differential chopper amplifier.

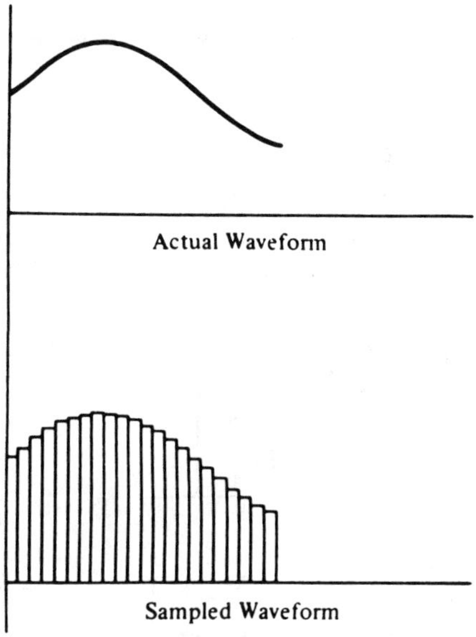

(c)

Figure 14-23 C) Chopped versus actual waveforms.

CARRIER AMPLIFIERS

A *carrier amplifier* is any type of signal-processing amplifier in which the signal carrying the desired information is modulated onto another signal (i.e., a *carrier*). The chopper amplifier is considered by many to fit this definition, but it is usually regarded as a type in its own right. The two principal carrier amplifiers are the DC-excited and AC-excited varieties.

Figure 14-24 shows a DC-excited carrier amplifier. The Wheatstone bridge transducer is excited by DC potential E. The output of the transducer, then, is a small DC voltage that varies with the value of the stimulating parameter. The transducer signal is usually of very low amplitude and is noisy. An amplifier builds up the amplitude, and a low-pass filter removes much of the noise. In some models the first stage is actually a composite of these two functions, being a filter with gain.

The signal at the output of the amplifier-filter section is used to amplitude-modulate a carrier signal. Typical carrier frequencies range from 400 Hz to 25 kHz, with 1 kHz and 2.5 kHz being very common. The signal frequency response of a carrier amplifier is a function of the carrier frequency and is usually considered to be one-fourth of the carrier frequency. A carrier of 400 Hz, then, is capable of signal frequency response to 100 Hz,

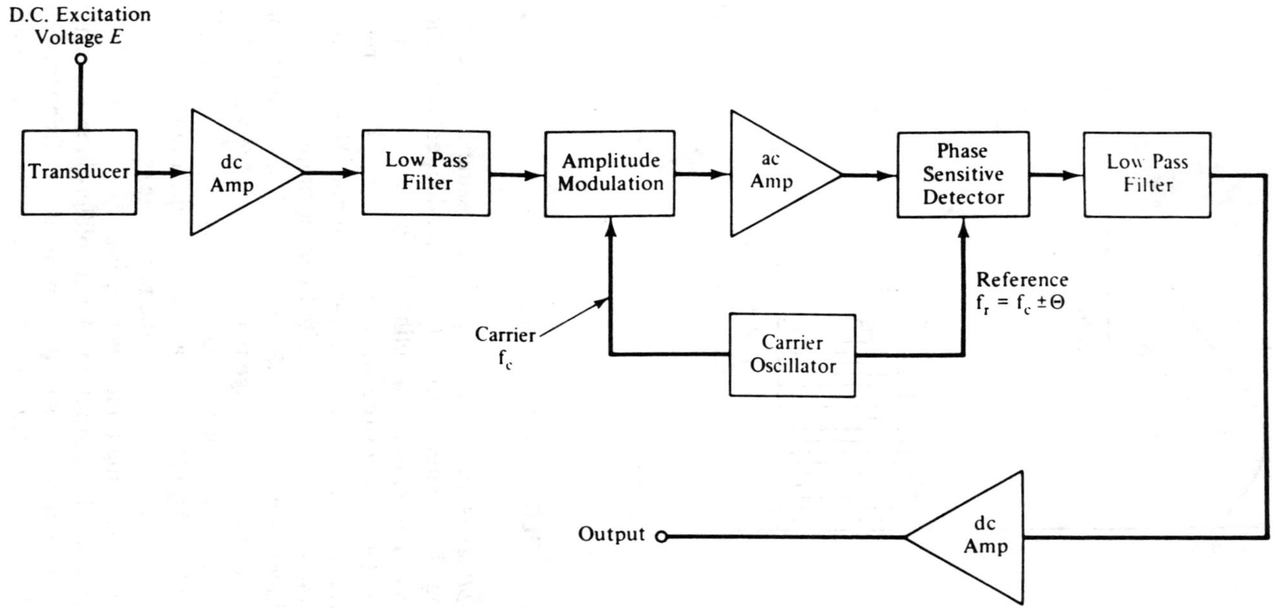

Figure 14-24 Carrier amplifier.

while the 25-kHz carrier will support a frequency response of 6.25 kHz. Further amplification of the signal is provided by an AC amplifier.

The key to the performance of any carrier amplifier worthy of the name is the phase sensitive detector (PSD) that demodulates the amplified AC signal. Envelope detectors, while very simple and of low cost, suffer from an inability to discriminate between the real signal and spurious signals.

Figure 14-25 shows a simplified PSD circuit. Transistors Q1 and Q2 provide a return path to ground for the opposite ends of the secondary winding of input transformer T1. These transistors are alternately switched into and out of conduction by the reference signal in such a way that Q1 is off when Q2 is on, and vice versa. The output waveform of the PSD is a full-wave rectified version of the input signal.

Other electronic switching circuits are also used in PSD design. All systems are designed using the fact that a PSD is essentially an electronic DPDT switch. The digital PSD circuit most often seen uses a CMOS electronic IC switch such as the CD4016/CD4066. These switches are toggled by the reference frequency in such a way that the output is always positive-going, regardless of the phase of the input signal.

The advantages of the PSD include the fact that it rejects signals not of the carrier frequency and certain signals that *are* of the carrier frequency. The PSD, for example, will reject even harmonics of the carrier frequency and those components that are out of phase with the reference signal. The

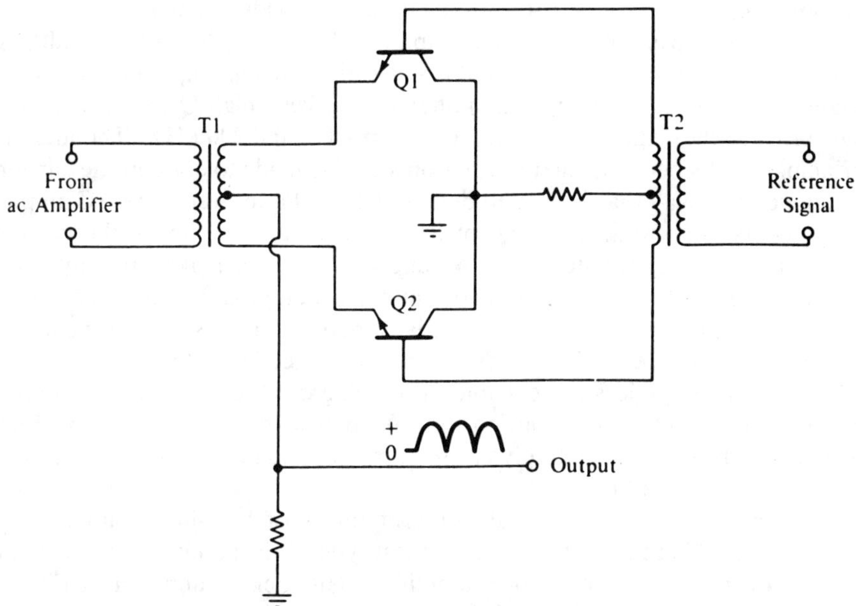

Figure 14-25 Phase-sensitive detector.

PSD will, however, respond to odd harmonics of the carrier frequency. Some carrier amplifiers seem to neglect this problem altogether. But in some cases, manufacturers will design the AC amplifier section to be a bandpass amplifier with a response limited to F_C +/− (F_C/4). This response will eliminate any third, or higher-order, odd harmonics of the carrier frequency before they reach the PSD. It is then necessary only to assure that purity of the reference signal.

An alternate, but very common, form of carrier amplifier is the AC-excited circuit shown in Figure 14-26. In this circuit the transducer is AC-excited by the carrier signal, eliminating the need for the amplitude modulator. The small AC signal from the transducer is amplified and filtered before being applied to the PSD circuit. Again, some designs use a bandpass AC amplifier to eliminate odd-harmonic response. This circuit allows adjustment of transducer offset errors in the PSD circuit instead of in the transducer by varying the phase of the reference signal.

LOCK-IN AMPLIFIERS

The amplifiers discussed so far in this chapter produce relatively large amounts of noise and will respond to noise present in the input signal. They suffer from shot noise, thermal noise, H-field noise, E-field noise, ground-loop noise, and so forth. The noise voltage or power at the output is directly proportional to the square root of the circuit bandwidth. The *lock-in amplifier* is a special case of the carrier amplifier idea in which the bandwidth is very narrow. Some lock-in amplifiers use the carrier amplifier circuit of Figure 14-26 but use an input amplifier with a very high Q bandpass. The carrier frequency may be anything between 1 Hz and 200 kHz. The lock-in principle works because the information signal is made to contain the carrier frequency in a way that is easy to demodulate and interpret. The AC amplifier accepts only a narrow band of frequencies centered about the carrier frequency. The narrowness of the bandwidth, which makes possible the improved signal-to-noise ratio, also limits the lock-in amplifier to very low frequency input signals. Even then it is sometimes necessary to time-average the signal for several seconds to obtain the needed data.

Lock-in amplifiers are capable of thinning out the noise and retrieving signals that are otherwise "buried" in the noise level. Improvements of up to 85 decibels (dP) are relatively easily obtained, and up to 100 dB is possible if the cost is no factor.

There are actually several different forms of lock-in amplifier. The type discussed here is perhaps the simplest type. It is merely a narrow-band version of the AC-excited carrier amplifier. The lock-in amplifier of Figure 14-27, however, uses a slightly different technique. It is called an autocorre-

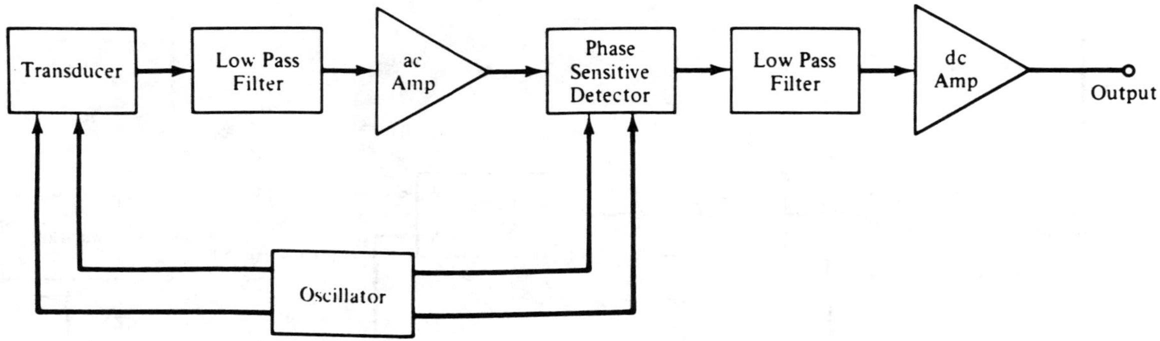

Figure 14-26 Lock-in amplifier.

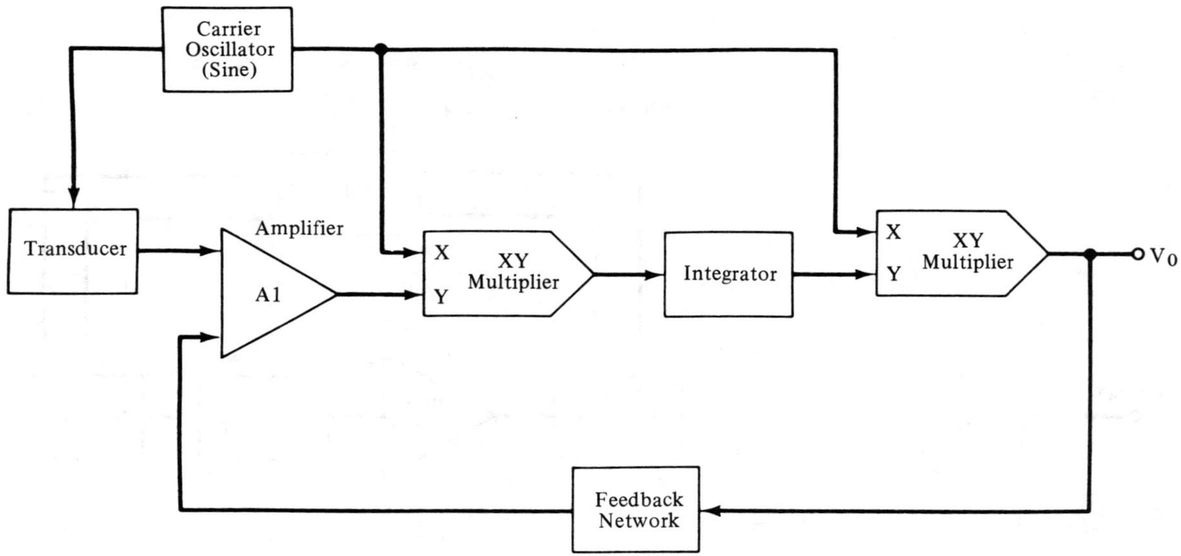

Figure 14-27 Autocorrelation amplifier.

lation amplifier. The carrier is modulated by the input signal and then integrated (i.e., time-averaged). The output of the integrator is demodulated in a product detector circuit. The circuit in Figure 14-27 produces very low output voltages for input signals that are not inphase with the reference signals, but it produces a relatively high output at the proper frequency.

15

SOME MISCELLANEOUS CIRCUITS

The purpose of this chapter is to tie up some loose ends, that is, cover subjects that warrant some attention but not a complete chapter to themselves and that do not fit easily into other chapters. We will take a look at sample and hold circuits (which are used with A/D converters), analog electronic switches, analog voltage and current reference sources, and auto-zero circuitry.

SAMPLE AND HOLD CIRCUITS

The sample and hold (S&H) circuit is designed to take a brief look—a sample—at an analog signal and hold its value indefinitely. Although certain practical design problems limit the term "indefinitely" to a very short duration, the S&H circuit can be used to hold a value at or close to its original value for a period of time after the signal itself has changed value . . . at least long enough to make a data conversion.

There seem to be two major uses of the S&H circuit in microcomputer data acquisition systems. It is used at the input of analog-to-digital converters where it keeps the analog signal constant during the data conversion process. Some A/D converters are particularly prone to substantial errors if the signal being converted changes during the conversion interval. Second, we sometimes see multiple S&H circuits used when a single A/D converter handles many channels. This arrangement allows us to take the digital value of several possibly related parameters. If we took the samples at sequential times, instead of identical times, then the correlation between related values in the different channels is reduced, and possibly destroyed altogether.

The use of S&H circuits in non-digital-computer applications is also widespread. We sometimes see S&H circuits used to cancel the drift in DC amplifiers by sampling first the signal in one S&H and then the drift component in a second S&H. This job is done with the amplifier input shorted, and it is performed during the hold period of the signal S&H. It is possible to subtract out the drift component from the signal by using an ordinary differential operational amplifier.

ANALOG SWITCHES

The heart of the S&H circuit is an analog electronic switch (see Figure 15-1). A typical circuit for a simple analog switch is shown in Figure 15-1A, while several possible circuit symbols are shown in Figure 15-1B. In this

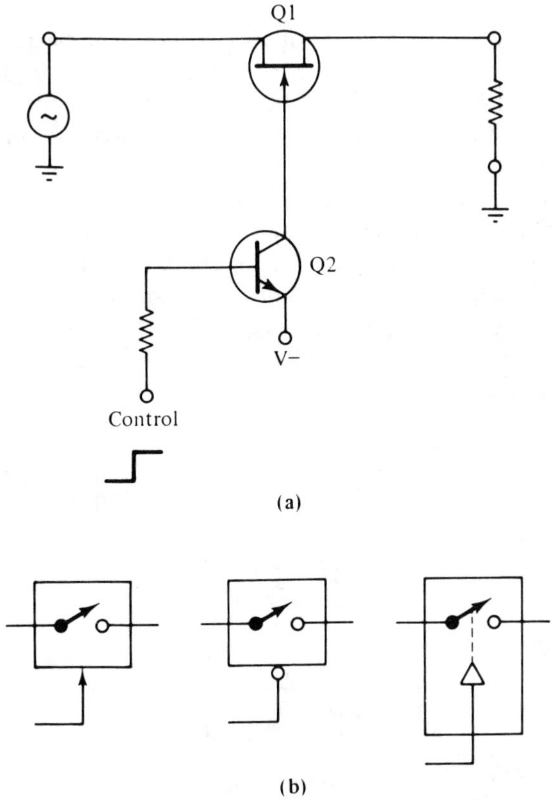

Figure 15-1 CMOS electronic switch: A) Circuit, B) Symbols.

example, the switching element is a single junction field effect transistor (JFET), while in others (perhaps most) it is either a single metal oxide semiconductor field effect transistor (MOSFET) or a complementary pair of N-channel and P-channel MOSFETs. In any case, the idea is that the transistor is connected in series with the signal path. When the transistor is biased fully off, then no signal easily passes from input to output sides. Typical off/on resistance ratio may be 1,000,000:1.

The usual field effect transistor analog switch contains a switch driver circuit (such as Q2) that will allow a TTL voltage signal to turn the switch on and off. In the example shown in Figure 15-1A, the driver transistor is used to control the V− bias to the gate of the JFET. When the base (control input) of Q2 is LOW, then the base is at a lower potential than the emitter, so the transistor will conduct. The gate of the JFET sees a high negative bias that shuts the switch off. Similarly, when the control line is HIGH, then the transistor is reverse-biased and the switch is turned on.

Simple S&H Circuit

Although there are a number of variations on the basic circuit, we see all of the necessary elements of the sample and hold circuit in Figure 15-2. The main components are an output buffer amplifier, sample capacitor, sample switch, and the input circuitry (which may or may not include an input buffer amplifier).

The analog sample switch (S1) closes during the sample period, allowing the input signal to charge the sample capacitor (C_H). The sample switch then opens up again, leaving the capacitor charged to the value of the analog input signal. The capacitor selected for C_H must be a very low leakage type,

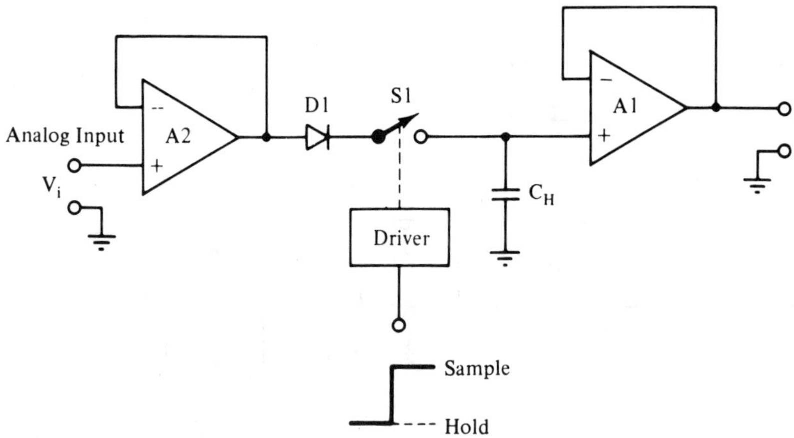

Figure 15-2 Sample and hold circuit.

or else the voltage will drop off ("droop") due to leakage current across the capacitor dielectric. Typical types are silver mica (not preferred), polycarbonate, polystyrene, or glass (considered obsolete for most uses). It is essential that the capacitor have a very low leakage dielectric.

The buffer amplifier is connected in a unity gain configuration. This operational amplifier circuit has a gain that is slightly less than unity (i.e., 0.99999). The open-loop gain (i.e., the gain without feedback) of the operational amplifier will determine how close to unity the gain will be; the higher A_{vol}, the closer to unity in the unity-gain noninverting circuit shown.

But it is not A_{vol} that is the most important property of the operational amplifier used for A1. After all, reasonably high gain (over 50,000) is available on even the cheapest operational amplifier devices. The most important property of the operational amplifier used as the output buffer is a high input impedance. The input impedance of the buffer amplifier represents a load across the sample capacitor that will tend to discharge the capacitor. Also, the input bias currents will tend to affect the charge on C_H. As a result, we must select an operational amplifier with an extremely high input impedance. Superbeta, BiMOS, or BiFET input operational amplifiers are a must. The RCA CA3140 device is a good selection because it has an input impedance of 1.5 teraohms ($1.5 \times 10^{12}\ \Omega$).

The isolation diode is used to keep the charge on the hold capacitor from leaking back through the input load, or the output terminal of the input operational amplifier (A2), if used. This diode must be selected to have a very low level of reverse leakage current, or else excessive droop will occur. It should also have a low value of forward voltage drop. This latter requirement, however, can be easily compensated for by using a gain circuit for either A1 or A2. In fact, there are numerous S&H applications where there will be a gain requirement for scaling purposes. In most cases, the input amplifier will have some gain in order to bring the input signal up to a level that will produce an in-range output signal. Besides, it is true that offset bias current errors in the charge of capacitor C_H are lessened in severity if a higher-level signal is applied. These errors tend to have a constant value and so will be a smaller percentage of the total if the input signal potential is higher.

Figure 15-3 shows the operation of a typical S&H circuit for a varying input signal, V_{in}. The dotted line represents the input signal, while the solid line represents the output signal V_0. For this example, the sample line must be LOW to sample and HIGH to hold. Initially, the line is HIGH so the output voltage will remain at the last valid level. In other words, the value is held. But, at time t_1 the S&H line drops LOW, so the input signal is sampled. Time $(t_2 - t_1)$ is the acquisition time. This period must be short enough that the input signal will not slew very far during the sample period; otherwise the data will contain an error. At time t_2 the S&H line snaps HIGH again, which causes the output voltage to remain at the last valid

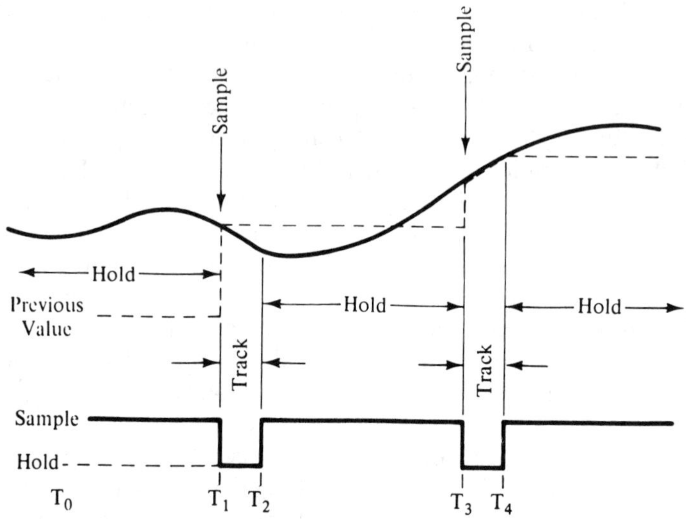

Figure 15-3 S&H circuit operation.

level. This cycle of sample and hold is repeated for different values of signal at time t_3.

Sample and Hold Errors

While our textbook circuits are always ideal and error-free, practical circuits usually contain substantial errors. Although the individual error in any one circuit is small, the cumulative error (i.e., the summation of all system errors) will be quite large in some cases. As a result, it is necessary for us to understand the errors peculiar to each circuit and their effect on the overall efficacy of the circuit or system that we are designing.

In sample and hold circuits there are at least eight forms of errors: these are aperture time, switching transients, settling time, S-to-H offset voltage, voltage droop, feedthrough, dielectric absorption, and gain.

We can define aperture time as the time period between the issuance of the command to sample and the actual closing of the switch (Figure 15-4A). We can attribute aperture time to two different classes of problems, switch actuation time and jitter. The switch actuation time is usually constant for any given type of switch, but this time can be substantial when dealing with very rapid S&H circuits. The primary cause of this problem is the charging of the capacitances in the base or gate structures of the switch driver circuits. All such circuits require some charging time. The problem of jitter is similar to the contact bounce problem of mechanical switches.

Aperture time becomes important because it limits the resolution of the circuit for any given input signal slew rate. Consider a hypothetical situation

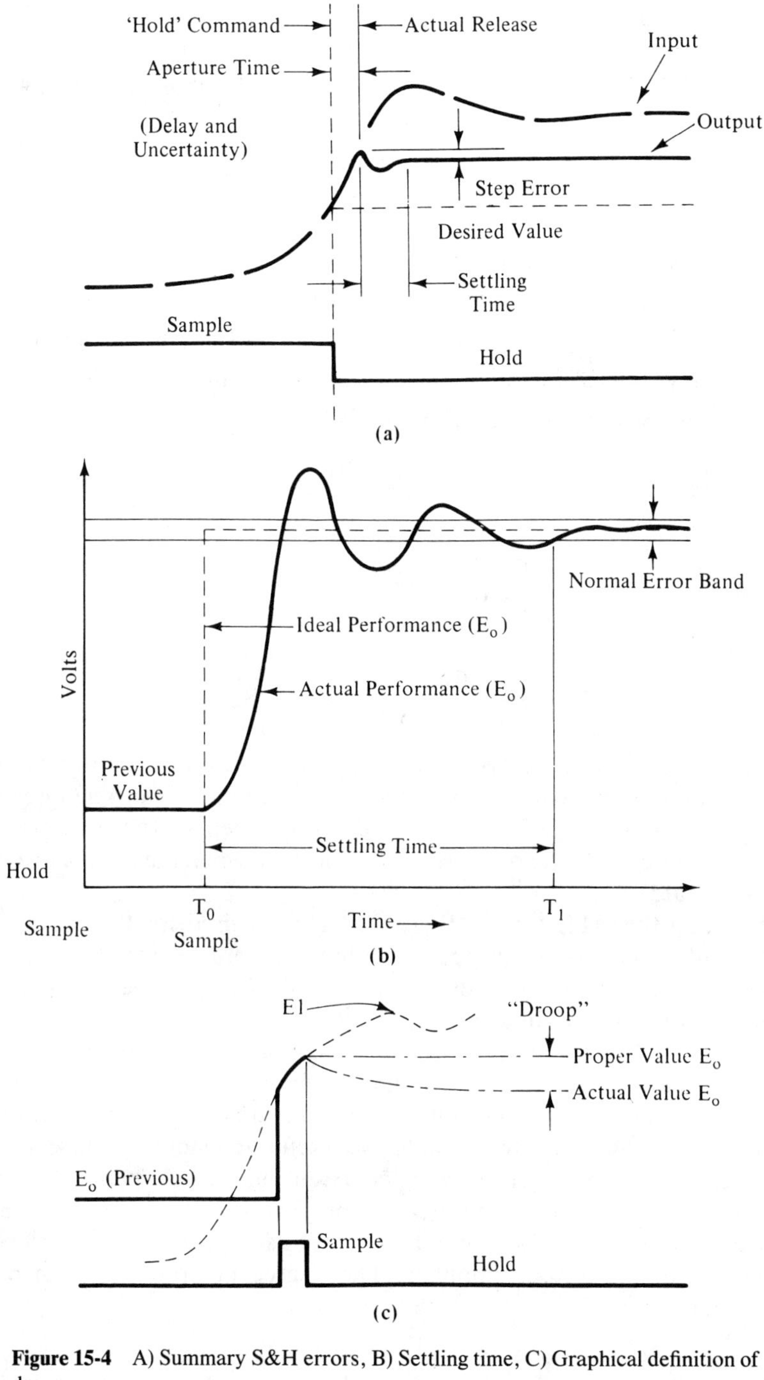

Figure 15-4 A) Summary S&H errors, B) Settling time, C) Graphical definition of droop.

in which we want to resolve the signal to 0.1 percent, and the full-scale input signal is 10 V. Also assume that the input signal slew rate is 2.5 V/μs. How do we find the minimum aperture time required for this signal? The equation is as follows:

$$T_a = \frac{V_{fs} r}{S_r} \tag{15-1}$$

where

T_a is the aperture time in seconds (s)

V_{fs} is the full-scale input signal in volts (V)

r is the resolution expressed as a decimal percent (i.e., 0.1 percent is 0.001)

S_r is the slew rate of the input signal in volts per microsecond (V/μs)

For the problem stated in the preceding text,

$$T_a = \frac{V_{fs} r}{S_r}$$

$$T_a = \frac{(10.0 \text{ V})(0.001)}{\dfrac{2.5 \text{ V}}{\mu s} \times \dfrac{10^6 \ \mu s}{S}}$$

$$T_a = \frac{0.01}{2.5 \times 10^{6/}/S} = 4 \times 10^{-9} S = 4 \text{ ns}$$

For the situation described above, then, we would require a very fast S&H circuit that was capable of an aperture time of less than 4 nanoseconds.

The second type of error is switching transients. These occur at the instant of switching and are of critical importance when the input signal has a high slew rate.

Setting time (Figure 15-4B) is the time required for the S&H circuit output voltage to permanently settle within the nominal error band specified for the device. This error is usually specified as a given percentage of the full-scale input signal voltage.

Also shown in Figure 15-4A is the sample-to-hold offset error. This error is due to the charge stored in the capacitances of the FET switch used to take the sample. At the instant that the hold mold is instituted, the charge in these capacitances is dumped into the circuit and stored in the capacitor.

Voltage droop (Figure 15-4C) occurs during the hold period and is due to a slight discharge of the hold capacitor. There will always be a leakage resistance across the capacitor and a shunt resistance that is due to the input impedance of the output amplifier. The degree of droop is given by the following equation:

$$\frac{dV_0}{dt} = \frac{I}{C_h} \tag{15-2}$$

where
V_0 is the output voltage in volts (V)
I is the capacitor leakage current from all resistance sources, both internal and external to the capacitor, in amperes (A)
C_h is the capacitance of the S&H capacitor in farads (F)

Output voltage droop is something we must live with, even though there are design steps that will mitigate the degree of droop in any given application. We can, for example, be careful to select low-leakage capacitors (e.g., polycarbonate and polyethylene), diodes with extremely low-leakage currents, and output amplifiers that have extremely high input impedance specifications.

Feedthrough is defined as the leakage of signal through the sampling switch. There are two factors to consider here: series resistance in the off state and the capacitance of the open switch (i.e., when in the off state). The presence of any substantial leakage resistance will permit the DC offset component of the input signal to affect the charge on the sampling capacitor. Similarly, the switch capacitance will permit alteration of the charge on the capacitor by AC signals of sufficient frequency.

Dielectric absorption is a property of capacitors that is often overlooked by capacitor users, except in cases such as S&H circuits and where the capacitor is used at higher voltages. The electrostatic field used to store energy in the dielectric of a capacitor causes the orbits of the electrons in the atoms of the dielectric to deform. This, in fact, is the manner in which the electrical energy is stored in the capacitor. When the capacitor is discharged the orbits return to their normal shapes. The energy lost in the dielectric by this process becomes the electrical current that flows in the external circuit. But not all electron orbits return to their correct shape at the same time. After the capacitor has discharged there will be a residual charge left in the capacitor. You can see this effect when working with high-voltage, oil-filled capacitors that have a high capacitance value. Charge the capacitor to several thousand volts. After a few seconds attempt to discharge the capacitor—note the spark. Now, take a voltmeter and measure the voltage across the terminals of the supposedly discharged capacitor; you may find over 100 volts remaining.

In the small capacitors used in instrumentation S&H circuits this voltage may be smaller than in the more spectacular example above, but it is nonetheless significant to the accuracy of the S&H output signal. This dielectric absorption voltage will be added to the output voltage of the next sample. The dielectric absorption factor is specified for capacitors and will be from 0.001 percent to 10 percent. Try to select as low a factor as possible; polystyrene is a good selection for most applications, as is Teflon.

Gain error is the error caused by problems with the gains of the operational amplifiers used in the S&H. Most operational amplifiers used in the noninverting unity gain follower configuration have negligible error. But when the amplifier is used in a gain circuit, then the tolerances of the resistors become important and gain errors pop up.

ANALOG REFERENCE VOLTAGE CIRCUITS

The data converters used in microprocessor-based instrumentation are devices that produce an output that is proportional to a reference voltage or current source. The digital-to-analog converter (DAC), for example, produces an output voltage (or current) that is proportional to both the applied binary word and an analog reference voltage (or current). The precision and accuracy of the data converter is determined in part by constraints placed on the reference source. Although the bit-length of the converter is normally thought to be the principle limiting factor that determines the resolution of the data, there is also a severe constraint placed on the reference source. For example, if the resolution of a DAC is 8 bits, then we would ordinarily expect that the 1-LSB (least significant bit) resolution of the device would be $1/2^N$, where N is the bit-length. In the case of an 8-bit DAC, then, we would expect the resolution to be the step voltage that is caused by 1-LSB, or $1/2^8$, which is 1/256 or 0.0039. Suppose that the full-scale output voltage of the DAC is 2.55 V. The 1-LSB voltage in that case would be 9.9 mV. The reference source should be accurate to 0.99 mV in this case. If it is not, then the 1-LSB resolution of the data converter is in doubt, and one cannot believe the results in that range.

It is quite possible to render a data converter no better than another data converter with two or three less bits of word length, just by failing to pay sufficient attention to the matter of the reference source. Consider a typical 10.000-V output 8-bit DAC. The 1-LSB voltage is 39 mV. If the DC reference supply is not adjusted with some precision, or if it drifts off to another voltage with changes in ambient temperature, then the actual 1-LSB voltage will be something different. Suppose the actual voltage is 10.24 V, instead of 10.000 V. This 8-bit DAC is incapable of resolving an output potential to more than 5 actual bits, rather than 8 because of the error in the reference potential.

Various types of reference voltage source are used. Another source, and the least desirable one, is the simple zener diode. Other sources, based on either zener diodes or precision versions of the zener such as the band-gap diode, will prove more useful in the development of microprocessor-based instruments. In the sections to follow, we will discuss some of the more popular circuits and IC devices used to make microprocessor data converter references.

ZENER DIODES

The zener (pronounced "zen-ner") diode is a special class of avalanche PN junction diode. Figure 15-5A shows the typical zener diode regulator circuit, while Figure 15-5B shows the response curve typical for a zener diode.

When a zener diode is forward-biased, it behaves very much like any other PN junction diode. Once the forward bias potential exceeds 0.6 to 0.7 V (its junction potential), the current-versus-voltage (I-vs.-V) curve becomes essentially linear. But in the reverse-bias condition, the behavior of the zener diode is markedly different from that of ordinary diodes. The reverse-bias current flow is limited to the normal leakage current, essentially zero or some extremely low value, until the reverse bias potential reaches a

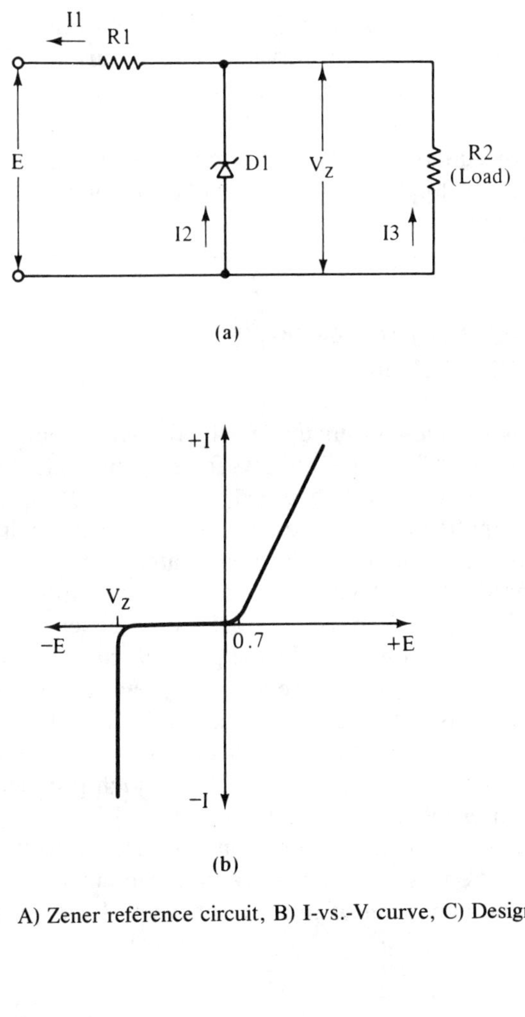

Figure 15-5 A) Zener reference circuit, B) I-vs.-V curve, C) Design equations.

critical point called the zener voltage, usually symbolized by V_Z. At that point the diode avalanches and a considerable reverse current will flow.

The circuit of Figure 15-5A shows a zener diode used as a shunt voltage regulator; that is, the zener is in parallel with the load. Resistor R1 serves to limit the current (I1) to a safe value, while R2 represents the load resistance.

Voltage regulation occurs if current I2 is considerably greater than I1; a common rule of thumb is to set I2 at approximately 10 times greater than I3. If this ratio is maintained, then fluctuations in I3 due to changing load conditions become an insignificant percentage of I1; therefore, V_z remains essentially constant.

There are three different circumstances in which the zener circuit of Figure 15-5A might be used, and all three have slightly different design equations, as follows:

1. Variable supply voltage (V), constant load current (I3)
2. Constant supply voltage (V), variable load current (I3)
3. Variable supply voltage (V), variable load current (I3)

The design equations for these conditions are summarized in Figure 15-5C. Examples using these equations will be included in the problem set at the end of the chapter.

PRECISION OPERATIONAL AMPLIFIER DC REFERENCE SOURCES

Most DC power supplies produce a specified nominal output voltage that is not very precise. A "12-V DC" supply, for example, might produce a stable output potential somewhere between 11.5 and 12.5 V and still be in tolerance. In many applications, however, a more precise voltage is required, such as the reference voltage used to calibrate meters.

The zener diode voltage tends to wander a little with temperature changes, so in a circuit such as Figure 15-6 the zener diode is placed in a controlled thermal environment. In the past this meant a component oven, but now "four-terminal" zeners are available that use a heating element and zener diode thermally coupled on the same IC substrate.

The manufacturer of the LM-199 device uses a zener diode nested among the transistors of a class-A amplifier, which produces constant heat dissipation under zero-signal conditions.

The circuit in Figure 15-6 uses a temperature-controlled zener diode to set the potential at one input of an operational amplifier. You should be able to understand the operation of this circuit if you have completed studying Chapter 12.

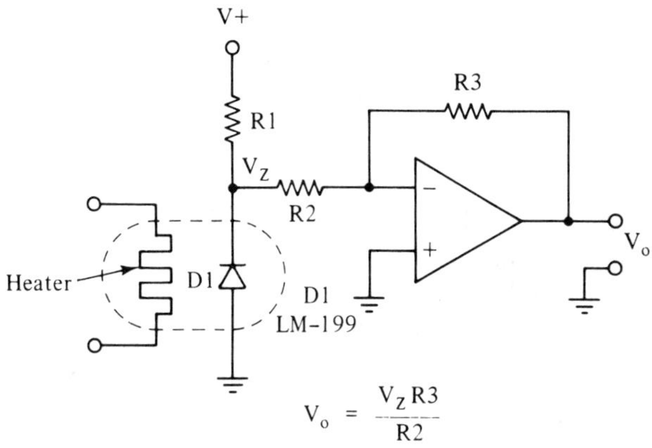

Figure 15-6 LM-199 reference source circuit.

The voltage at the output terminal of the circuit in Figure 15-6 is given by the normal transfer function of the inverting follower operational amplifier, as follows:

$$V_0 = -V_z \times \frac{R3}{R2} \qquad (15\text{-}3)$$

The zener voltage and the desired output voltage are usually known, so we will need to rearrange Equation (15- 3) to find either the ratio R3/R2 or, if one or the other resistor is selected for convenience sake, the value of the remaining resistor. Also, note that this circuit produces a negative output voltage, so it can be used only with data converters that either need or allow a negative reference source (e.g., DAC-08 or MC1408-8 devices).

The circuit for a positive output operational amplifier reference source is shown in Figure 15-7. In this case, the operational amplifier is connected in the noninverting follower configuration. The zener reference diode (which could also be an LM-199 or similar device) is connected with its series current-limiting resistor to the noninverting input of the amplifier. This connection makes the input voltage of the amplifier equal to the zener potential. The transfer equation for this circuit is

$$V_0 = V_z \times \frac{R_a}{R1} + 1 \qquad (15\text{-}4)$$

where R_a is the series combination R2 + R3, and the other factors bear their normal meanings.

The potentiometer in the circuit of Figure 15-7 is used to trim the output voltage to a precise value. For this reason, it should have a maximum value of approximately 10 percent of R_a. Further, the potentiometer

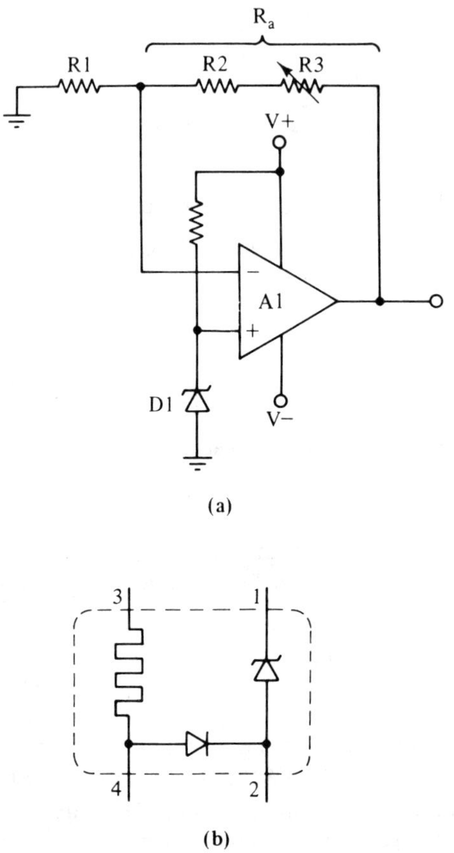

Figure 15-7 A) Adjustable LM-199 reference source circuit, B) LM-199.

should be a 10- to 20-turn "trimpot" type in order to increase the resolution per turn. All of the resistors in this circuit should be precision types of one percent or better tolerance. The purpose in selecting precision resistors is not as much for the precise setting of the amplifier gain (the potentiometer does that job), but to acquire low temperature coefficient resistors. It does little good to specify a 20-turn potentiometer for R3, only to use carbon composition resistors for R1 and R2, thereby introducing more drift than the potentiometer could compensate. The low temperature coefficient resistors will keep the gain of the amplifier reasonably stable even after warm-up of the circuit.

In some applications, the circuits of Figures 15-6 and 15-7A can use an ordinary zener diode, especially if the ambient temperature remains relatively constant. But for most instrumentation and control applications, a premium reference diode must be used in place of the zener. Recommended are the LM-199, which keeps the buried zener at a constant temperature, or one of the band-gap zener reference diodes. The LM-199 (and the lower-grade LM-299 and LM-399 devices in the same series) offer good temperature-tracking specifications because the zener element is buried on the same semiconductor die with a class-A amplifier that has its input terminals shorted together. The effect of this configuration is that the class-A amplifier transistors dissipate a constant amount of power, thus producing a constant temperature environment on the IC die. Burying the zener also produces a lower noise operation and provides a 20-ppm long-term stability and short-term stability of 1 ppm. The rated terminal voltage of the LM-199 is +/− 2 percent, but this is corrected by the operation of potentiometer R3 in Figure 15-7A.

The operation of the LM-199 (and others of the series) is such that the diode-side connections are identical to zener diode connections. The heater terminals, however, must be connected to a DC source of 9 to 40 V. A diagram of the LM-199 circuit is shown in Figure 15-7B.

An example of a band-gap zener diode is shown in Figure 15-8. This device is the Ferranti Semiconductor, Inc., ZN-458A/B device, but it is representative of the products of several manufacturers. The band-gap zener diode is on the same substrate with an amplifier and output transistor that will allow the device to pass up to 120 mA of current. Note, however, that best performance occurs when the diode is operated at a lower current, in the 2 to 15 mA range. The Ferranti devices are available in several differ-

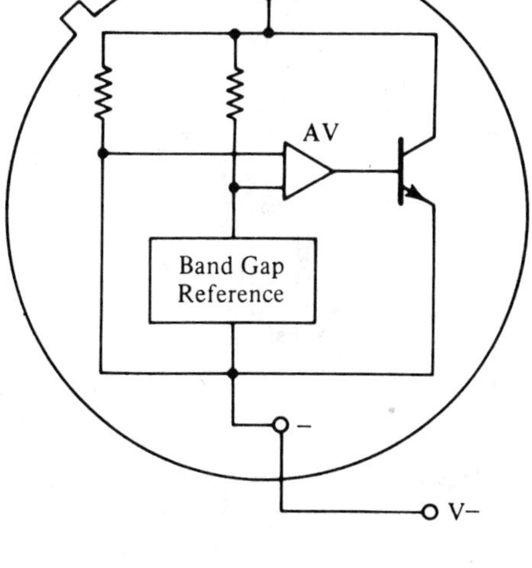

Figure 15-8 Band gap reference diode circuit.

ent voltages and with different stability ratings. Most of the devices in this line offer either 1.26 V or 2.45 V, with stability ratings to 10 ppm, at a temperature coefficient of 0.003 percent per degree centigrade. The series current-limiting resistor is selected according to the same rules as those for ordinary zener diode regulators.

A final example of the band-gap reference diode is shown in Figure 15-9A, the Intersil ICL8069 device. The specifications for various grades of

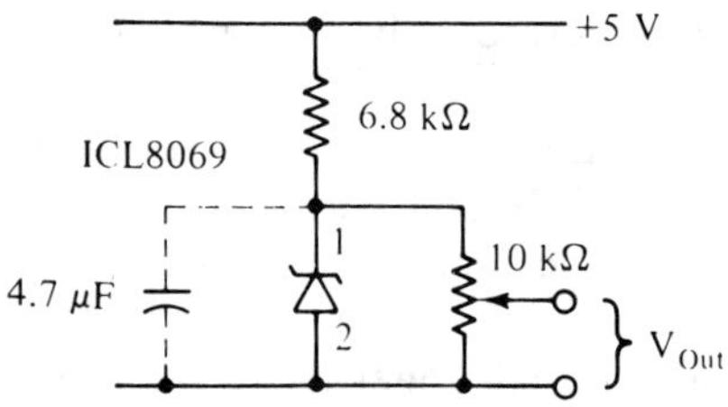

(a)

Max Temperature Coefficient of V_{Ref}	Temp Range	Order Part #
0.001%/°C	0°C to +70°C	ICL8069ACQ
0.0025%/°C	0°C to +70°C	ICL8069BCQ
0.005%/°C	−55°C to +125°C	ICL8069CMQ
0.005%/°C	0°C to +70°C	ICL8069CCQ
0.01%/°C	−55°C to +125°C	ICL8069DMQ
0.01%/°C	0°C to +70°C	ICL8069DCQ

(b)

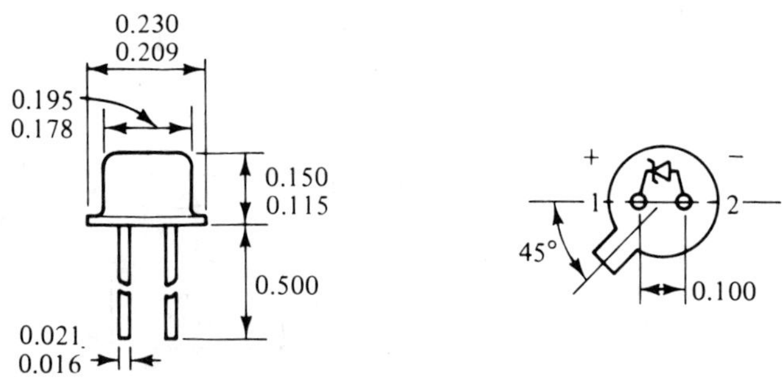

(c)

Figure 15-9 ICL-8069 reference diode: A) Circuit, B) Table of performance, C) Package.

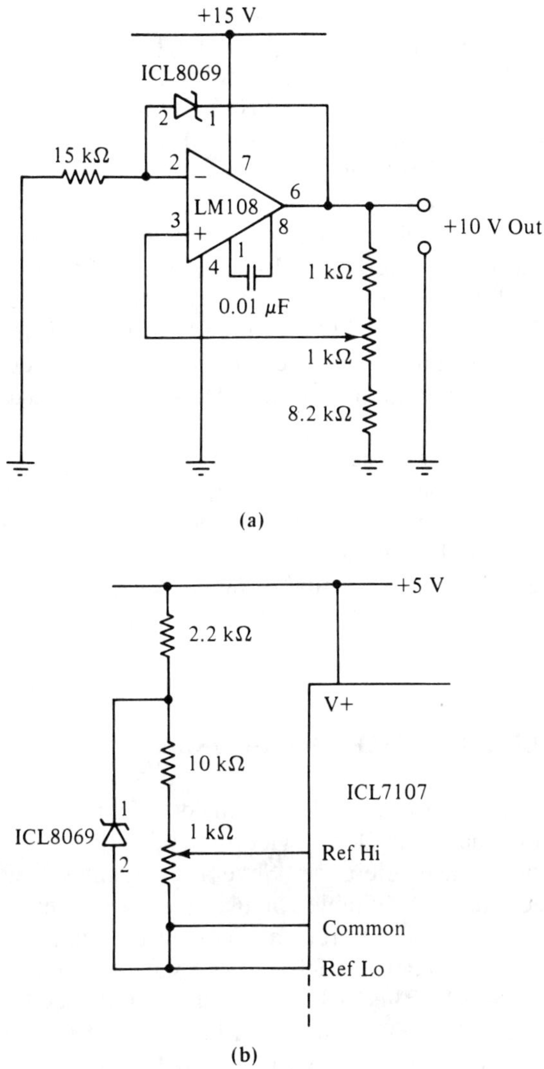

Figure 15-10 ICL-8069 circuits: A) Buffered, B) Nonbuffered.

this diode are shown in Figure 15-9B, while the case is shown in Figure 15-9C. The device is available in both commercial (O° C to +70° C) and military (−55° C to +125° C) temperature ranges, with temperature coefficients from 0.001%/°C to 0.01%/°C.

Circuits using the ICL8069 device are shown in Figure 15-10. The circuit on Figure 15-10A shows the use of an operational amplifier with the

ICL8069 to produce a 10.000-V reference source. In Figure 15-10A, the reference diode is connected in the negative feedback path of the operational amplifier, while a resistor feedback network is used to the noninverting input. The potentiometer in this circuit is selected to have a value of approximately 10 percent of the total, and it should be a 10- to 20-turn type. All resistors in this circuit should be precision types, again to take advantage of the temperature coefficient.

The operational amplifier selected for a reference source should be able to produce low drift characteristics. In general, the lowest-cost operational amplifiers, such as the 741 device, are not suitable. A premium device such as the LM108, LM156, or CA3140 is required. It makes little sense to specify a premium reference diode and then connect it into a circuit in which the drift factor and offset potential/currents negate the advantages of the reference diode.

In Figure 15-10B we see the use of the ICL8069 device in a data converter without the benefit of the operational amplifier. In this case, a series combination of a fixed resistor and a potentiometer is used to trim the fixed output of the ICL8069. Note that the drift/stability factors of these diodes are quite good, but the actual terminal voltage is nominal, so it may be as much as two percent in error. The potentiometer allows us to cancel some of that error.

INTEGRATED CIRCUIT REFERENCE SOURCES

Integrated circuit technology has allowed the development of a large number of special-function devices. One device that takes good advantage of the technology is the voltage reference. Several companies now offer voltage reference devices that are suitable for use in data converter applications. Although there are several different devices on the market, we will cover only one of the more popular: the REF-01/02 by Precision Monolithics, Inc., shown in Figure 15-11A. The REF-01 is a 10.000-V device (Figure 15-11B), while the REF-02 is a 5.000-V device (Figure 15-11B). The REF-02 also offers a temperature output (pin number 3 in Figure 15-11B) voltage that produces a slope of 2.1 mV/°C. This feature makes the REF-02 useful in temperature measurement applications and also provides a signal that can be used to trim circuits according to temperature changes.

The basic circuit of the REF-01 and REF-02 is shown in Figure 15-11C. A sample of the output voltage is fed back to the device through a potentiometer that can be used to trim the output voltage to within a few millivolts of the rated output voltage. These devices will sink up to 20 mA of current, but one is advised to use as little current as possible in reference sources.

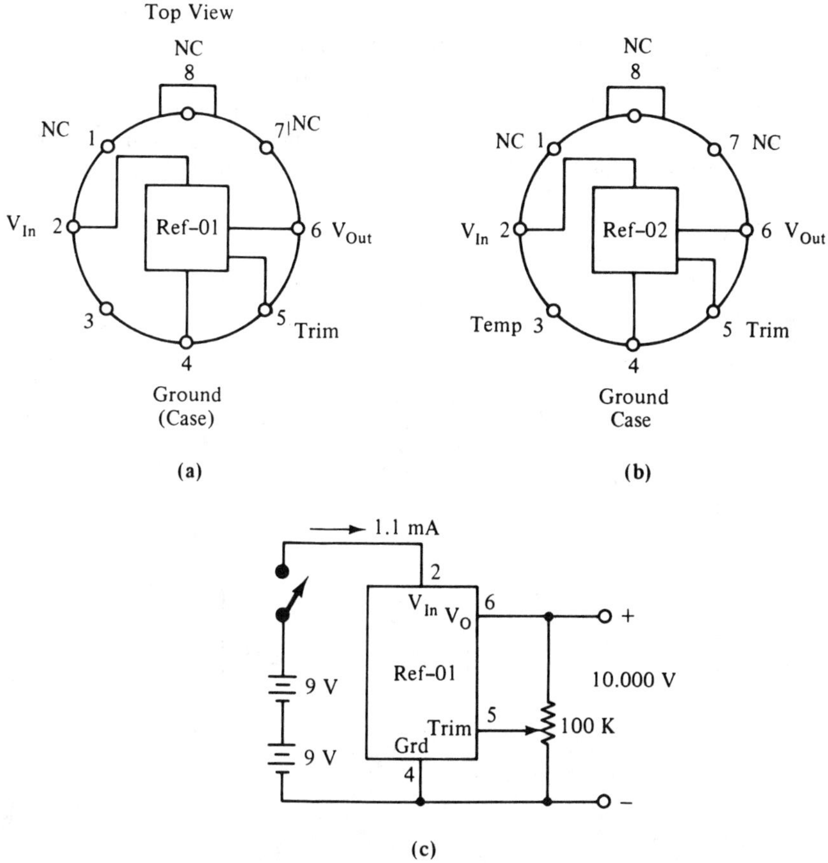

Figure 15-11 A) REF-01, B) REF-02, C) Typical circuit.

CURRENT REFERENCE SOURCES

Some data converter circuits require a current reference instead of a voltage reference. In most of those cases, we can create a current reference by using one of the available reference voltage sources and a resistor—Ohm's law has not yet been repealed by Congress! The simple tactic is to apply a reference voltage to one end of a precision resistor connected to the reference current input of the DAC. The reference current I_{ref} is then simply the quotient of the reference voltage over the resistance. But sometimes it is also desirable to use a separate current reference source, and we have several methods for accomplishing this job.

One way to provide a current reference is to purchase a reference current "diode." These devices will be rated at some nominal current and produce relatively consistent results. But the device is actually little more than a JFET that is connected with the gate and source terminals shorted together. Although the connection is made internally, making the device look like a two-element diode, it is actually similar to Figure 15-12A. All

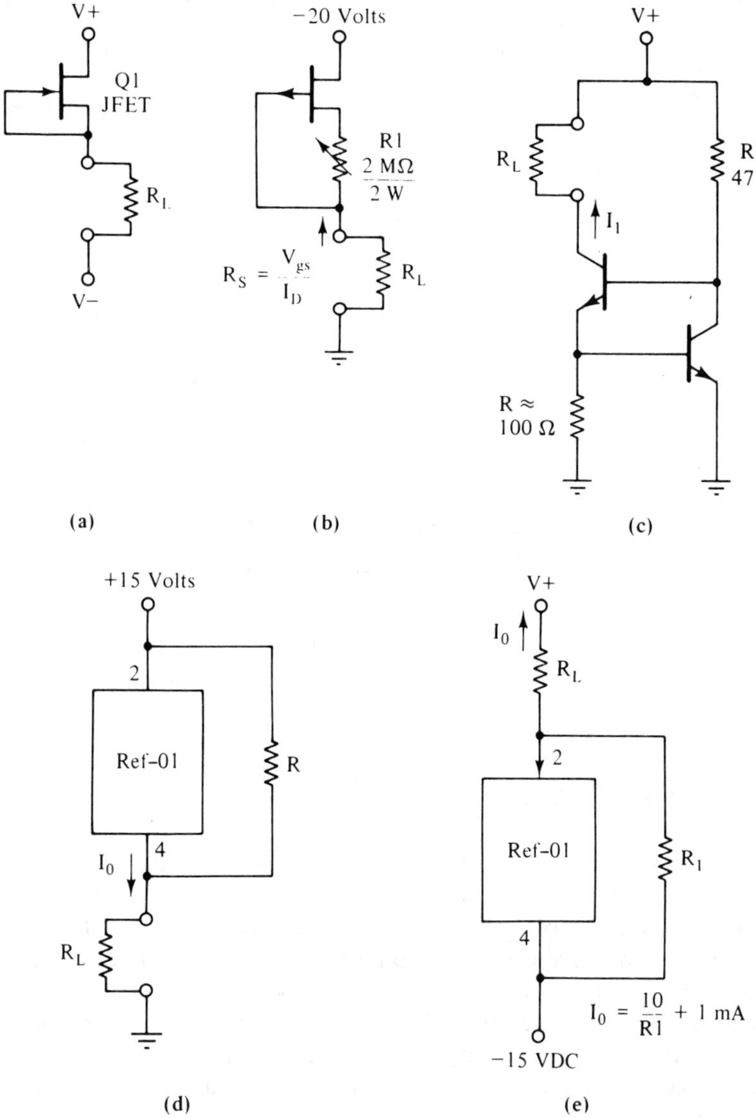

Figure 15-12 Current source circuits.

JFETs have a certain knee current at which the device becomes saturated. Applied voltages higher than that required to saturate the device will not produce a higher source-drain current. We can, therefore, use the JFET in that region to produce a constant reference current. A method for making the current variable over the range 6 μA to 2 mA is shown in Figure 15-12B. Here we have a potentiometer connected as a rheostat in series with the source of the JFET, while the gate continues to be connected to the source.

The use of a pair of bipolar NPN transistors as a constant current source is shown in Figure 15-12C. This circuit uses slight regeneration to servo out changes on the load resistance, hence keeping the load current I1 constant. The derivation of the equation for I1 is left as an exercise for the reader.

The REF-01 and REF-02 devices can be used as a constant current source by using the circuits of Figures 15-12D and 15-12E. The circuit in Figure 15-12D is a current source, while that of Figure 15-12E is a current sink. In both cases, the output current is given by

$$I_{out} = \frac{10.0\ V}{R} + 1\ mA \qquad (15\text{-}5)$$

where

I_{out} is the current in milliamperes (mA)

R is the resistance in kiloohms (kΩ)

V is the potential 10.000 V

AUTO-ZERO AND DRIVE TO VALUE CIRCUITRY

In another chapter it is stated that transducers are rarely perfectly balanced. Even when the individual arm elements have the same nominal at-rest resistance (R), an offset error may exist due to the fact that the bridge arm resistances each have a certain tolerance, and that causes a difference in the actual values of resistance. Because of this problem, the output voltage of the Wheatstone bridge transducer will be non-zero when there is no stimulus applied.

Another problem that sometimes needs solution in computer-based experiments or instruments is the *drive to preset value* problem. Consider a transducer used to weigh some sample or another. The container has a certain weight that must be subtracted from the measured weight in order to arrive at the weight of the material in the container. This is called the *tare weight* of the container. There are two popular solutions: 1) subtract the tare weight in software (certainly viable), and 2) drive the measuring sensor to a preset offset that represents the tare weight. That is, with the offset bias present, setting the container on the scale will result in a zero reading.

Monitoring the output of the bridge amplifier in Figure 15-7 is a voltage comparator (A2). When V_0 is zero, the A2 input is zero, so the output is LOW. Alternatively, when the amplifier (A1) output voltage is nonzero, the A2 output is HIGH. The DAC binary inputs are connected to the digital outputs of a binary counter that is turned on when the enable (EN) line is HIGH. Circuit operation is as follows:

1. The operator sets the transducer to zero stimulus (e.g., for pressure transducers the valve to atmosphere is opened).

2. The ZERO button is pressed, thereby triggering the one-shot to produce an output pulse of time T, which makes one input of NAND gate G1 HIGH.

3. If voltage V_0 is nonzero, then the A2 output is also HIGH, so both inputs of G1 are HIGH—making the G1 output LOW, thereby turning on the binary counter.

4. The binary counter continues to increment in step with the clock pulses, thereby causing the DAC output to rise continously with each increment. This action forces the bridge output towards null.

5. When V_0 reaches zero, the A2 output turns off, stemming the flow of clock pulses to the binary counter and stopping the action. The DAC output voltage will remain at this voltage level.

16

ACTIVE FILTER CIRCUITS

A frequency-selective filter is an electronic circuit that favors some frequencies and discriminates against certain other frequencies (or bands of frequencies). In other words, a filter circuit will pass some frequencies and reject or sharply attenuate others. Frequencies that pass through the filter with little attenuation are said to be in the passband, while frequencies that are heavily attenuated are said to be in the stopband. Filter circuits can be classified several ways: passive versus active, analog versus digital, hardware versus software, by frequency range (e.g., audio, RF, or microwave), or by passband characteristic. Passive filters are made of combinations of passive components such as resistors (R), capacitors (C), and inductors (L). In general, passive filters are lossy and not very flexible. An active filter, on the other hand, is one based on an active device such as a transistor or an operational amplifier along with passive components that determine frequency. In most cases, the passive components are resistors and capacitors (although a few with inductor-based circuits are known).

Active filters use linear circuit techniques such as those found throughout this book. Digital filters use digital IC devices, and are often based on switching techniques. Software filters implement solutions to frequency-selective equations using computer programming techniques.

Filters can also be classified by frequency range. ''Audio'' filters operate from the subaudio to ultrasonic ranges (near-DC to about 20 kHz). RF filters operate at frequencies above 20 kHz, up to about 900 MHz. Microwave filters operate at frequencies above 900 MHz. These range designations are not absolute, but they do serve to indicate approximate points at which a change of design technique usually takes place. For example, filters can be made frequency-selective using inductors. But in the audio range the inductance values are typically very large, so the inductors are bulky, costly, and lossy. In addition, inductors also produce stray magnetic fields that can interfere with other nearby circuits. On the other hand, inductors are the elements of choice in the RF region. But once frequencies approach several hundred megahertz, the inductance values required become too low for practical use, so other techniques are required. In the microwave and high-UHF region, transmission line and cavity techniques are used.

Finally, filters may be classified by their frequency response characteristic. This method of categorizing filters takes note of the filter's passband. In this chapter we will examine low-pass filters, high-pass filters, bandpass filters, and stopband filters. We will also examine a related circuit called the all-pass phase shifter.

FILTER CHARACTERISTICS

Figure 16-1A shows in general terms the characteristics of theoretically ideal filters. These curves will be discussed in greater detail later. A low-pass filter has a passband from DC to a specified cutoff frequency (F1). All frequencies above the cutoff frequency are attenuated, so they are in the stopband. A bandpass filter has a passband between a lower limit (F2) and an upper limit (F3). All frequencies lower than F2 or greater than F3 are in the stopband. A high-pass filter has a stopband from DC to a certain lower limit (F4). All frequencies greater than F4 are in the passband.

A stopband filter response is shown in Figure 16–1B. This filter severely attenuates frequencies between lower and upper limits (F5 to F6) but passes all others. When the stopband is very narrow, the stopband filter is called a notch filter. Such filters are often used to remove a single, unwanted frequency. An example of such an application is removal of unwanted 60-Hz interference caused by local power lines.

"Ideal" versus Practice Filter Response Curves

The response curves shown in Figure 16-1 are unrealistic "ideal" generalizations. It might seem that a step function cutoff is desirable, but in actual practice such a response is neither attainable nor desirable. The reason it is undesirable is that the "ideal" response actually causes a problem because the filter may ring when a fast rise-time signal is applied. This phenomenon is especially likely to occur on narrow bandpass filters. While sinusoidal signals do not usually pose a problem, noise impulses or transient step functions can easily cause unwanted ringing.

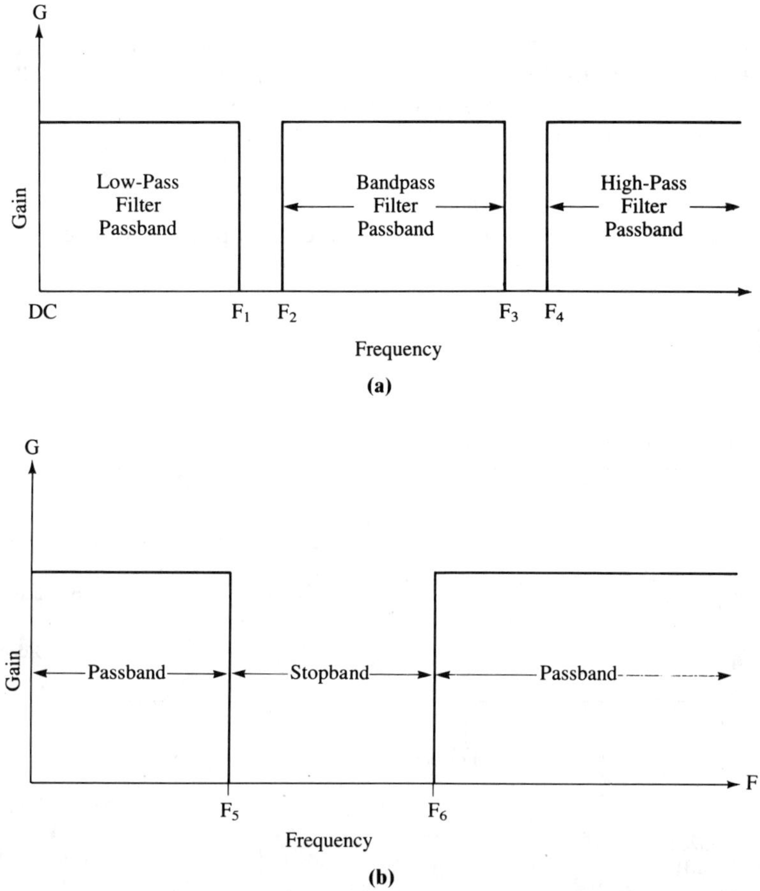

Figure 16-1 A) Respective characteristics for low-pass, bandpass, and high-pass filters, B) Bandstop filter characteristic.

There are several common filter responses: Butterworth, Chebychev, Cauer (also sometimes called elliptic), and Bessel. The Butterworth response is shown in Figure 16-2A. The noteworthy properties of the Butterworth filter are that both the passband and stopband are relatively flat, as is the transition region slope between them.

It is standard practice in filter design to specify the passband between points where the response falls off −3 dB. Therefore, for the low-pass filter shown in Figure 16-2A the cutoff frequency is at the point where gain (Al) falls off to 0.707 times the low-frequency gain (A1), i.e. the −3-dB point.

At frequencies $f > f_c$, the gain falls off linearly at a rate that depends on the order of the filter. The slope (S) of the falloff is measured in either decibels per octave (a 2:1 frequency change) or decibels per decade (a 10:1

frequency change). Note that these two systems can be scaled: −6 dB/octave is the same slope as −20 dB/decade. The slopes shown in Figure 16–2A cover three Butterworth cases. A first-order filter offers a roll-off of −20 dB/decade, a second-order filter offers a roll-off of −40 dB/decade, and a third-order filter rolls off of −60 dB/decade. These correspond to 6, 12, and 18 dB/octave, respectively.

On first glance, it might appear that only third-order filters would be used, because they transition from passband to stopband more rapidly. But

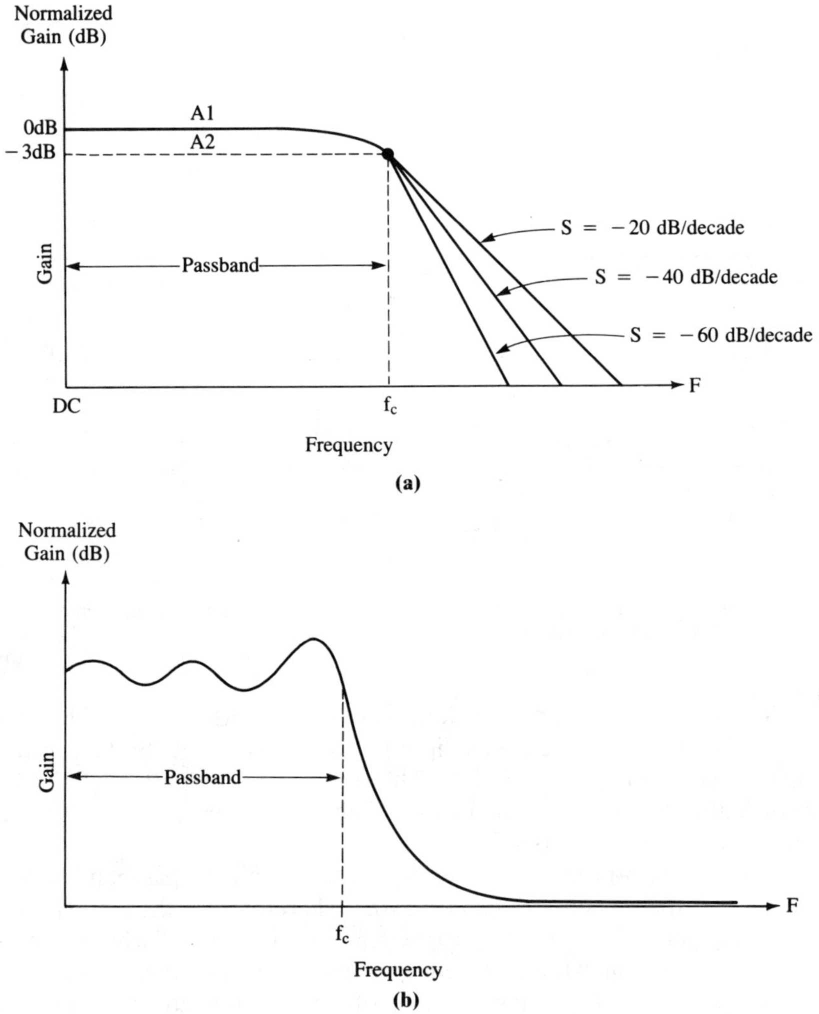

Figure 16-2 A) Butterworth characteristic, B) Chebychev characteristic, C) Cauer/elliptical characteristic, D) Bessel characteristic.

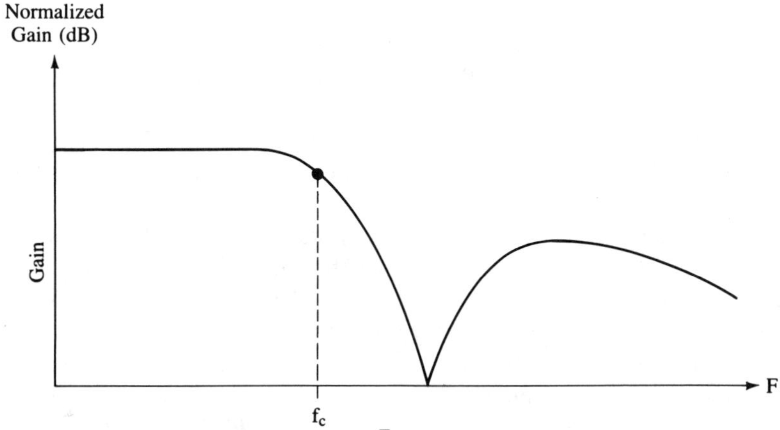

(c)

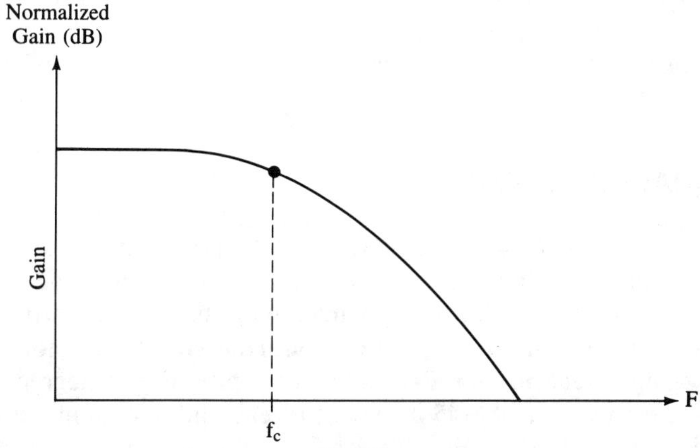

(d)

higher-order response is obtained at the cost of more complexity, greater component value error sensitivity, and more difficult design procedures. Some higher-order filter designs are also more likely to oscillate than their lower-order equivalents. The selection of filter "order" is a trade-off between system requirements and complexity. If a large, undesired signal is expected at a frequency up to, say, the third harmonic of f_c, then a higher-order response might be indicated. Conversely, weaker stopband signals in that region may permit the use of a lesser-order filter.

The steepness and shape of the roll-off curve is a function of the filter damping factor. As a class, Butterworth filters tend to be heavily damped, which explains the gradual roll-off in the response curve. The Chebychev filter response (Figure 16-2B) is lightly damped, so it has a variation (or "ripple") within the passband. The Chebychev filter offers a generally faster roll-off than does the Butterworth filter, but at the cost of less flatness within the passband.

The Cauer or elliptic filter response curve (Figure 16-2C) offers the fastest roll-off for frequencies close to the cut-off frequency, as well as relatively good flatness within the passband. Notches of −40 dB to −60 dB can be achieved close to f_c, but only at a cost of less attenuation further into the stopband. A typical Cauer filter will have a deep notch close to f_c, rise to a peak at some high frequency removed from f_c, and then gradually fall off at even higher frequencies at a rate of −20 dB to −40 dB/decade.

The response of the Bessel filter is shown in Figure 16-2D. Although it appears similar to the Butterworth response, it is not maximally flat within the passband, and it falls off somewhat less rapidly. The benefit of the Bessel filter is a flat phase response across the passband. In some applications the phase response is the overriding consideration, so the Bessel response is preferred.

FILTER PHASE RESPONSE

Most frequency-selective filter circuits exhibit a phase change over the passband. The responses for two different filters, Butterworth and Bessel, are shown in Figure 16-3. Note that the maximally flat Butterworth exhibits a decidedly nonlinear phase response in both passband and stopband. The frequency-dependent phase shift of a low-pass filter is −45 degrees at f_c, and it increases by a factor of −45 degrees for each additional increase of −20 dB/decade in the roll-off slope. Thus a first-order low-pass filter phase shift is −45 degrees, while the second-order phase shift (for −40 dB/decade roll-off) is −90 degrees, and for a third-order filter (−60 dB/decade) it is −135 degrees. The high-pass response is of the same magnitude for each order of filter, but the sign is opposite. Thus a first-order high-pass filter shows a +45 degree phase shift, a second-order filter shows +90 degrees, and a third-order high-pass filter shows +135 degrees.

The Bessel filter also shows a phase shift over the passband, but it is nearly linear. A useful feature of this characteristic is that it allows a uniform time delay all across the passband. As a result, the Bessel filter offers the ability to pass transient pulse waveforms with minimum distortion. For the Bessel filter, the phase shift maximum occurs at:

$$\Delta\Phi_{max} = \frac{-n\,\pi}{2} \tag{16.1}$$

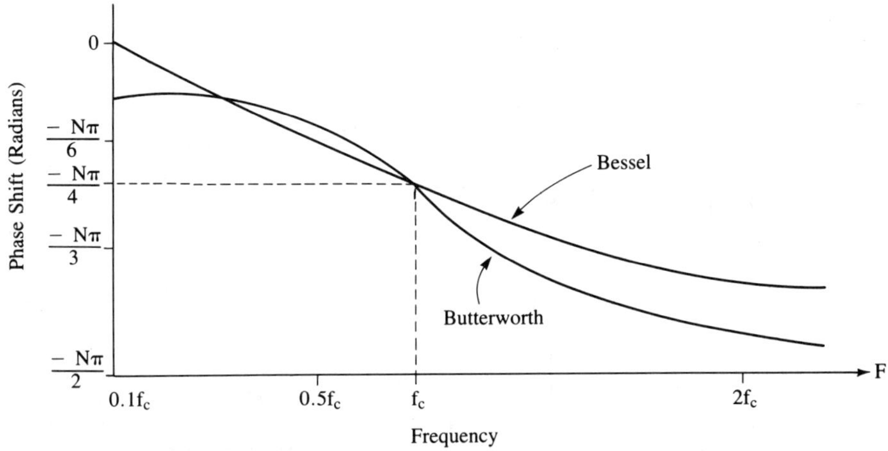

Figure 16-3 Phase shift performance of Bessel and Butterworth filters.

where

Φ_{max} is the maximum phase shift

n is the order of the filter (i.e., number of poles)

In a properly designed Bessel filter, the cutoff frequency, f_c, occurs at a point where the phase shift is half the maximum phase shift, or

$$\Delta\Phi_{fc} = \frac{-n\,\pi}{4} \tag{16.2}$$

The Bessel filter is said to work best at the frequency where $f = f_c/2$.

LOW-PASS FILTERS

In this section and the sections to follow, practical design of active filters will be discussed. The model for the filter is shown in Figure 16-4. This filter is called the voltage-controlled voltage source (VCVS) filter or Sallen-Key filter. The basic configuration is a noninverting follower operational amplifier (A1). The op-amp selected should have a high-gain bandwidth product, relative to the cutoff frequency, in order to permit the filter to operate properly. The gain of the circuit is given by

$$A_v = \frac{R_f}{R_{in}} + 1 \tag{16.3}$$

So, if $R_{in} = R$, then we may deduce

$$R_f = R(A_v - 1) \tag{16.4}$$

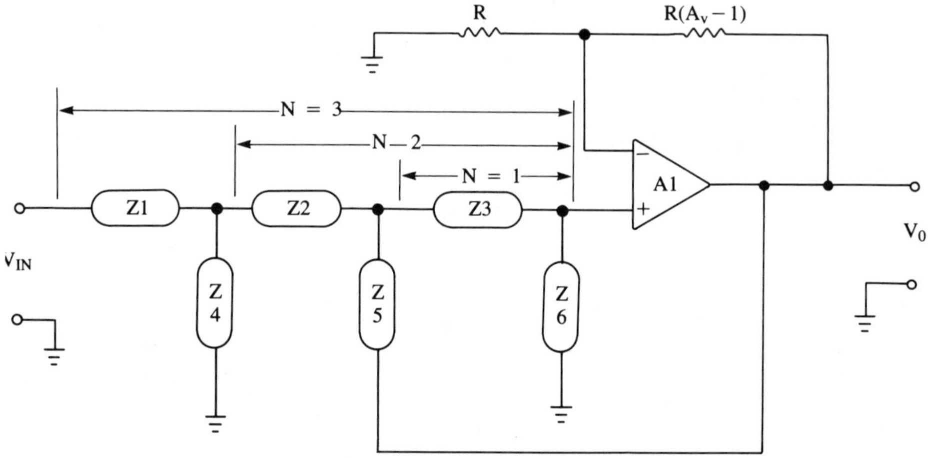

Figure 16-4 Model for a multisection filter.

In some other circuits, the gain may be unity. In those circuits the resistor voltage divider feedback network is replaced with a single connection between output and the inverting input.

The input circuitry of the generic VCVS filter consists of a network of impedances labeled Z1 through Z6. Each of these blocks will be either a resistance (R) or a complex capacitive reactance ($-jX_c$). Which element becomes which type of component is determined by whether the filter is a low-pass or high-pass type.

The order of the filter, denoted by n, refers to the number of poles in the design, or in practical terms, the number of RC sections. A first-order filter ($n = 1$) consists of Z3 and Z6, a second-order filter ($n = 2$) consists of Z2, Z3, Z5, and Z6 a third-order filter ($n = 3$) consists of all six impedances (Z1–Z6). Higher-order filters ($n > 3$) can also be built but are not discussed here.

In a low-pass filter, Z1 through Z3 are resistances, while Z4 through Z6 are capacitances. The component roles are reversed in high-pass filters. Now that we have laid a foundation, let's review the properties of the low-pass filter and then learn to design first-, second-, and third-order filters.

By way of review, Figure 16-5 shows the low-pass Butterworth filter response curve. This type of filter is maximally flat within the passband, and passes all frequencies below a certain critical frequency (F_c). The breakpoint between the passband and the stopband is the point at which the gain of the circuit has dropped off −3 dB from its lower frequency value. Above the critical frequency the gain falls off at a certain slope. The steepness of the slope is usually given in terms of decibels (dB) of gain per octave of

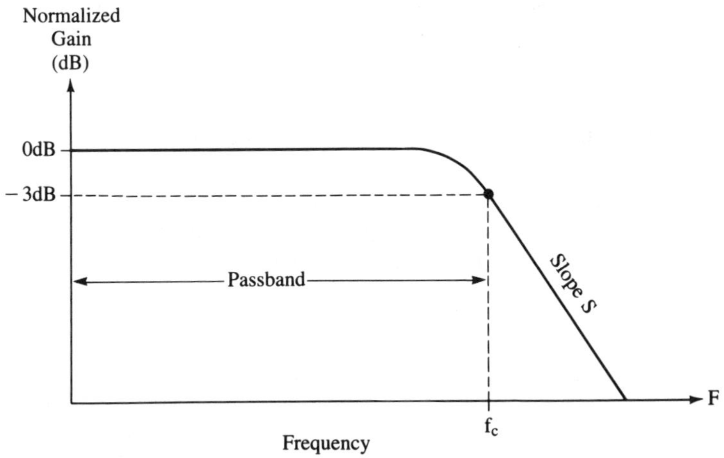

Figure 16-5 Low-pass characteristic.

frequency (an octave is a 2:1 change in frequency); alternatively, dB/decade (a decade is a 10:1 change in frequency) is sometimes used.

Frequency sweeps of a set of low-pass filters are shown in Figure 16-6. These oscilloscope traces were created using a sweep generator that varies a sine-wave signal from 1 kHz to 10 kHz. The −3 dB frequency of the filters was 3 kHz. The Y-input of the oscilloscope displays the output amplitude of the filter, while the X-input was externally swept using the same sawtooth that was used to sweep the signal generator. Thus the X-Y oscilloscope trace is a plot of the frequency response of the filter under test. Figure 16-6A shows the response of a first-order filter in the area from just below f_c up to 10 kHz. The response of the second-order filter is shown in Figure 16-6B,

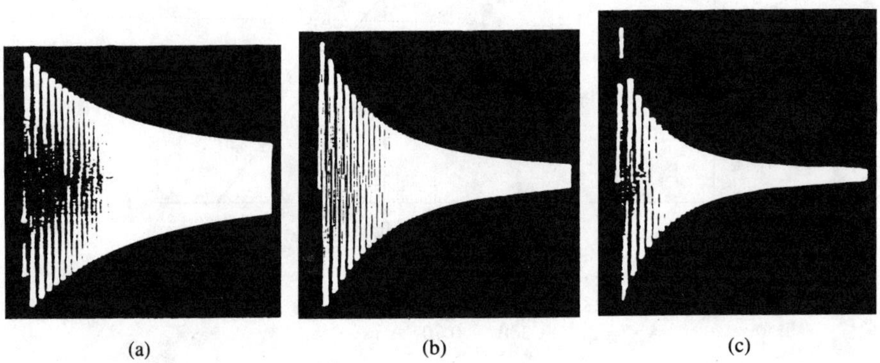

Figure 16-6 Effect of increasing roll-off on swept frequency: A) 6 dB/octave, B) 12 dB/octave, C) 18 dB/octave.

the response of the third-order filter is in Figure 16-6C. All three traces were taken under the same conditions, so one can see how the attenuation of frequencies greater than f_c is greater in the higher-order filters.

First-Order Low-Pass Filters (−20 dB/Decade)

The first-order low-pass filter is shown in Figure 16-7A, and its response curve is shown in Figure 16-7B. The filter consists of a single-section RC low-pass filter driving the noninverting input of an operational ampli-

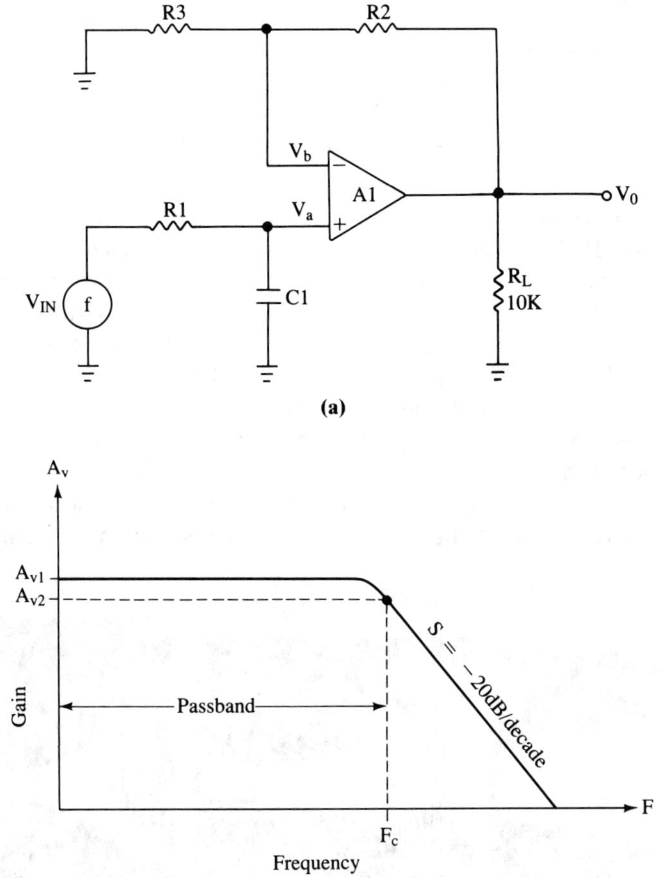

$$A_{v1} = \frac{R2}{R3} + 1$$

$$A_{v2} = A_{v1} - 3dB = 0.707\ A_{v1}$$

(b)

Figure 16-7 First-order low-pass filter: A) Circuit, B) Characteristic.

fier. The gain of the op-amp is [(R2/R3) + 1]. The high input impedance of A1 prevents loading of the RC network. The general form of the transfer equation for the amplitude versus frequency response for the first-order filter is as follows:

$$A_{dB} = 20 \log (A_v) - 20 \log [(1 + (\omega_0)^2)]^{1/2} \qquad (16.5)$$

Where:

A_{dB} is the gain of the circuit in decibels

A_v is the voltage gain within the passband

log denotes the base-10 logarithms

ω_0 is the ratio of the input frequency to the cutoff frequency ($f_0 = f/f_c$)

The voltage at the output of the RC network (V_a) is found from the voltage divider equation:

$$V_a = \frac{-jX_c V_{in}}{R - jX_c} \qquad (16.6)$$

Where:

$-jX_c = 1/j\pi fC$ and j is the imaginary operator ($[-1]^{1,2}$)

Substituting the value for $-jX_c$:

$$V_a = \frac{\dfrac{V_{in}}{j2\pi fC}}{R + \dfrac{1}{j2\pi fC}} \qquad (16\text{-}7)$$

which simplifies to

$$V_a = \frac{V_{in}}{1 + 2\pi fCR} \qquad (16.8)$$

If the transfer function of the noninverting follower is

$$V_0 = V_{in}\left[\frac{R2}{R3} + 1\right] \qquad (16.9)$$

and since $V_{in} = V_a$ [Equation (16.7)], then

$$V_0 = \left[\frac{V_{in}}{1 + 2\pi fCR}\right]\left[\frac{R2}{R3} + 1\right] \qquad (16.10)$$

Equation (16.9) can be put into a more generic transfer equation of the form

$$\frac{V_0}{V_{in}} = \frac{A_v}{1 + j(f/f_c)} \qquad (16.11)$$

where
V_0 is the output signal voltage
V_{in} is the input signal voltage
A_v is the passband gain [(R2/R3) + 1]
f is the signal frequency
f_c is the −3-dB frequency (1/2#RC)

The filter parameters are required to define the operation of any particular circuit. The gain magnitude and phase shift are found from the equations for gain magnitude:

$$\left|\frac{V_0}{V_{in}}\right| = \frac{A_v}{[1 + (f/f_c)^2]^{1/2}} \tag{16.12}$$

and phase shift angle (in radians):

$$\phi = -\tan^{-1}(f/f_c) \tag{16.13}$$

Because the filter characterization depends in part on the ratio f/f_c, the equations take different forms at different values of f and f_c. These can be reduced as follows. At low frequencies well within the passband ($f < f_c$):

$$\left|\frac{V_0}{V_{in}}\right| = A_v = \frac{R2}{R3} + 1 \tag{16.14}$$

At the −3-dB cutoff frequency ($f = f_c$):

$$\left|\frac{V_0}{V_{in}}\right| = 0.707A_v \tag{16.15}$$

At a high frequency well above the −3-dB cutoff frequency ($f > f_c$):

$$\left|\frac{V_0}{V_{in}}\right| < A_v \tag{16.16}$$

Design Procedure for a First-Order Low-Pass Filter

There are two ways to design a low-pass filter: from the ground up and by frequency scaling. In this section the ground-up method is discussed.

PROCEDURE

1. Select the −3-dB cutoff frequency (f_c) from consideration of the circuit requirements and applications.
2. Select a standard value capacitance ($C <= 1\mu F$).
3. Calculate the required resistance from

$$R1 = \frac{1}{2\pi f_c C}$$

4. Select the passband gain for $f < f_c$.
5. Select a value for resistor R.
6. Calculate R_f as follows:

$$R_f = R(A_v = 1)$$

Example

A low-pass filter is needed for a medical blood pressure transducer. The cutoff frequency should be 100 Hz, and the gain should be 5.

Solution:

1. $F_c = 100$ Hz
2. Select trial value for C1: 0.1 μF
3. Calculate R1:

$$R1 = \frac{1}{2\pi f_c C1}$$

$$R1 = \frac{1}{(2)(3.14)(100 \text{ Hz})(0.1 \times 10^{-6} \text{ F})}$$

$$R1 = 1/6.28 \times 10^{-5}\ \Omega = 15{,}923\ \Omega$$

4. Select a trial value for R1: 10 kΩ
5. Calculate R_f:

$$R_f = R(A_v - 1)$$
$$R_f = (10\text{ k}\Omega)(5 - 1)$$
$$R_f = (10\text{ k}\Omega)(4) = 40\text{ k}\Omega$$

Design by Normalizing Model

The filter design can be simplified by using a normalized model. First, design a filter for a standardized frequency (e.g., 1 Hz, 10 Hz, 100 Hz, 1000 Hz, or 10,000 Hz) and list the component values. The values for any other frequency can then be computed for any other frequency by a simple ratio and proportion. An example of a first-order low-pass Butterworth filter is shown in Figure 16-8. The component values shown in Figure 16-8 are normalized for 1 kHz. The actual required component values (R1′ and C1′) are found by dividing the normalized values shown by the desired cutoff frequency in kilohertz.

$$C1' = \frac{(C1)(1\text{ kHz})}{F} \tag{16.17}$$

or

$$R1' = \frac{(R1)(1\text{ kHz})}{F} \tag{16.18}$$

Leave one of the values alone, and calculate the other. In general, it is easier to obtain precision resistors in unusual values, so it is common practice to select a standard capacitance and calculate the new resistance.

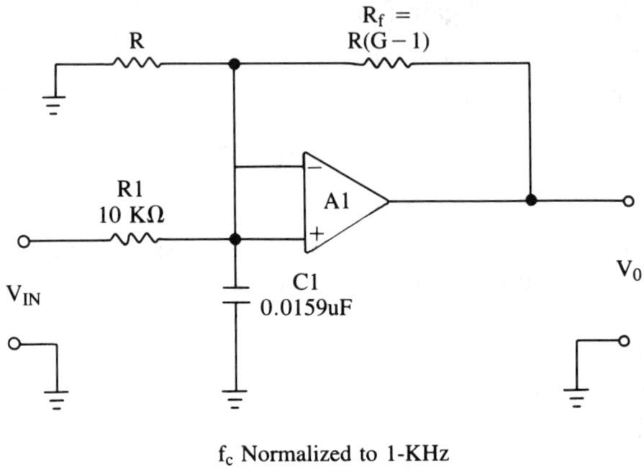

Figure 16-8 First-order LPF normalized to 1 Hz.

Example

Change the frequency of the normalized 1-kHz filter to 60 Hz (i.e., 0.06 kHz).

Solution:

$$C1' = \frac{(C1)(1\ \text{kHz})}{F}$$

$$C1' = \frac{(0.0159\ \mu F)(1\ \text{kHz})}{0.06} = 0.265\ \mu F$$

A generalized form of the equations is given below:

$$A' = A(f_c/f) \tag{16.19}$$

where

A' is the new component value for either C1 or R1

A is the original component value for either C1 or R1

f_c is the filter −3-dB cutoff frequency

f is the new design frequency

Second-Order Low-Pass Filters (−40 dB/Decade)

The circuit for a second-order low-pass filter is shown in Figure 16-9A, while the response curve is shown in Figure 16-9B. Note that this circuit is similar to the first-order filter but with an additional RC network in the frequency-selective portion of the circuit.

The particular version of this circuit shown in Figure 16-9A is connected in the unity gain configuration. The purpose of R3 is to help counteract the DC offset at the output of the operational amplifier that is created by input bias currents charging the capacitors in the frequency-selective network. The value of R3 in the unity gain situation is 2R, where R is the value of the resistors in the frequency-selective network. In cases where DC offset is not a problem, resistor R3 can be replaced with a short ciruit between the op-amp output and the inverting input. If passband gain is required, then resistors R3 and R4 are used.

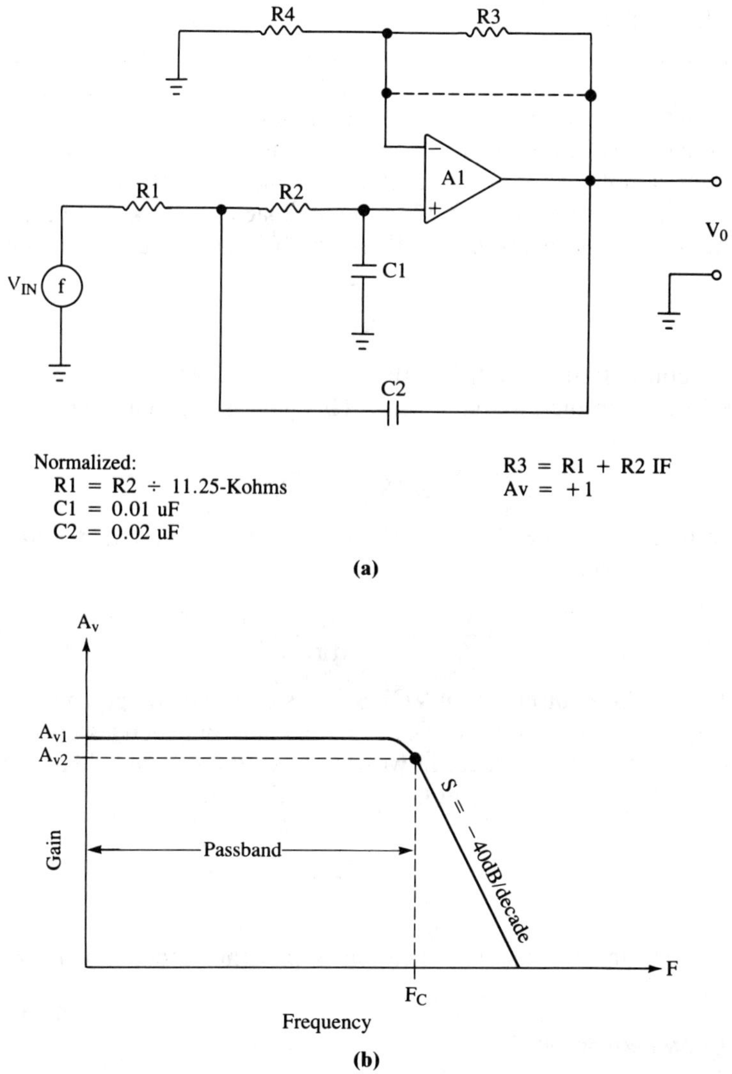

Figure 16-9 A) Second-order low-pass filter with 1000-Hz normalized values, B) Characteristic.

The second-order VCVS filter is by far the most commonly used type. Its −40-dB/decade roll-off, coupled with a high degree of stability, results in a generally good trade-off between performance and complexity.

The general form of the second-order filter transfer equation is similar to the expression for the first-order filter:

$$A_{dB} = 20 \log (A_v) - 20 \log [(\omega_0^4) + (a^2 - 2)(\omega_0)^2 + 1]^{1/2} \qquad (16.20)$$

Where a is the damping factor of the circuit, and other terms are as defined earlier for the first-order case.

The damping factor term (a) is determined by the form of filter circuit. For the Butterworth design, which is used in most of the examples of filters in this chapter, the value of $a = [2]^{1/2}$, or 1.414.

The passband gain for this circuit is the normal gain for any noninverting follower/amplifier. If the output is strapped directly to the inverting input, or if R3 (but not R4) is used in the feedback network, then the gain is unity ($A_v = +1$). For gains greater than unity ($A_v > 1$), the following is true:

$$A_v = \frac{R3}{R4} + 1 \qquad (16.21)$$

The cutoff frequency (f_c) is the frequency at which the voltage gain drops −3 dB from the passband gain. This gain is found as follows:

$$A_v = \frac{1}{2\pi[R1R2C1C2]^{1/2}} \qquad (16.22)$$

The gain magnitude (ABS(V@-{o}/V@-{in}) is found in a manner similar to the first-order case:

$$\frac{V_0}{V_{in}} = \frac{A_v}{[1 + (f/f_c)^4]^{1/2}} \qquad (16.23)$$

There is no requirement in VCVS filters that like components (R or C) in the frequency-selective network be made equal, but such a step simplifies the design procedure. It does, however, make the design other than Butterworth. If R1 = R2 = R, and C1 = C2 = C, then

$$f_c = \frac{1}{2\pi RC} \qquad (16.24)$$

A constraint on this simplification is that the response is guaranteed only if $A_v <= 1.586$.

DESIGN PROCEDURE

1. Select the −3-dB cutoff frequency (f_c) from consideration of the circuit requirements and applications.

2. Select a standard value capacitance (30 pF <= C <= 1 μF).

3. Calculate the required resistance:

$$R1 = \frac{1}{2\pi f_c C} \tag{16.25}$$

4. Select the passband gain for $f < f_c$.

5. Select a value for resistor R4.

6. Calculate R3:

$$R3 = R4(A_v - 1 \tag{16.26}$$

Example

Design a 1-kHz second-order low-pass filter with a unity gain based on Figure 16-9.
Solution:
1. Select f_c: 1 kHz (given)
2. Select a trial value for C1 and C2: C = 0.0056 μF
3. Calculate R1 = R2 = R:

$$R = \frac{1}{2\pi f_c C}$$

$$R = \frac{1}{(2)(3.14)(1000\ \text{Hz})(0.0056 \times 10^{-6}\ \text{F})}$$

$$R = \frac{1}{3.52 \times 10^{-5}}\ \Omega = 28{,}435\ \Omega$$

4. Select R3:

$$R3 = 2R = (2)(28{,}435\ \Omega) = 56{,}870\ \Omega$$

The normalized 1-kHz trial values for doing scaling design of the second-order low-pass filter are shown in Figure 16-9 as an inset. The design here is based on a more complex arrangement whereby C2 = 2C1. Some authorities maintain that this is the superior design. The same scaling rule is applied to the second-order filter as was used in the first-order.

Third-Order Low-Pass Filters (−60 dB/Octave)

A third-order filter has a frequency roll-off slope of −60 dB/decade, or −18 dB/octave. There are two main forms of third-order filter. One type is similar to the first- and second-order filters, except there is an extra low-pass RC filter in the frequency-selective network. The other method is to cascade first- and second-order filters. Figure 16-10A shows the circuit for the former type, along with the response curve shown in Figure 16-10B. This circuit is the normalized version. In past examples, filters were normalized using frequency f_c as the determining factor, so in this example we will use the radian form in which frequency ω_c is the cutoff frequency in radians per

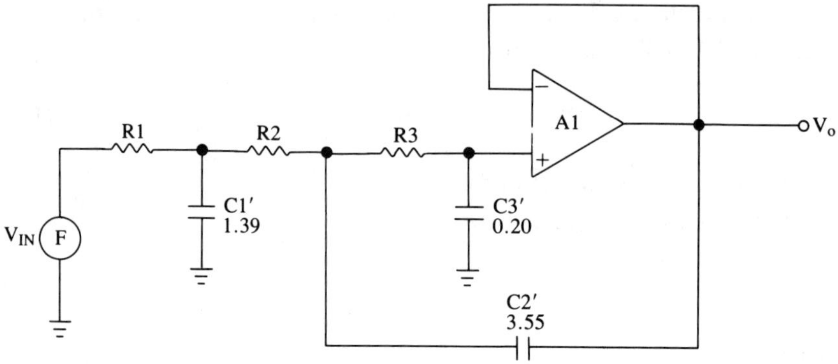

Normalized to 1 radian/second (1-Hz)
R1 = R2 = R3 = R

(a)

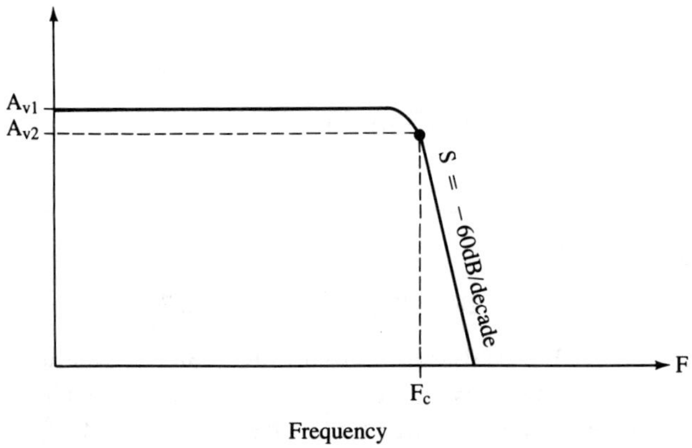

(b)

Figure 16-10 Third-order low-pass filter: A) Circuit, B) Characteristic.

second and is equal to $2\pi f_c$. The filter circuit shown in Figure 16-10A is normalized to one radian per second.

The values assigned to the capaci*or values in Figure 16-10A are parametric values and are used in a two-step process for finding the actual final values C1, C2, and C3. The parameters are C1′ = 1.39, C2′ = 3.55, and C3 = 0.20 (for Butterworth filters).

Example

Design a −60 dB/octave low-pass filter for a frequency of 100 Hz. Use the scaling method and the normalized values shown in Figure 16-10A.

Solution

1. Calculate ϕ_c:

$$\phi_c = 100 \text{ Hz} \times 6.28 = 628 \text{ radians/second}$$

2. State the parameter values:

$$C1' = 1.39$$
$$C2' = 3.55$$
$$C3' = 0.20$$

3. Perform frequency scaling:

$$C1a = \frac{C1'}{\omega_c} = \frac{1.39}{628} = 2.2 \times 10^{-3}$$

$$C2a = \frac{C2'}{\omega_c} = \frac{3.55}{628} = 5.65 \times 10^{-3}$$

$$C3a = \frac{C3'}{\omega_c} = \frac{0.20}{628} = 3.19 \times 10^{-4}$$

4. Select a trial value for C2: 0.1 μF (i.e., 1×10^{-7} F).
5. Solve for the value of R1 = R2 = R3 = R:

$$R = \frac{C2a}{C2} = \frac{5.65 \times 10^{-3}}{1 \times 10^{-7}} = 56{,}500\ \Omega$$

$$C1 = \frac{C1a}{R} = \frac{2.2 \times 10^{-3}}{56{,}500}$$

$$= 3.9 \times 10^{-8} \text{ farads} = 0.0039\ \mu\text{F}$$

$$C3 = \frac{C3a}{R} = \frac{3.19 \times 10^{-4}}{56{,}500}$$

$$= 5.7 \times 10^{-9} \text{ F} = 0.0057\ \mu\text{F}$$

A second method for designing a third-order low-pass filter is to cascade first-order and second-order filters of the same cutoff frequency (Figure 16-11). The roll-off of the first-order filter is −20 dB/decade, and for the second-order filter it is −40 dB/decade. The combined roll-off is −(20 + 40) = −60 dB/decade. The components in this circuit have the following relationships:

$$R1 = R2 = R3 = R$$

$$R = \frac{1{,}000{,}000}{2\pi f_c C3_{\mu F}} \tag{16.27}$$

C3 is selected for a convenient value:

$$C1 = \frac{C3}{2} \tag{16.28}$$

$$C2 = 2 \times C3 \tag{16.29}$$

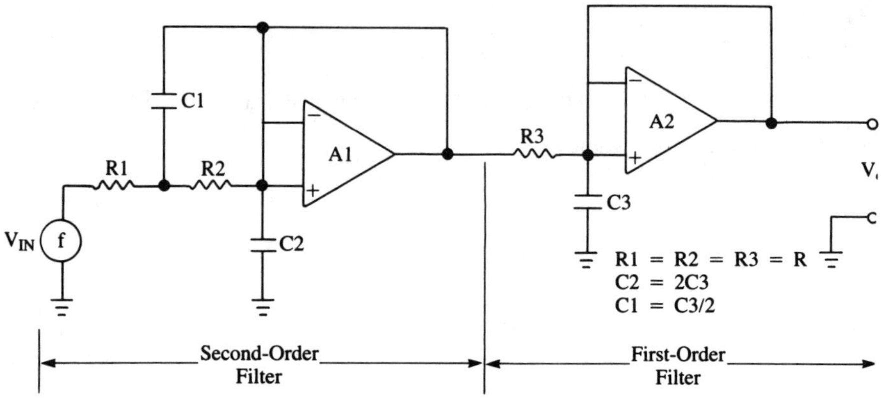

Figure 16-11 Two-stage third-order loss-pass filter.

The design procedure calls for selecting a reasonable trial value for C3, and then calculating C1 and C2. If these values are not standards, then set another trial value of C3 and try again. Repeat the procedure until a good set of values is obtained. Finally, calculate R.

Example

Design a −60 dB/decade cascade filter for a frequency of 500 Hz.

Solution:

1. Select frequency f_c: 500 Hz
2. Select C3: 0.01 μF
3. Calculate C1:

$$C1 = C3/2 = 0.01\ \mu F/2 = 0.005\ \mu F$$

4. Calculate C2:

$$C2 = 2 \times C3 = (2)(0.01\ \mu F) = 0.02\ \mu F$$

5. Calculate R1 = R2 = R3 = R:

$$R = \frac{1{,}000{,}000}{2\pi f_c C3_{\mu F}}$$

$$R = \frac{1{,}000{,}000}{(2)(3.14)(500\ Hz)(0.01\ \mu F)}$$

$$R = \frac{1{,}000{,}000}{31.4} = 31{,}850\ \Omega$$

HIGH-PASS FILTERS

The high-pass filter is the inverse of the low-pass filter, so one can reasonably expect its frequency response characteristic to mirror that of the low-pass filter response. Figure 16-12 shows the basic high-pass filter response with roll-off slopes of −20, −40, and −60 dB/decade. The passband of the high-pass filter are all frequencies above the cutoff frequency f_c. As in the low-pass case, f_c is the frequency at which passband gain drops −3 dB; that is, $A_{vc} = 0.707A_v$.

The cutoff frequency phase shift in a high-pass filter has the same magnitude as the low-pass case, but the sign is opposite. At f_c, the high-pass filter exhibits a phase shift of +45 degrees per 20 dB/decade of roll-off. Put another way, the phase shift is (n × 45 degrees), where n is the order of the filter.

The high-pass versions of the VCVS filters are of the same form as the low-pass filter. That form was laid out in Figure 16-4. In the case of the high-pass filter, however, impedances Z2 through Z3 are capacitances, while Z4 through Z6 are resistances. In the high-pass filter the roles of the resistors and capacitors are reversed.

First-Order High-Pass VCVS Filters (−20 dB/Decade)

The circuit for a first-order high-pass filter is shown in Figure 16-13. This circuit is identical to the first-order low-pass filter in which the roles of C1 and R1 are interchanged. The filter shown here is normalized for 1 kHz. Passband gain of this circuit is as follows:

$$A_V = \frac{R_f}{R_{in}} + 1 \qquad (16.30)$$

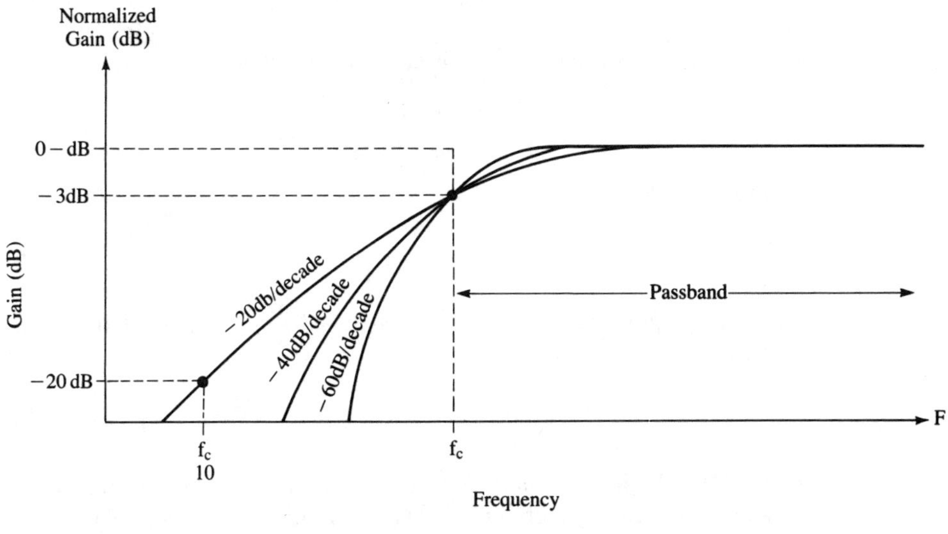

Figure 16-12 High-pass filter characteristics for various roll-off rates.

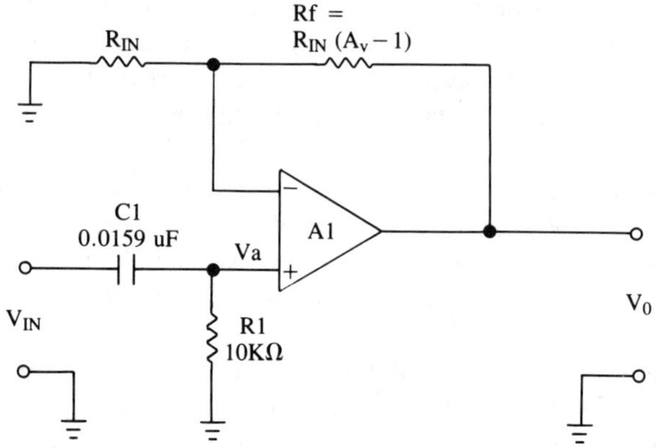

Figure 16-13 First-order high-pass filter.

The voltage at the noninverting input of the operational amplifier (V_a) is developed across resistor R1 and is given by

$$V_a = \frac{j2\pi fR1C1V_{in}}{1 + j2\pi fR1C1} \tag{16.31}$$

The transfer equation for the circuit is

$$V_0 = A_vV_a \tag{16.32}$$

$$V_0 = \left[\frac{R_f}{R_{in}} + 1\right]\left[\frac{j2\pi fR1C1V_{in}}{1 + j2\pi fR1C1}\right] \tag{16.33}$$

And, in the traditional form, the equation becomes

$$\frac{V_0}{V_{in}} = \frac{A_vj(f/f_c)}{1 + j(f/f_c)} \tag{16.34}$$

As in the previous cases, the cutoff frequency f_c is found from

$$f_c = \frac{1}{2\pi R1C1} \tag{16.35}$$

The gain magnitude of this circuit is the absolute value of the traditional form of the transfer equation:

$$\left|\frac{V_0}{V_{in}}\right| = \frac{A_v(f/f_c)}{[1 + (f/f_c)^2]^{1/2}} \tag{16.36}$$

The VCVS high-pass filter shown in Figure 16-12 is normalized to 1 kHz. The same scaling technique is used for this circuit as was used for the low-pass filters discussed earlier.

Second-Order High-Pass Filter (−40 dB/Decade)

The second-order high-pass filter offers a roll-off slope of −40 dB/decade. This VCVS filter circuit is, like its low-pass counterpart, probably the most commonly used form of high-pass filter. The circuit is similar to the low-pass design except for a reversal of the roles of capacitors and resistors. The cutoff frequency is the frequency at which gain falls off −3 dB, and it is found from

$$f_c = \frac{1}{2\pi[R1R2C1C2]^{1/2}} \tag{16.37}$$

Or, in the case where R1 = R2 = R and C1 = C2 = C:

$$f_c = \frac{1}{2\pi RC} \tag{16.38}$$

The gain magnitude of the circuit is found from

$$\left|\frac{V_0}{V_{in}}\right| = \frac{A_v}{[1 + (f_c/f)^4]^{1/2}} \tag{16.39}$$

Example

Calculate the cutoff frequency of a filter such as Figure 16-14 in which C1 = C2 = 0.0056 μF and R1 = R2 = 22 kΩ.

Solution:

$$f_c = \frac{1}{2\pi RC}$$

$$f_c = \frac{1}{(2)(3.14)(22{,}000\ \Omega)(0.0056 \times 10^{-6}\ \text{F})}$$

$$f_c = \frac{1}{7.74 \times 10^{-4}} = 1{,}293\ \text{Hz}$$

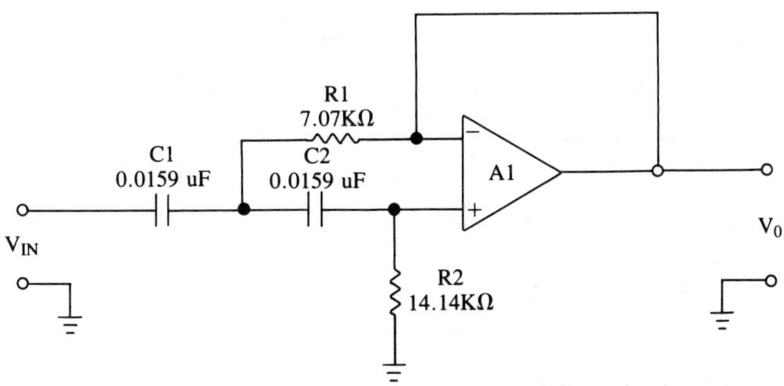

Figure 16-14 Second-order high-pass filter.

Multiple-Feedback Path Filters

A different form of active filter, the multiple-feedback path (MFP) circuit, is shown in Figure 16-15. The low-pass version is shown in Figure 16-15A, and the high-pass version is in Figure 16-15B. The values are normalized for 1 kHz, and we find the actual values in the manner described previously. Change either the capacitor or the resistor values, but not both.

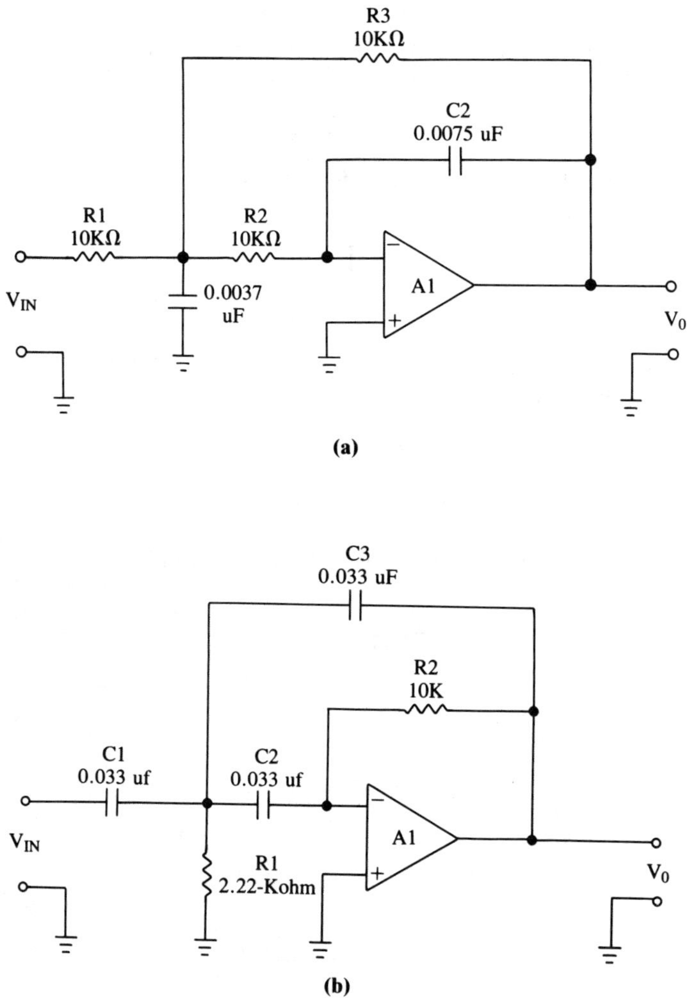

Figure 16-15 A) Multiple feedback path low-pass filter, B) High-pass version.

BANDPASS FILTERS

The bandpass filter is a circuit that has a passband between an upper limit and a lower limit. Frequencies above and below these limits are in the stopband. There are two forms of bandpass filter: wide bandpass and narrow bandpass. These two types are sufficiently different that they offer different responses. The wide bandpass filter may have a passband that is wide enough to be called a bandpass amplifier rather than a filter. The wide bandpass filter response is shown in Figure 16-16A, while the narrow re-

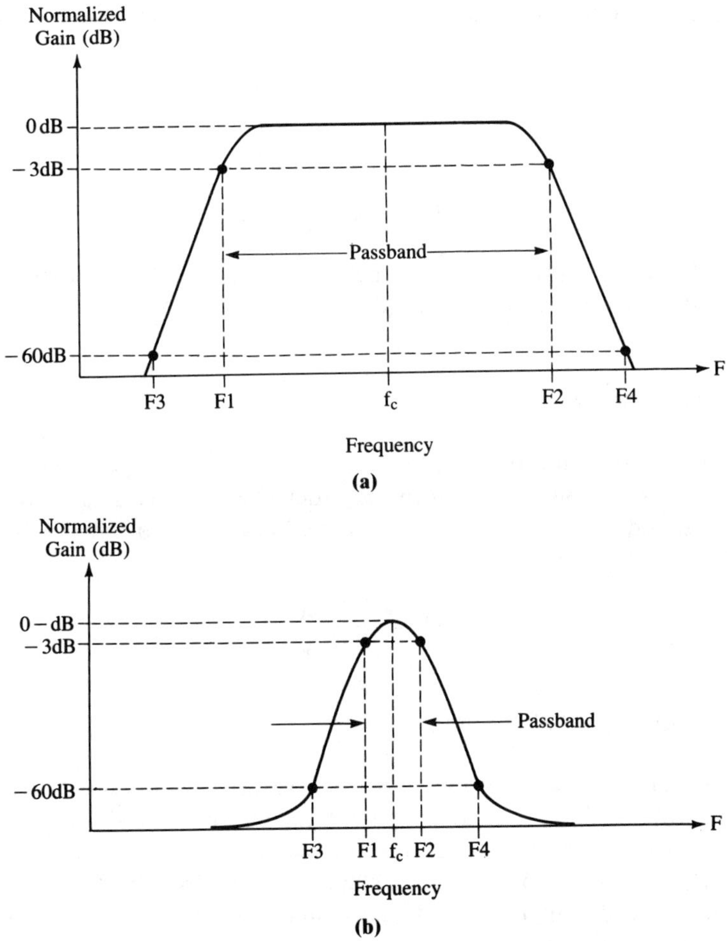

Figure 16-16 A) Wide passband filter characteristic (low Q), B) Narrow passband filter (high Q).

sponse is shown in Figure 16-16B. The passband is defined as the frequency difference between the upper −3 dB point (F2), and the lower −3 dB point (F1). The bandwidth BW is

$$BW = F2 - F1 \tag{16.40}$$

The center frequency f_c of the bandpass filter is usually symmetrically placed between F1 and F2, or (F2 − F1)/2. If the filter is a very wide band type, however, the center frequency is:

$$f_c = [F1F2]^{1/2} \tag{16.41}$$

Bandpass filters are sometimes characterized by the figure of merit, or Q. The Q is a factor that describes the sharpness of the filter and is found as follows:

$$Q = \frac{f_c}{BW} \tag{16.42}$$

$$Q = \frac{f_c}{F2 - F1} \tag{16.43}$$

The Q of the filter tells us something of the passband characteristic. Wideband filters generally have a $Q < 10$, while narrow band filters have a $Q > 10$.

Shape Factor

The shape factor of the filter characterizes the slope of the roll-off curve, so it is obviously related to the order of the filter. The shape factor (SF) is defined as the ratio of the −60-dB bandwidth to the −3-dB bandwidth:

$$SF = \frac{BW_{-60dB}}{BW_{-3dB}} \tag{16.44}$$

First-Order VCVS Bandpass Filters

A wideband first-order bandpass filter response is obtained by cascading first-order high-pass and low-pass filter circuits, as shown in Figure 16-17A. This arrangement overlays, or superimposes, the frequency response characteristics of both filter stages. Figure 16-17B shows the situation when cascaded high- and low-pass filters are used. The low-pass filter response (solid line) is from DC to the −3-dB point at F2. The high-pass filter response is from the highest possible frequency within the range of the circuit down to the −3-dB point at F1. The passband is the intersection of

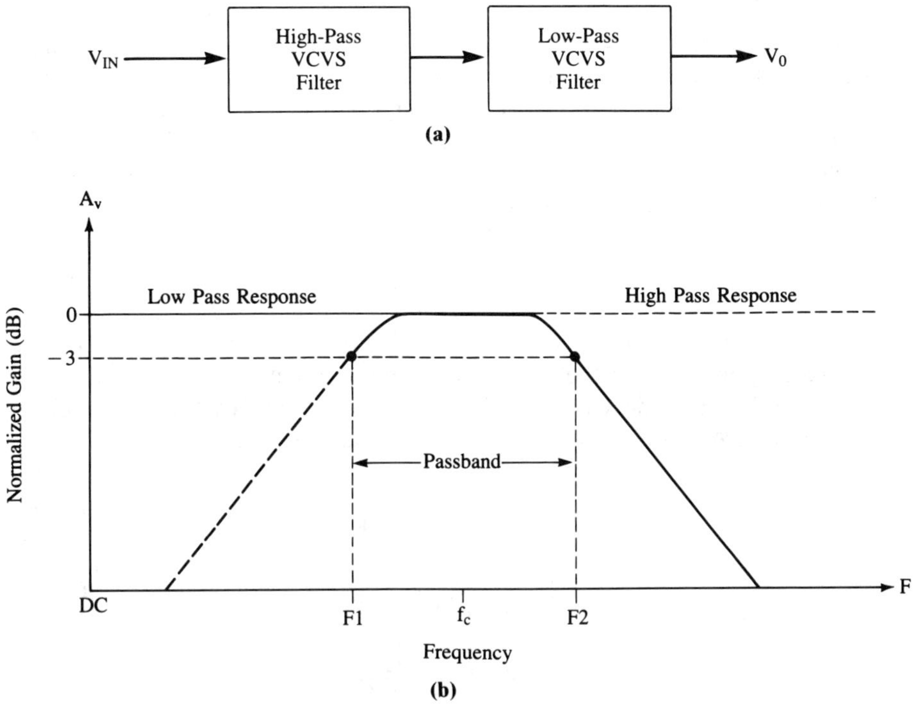

Figure 16-17 Two-step approach to bandpass filter design: A) Block diagram, B) Overlapping high- and low-pass characteristics.

the two sets (high- and low-pass characteristics), which falls between F1 and F2.

The gain of the overall bandpass filter within the passband is the product of the two individual gains: $A_{vt} = A_{Vl} \times A_{vh}$. The gain magnitude term of this form of filter is found as follows:

$$\left|\frac{V_0}{V_{in}}\right| = \frac{A_{vt}(f/F1)}{[(1 + (f/F1))^2(1 + (f/F2)^2]^{1/2}} \tag{16.45}$$

where

V_0 is the output signal voltage

V_{in} is the input signal voltage

f is the applied frequency

F1 is the lower −3-dB point frequency

F2 is the upper −3-dB point frequency

A_{vt} is the total cascade gain of the filter

Cascading low- and high-pass filter sections can be used to make wide-band filters, but because of component tolerance and problems it becomes less useful as Q increases above about 10 or so. For narrow band filters a MFP filter circuit such as Figure 16.18 can be used. This filter offers first-order performance and relatively narrow bandpass. The circuit will work for values of $10 <= Q <= 20$ and gains up to about 15. The center frequency of the MFP bandpass filter is calculated as follows:

$$f_c = \frac{1}{2\pi}\left[\frac{1}{R3C1C2}\left(\frac{1}{R1} + \frac{1}{R2}\right)\right]^{1/2} \tag{16.46}$$

In order to calculate the resistor values it is necessary to first set the passband gain (A_v) and the Q. It is the general practice to select values for C1 and C2, and then calculate the required resistances for the specified values of f_c, A_v, and Q. The resistor values are as follows:

$$R1 = \frac{1}{2\pi A_v C2 f_c} \tag{16.47}$$

$$R2 = \frac{1}{2\pi f_c(2Q^2 - A_v)C2} \tag{16.48}$$

$$R3 = \frac{2Q}{2\pi f_c C2} \tag{16.49}$$

and the gain is

$$A_v = \frac{R3}{R1\left[1 + \frac{C2}{C1}\right]} \tag{16.50}$$

These equations can be simplified if the two capacitors are made equal ($C1 = C2 = C$), and assuming that $Q > (A_v/2)^{1/2}$:

$$R1 = \frac{Q}{2\pi f_c A_v C} \tag{16.51}$$

$$R2 = \frac{Q}{2\pi f_c C(2Q^2 - A_v)} \tag{16.52}$$

$$R3 = \frac{2Q}{2\pi f_c C} \tag{16.53}$$

$$A_v = \frac{R3}{2R1} \tag{16.54}$$

Example

Design an MFP bandpass filter with a gain of 5 and a Q of 15 when the center frequency is 2,200 Hz. Assume that $C1 = C2 = 0.01\ \mu F$

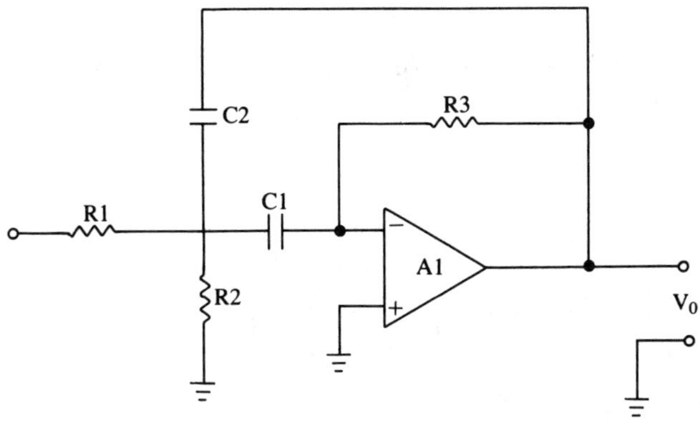

Figure 16-18 MFP high-pass first-order filter circuit.

Solution:

$$R1 = \frac{Q}{2\pi f_c A_v C}$$

$$R1 = \frac{15}{(2)(3.14)(2200\ \text{Hz})(5)(0.01 \times 10^{-6}\ \text{F})}$$

$$R1 = \frac{15}{0.00069} = 217{,}140\ \Omega$$

$$R2 = \frac{Q}{2\pi f_c C(2Q^2 - A_v)}$$

$$R2 = \frac{15}{(2)(3.14)(2200\ \text{Hz})(0.01 \times 10^{-6}(2(15)^2 - (5))}$$

$$R2 = \frac{15}{0.062} = 240\ \Omega$$

$$R3 = \frac{Q}{2\pi f_c C}$$

$$R3 = \frac{(2)(15)}{(2)(3.14)(2200\ \text{Hz})(0.01 \times 10^{-6}\ \text{F})}$$

$$R3 = \frac{30}{0.000138} = 217{,}140\ \Omega$$

The MFP bandpass filter is capable of being tuned using only one of the resistors. If R2 is varied, then the center frequency will shift, but the bandwidth, Q, and gain will remain constant. To scale the circuit to a new center frequency using only R2 as the change element, select a new value of R2 according to

$$R2' = R2 \left[\frac{f_c}{f_{c'}}\right]^2 \qquad (16.55)$$

Example

Calculate the new value of R2 that will force the MFP filter in the previous example from 2,200 Hz to 1,275 Hz.

Solution:

$$R2' = R2 \left[\frac{f_c}{f_{c'}}\right]^2$$

$$R2' = (240\ \Omega)\left[\frac{2200\ \text{Hz}}{1275\ \text{Hz}}\right]^2$$

$$R2' = (240\ \Omega)(1.726) = 414\ \Omega$$

BAND REJECT (NOTCH) FILTERS

A band reject or notch filter is used to pass all frequencies except a single frequency or small band of frequencies. An application for this circuit is to remove 60-Hz interference from electronic instruments. The medical electrocardiograph machine, for example, often suffers 60-Hz interference because the input leads are unshielded at the tips. These machines often include a switch-selectable 60-Hz notch filter to remove the 60-Hz artifact that could result.

Figure 16-19A shows a typical active notch filter, while Figure 16-19B shows the frequency response for the circuit. Note that the gain is constant throughout the frequency spectrum except in the immediate vicinity of f_c. The depth of the notch is infinite in theory, but in practical circuits precision-matched components will offer −60 dB of suppression, while ordinary "bench run" components (not precision) can offer −40 to −50 dB of suppression. The resonant frequency of this notch filter is found from

$$f_c = \frac{1}{2\pi RC} \tag{16.56}$$

The gain of the circuit is unity, but the Q can be set according to the following equations:

$$Q = \frac{R_a}{2R} \tag{16.57}$$

or

$$Q = \frac{C}{C_a} \tag{16.58}$$

Example

Design a notch filter with a Q of 8 for a frequency of 60 Hz.

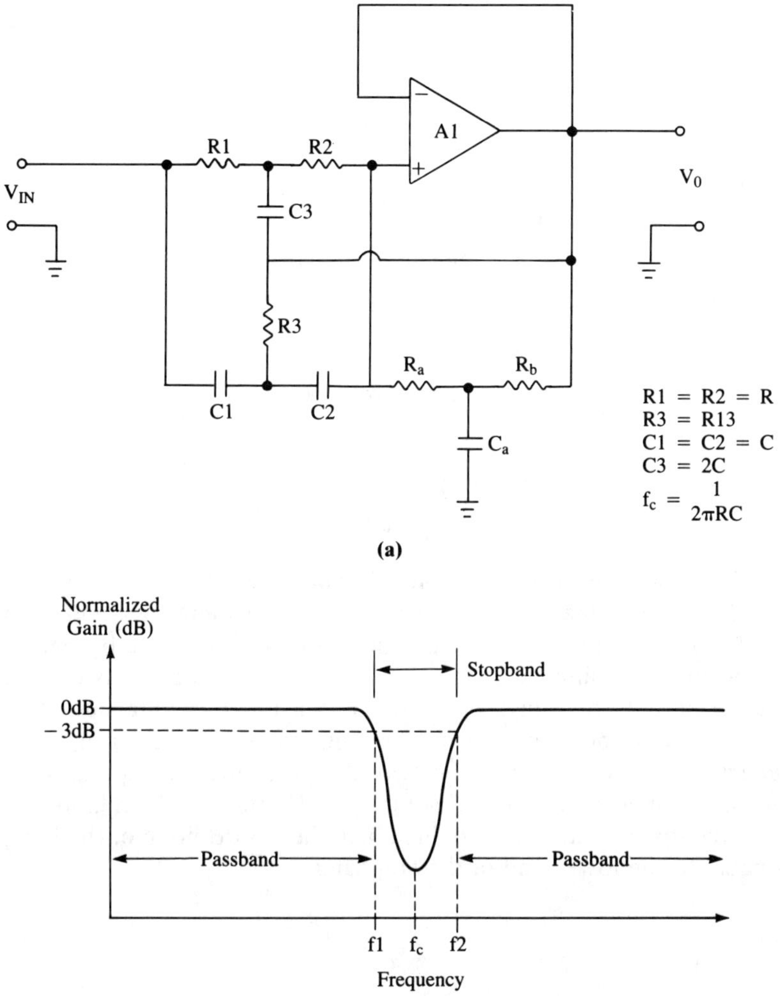

(b)

Figure 16-19 A) Notch filter circuit, B) Characteristic.

Solution:

1. Select a trial value of C1 = C2 = C: 0.01 μF
2. Calculate the value of R1 = R2 = R:

$$R = \frac{1}{2\pi fC}$$

$$R = \frac{1}{(2)(3.14)(60\text{ Hz})(0.01 \times 10^{-6}\text{ F})}$$

$$R = 1/3.768 \times 10^{-6} = 265{,}392\ \Omega$$

3. Select R3:

$$R3 = R/2$$

$$= \frac{265{,}392\ \Omega}{2} = 132{,}696\ \Omega$$

4. Select C3:

$$C3 = 2C$$

$$C3 = (2)(0.01\ \mu F) = 0.02\ \mu F$$

5. Select R_a:

$$R_a = 2QR$$

$$R_a = (2)(8)(265{,}392) = 4.24\ M\Omega$$

6. Select C_a:

$$C_a = C/Q$$

$$C_a = 0.01\ \mu F/8 = 0.0013\ \mu F$$

A bandstop filter is an example of a notch filter with a wide stopband. Just as the windbandpass filter can be made by cascading high- and low-pass filters, the wideband notch (or stopband) filter can be made by paralleling high- and low-pass filter sections. Figure 16-20 shows a bandstop filter in which the outputs of a high-pass filter and a low-pass filter are summed together in a two-input unity gain inverting follower amplifier circuit. The frequency response curves of the two filter sections are superimposed to eliminate the undesired band. Make the −3-dB point of the high-pass filter equal to the upper end of the stopband and the −3-dB point of the low-pass filter equal to the lower end of the stopband.

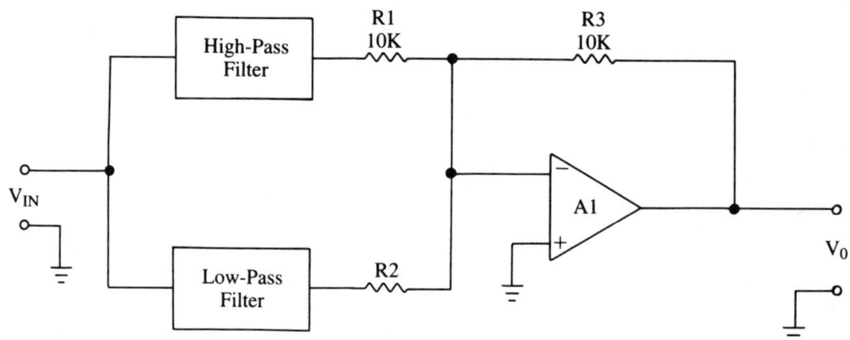

Figure 16-20 Summation notch filter circuit.

ALL-PASS PHASE-SHIFT FILTERS

The all-pass phase-shifter (APPS) is a special category of filter in which all frequencies (within the ability of the op-amp) are passed but are shifted in phase a specified amount. Figure 16-21A shows the circuit for an APPS that will exhibit phase shift between input and output of −180 to 0 degrees, while Figure 16-21B shows the timing waveforms. If the roles of R1 and C1 are reversed, then the phase shift will be −360 to −180 degrees. The gain re-

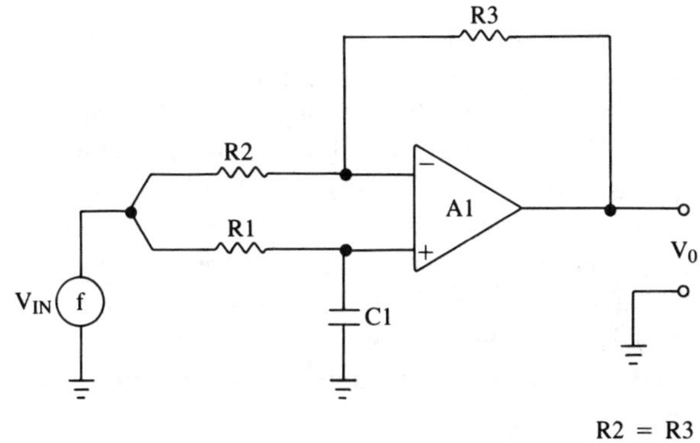

(a)

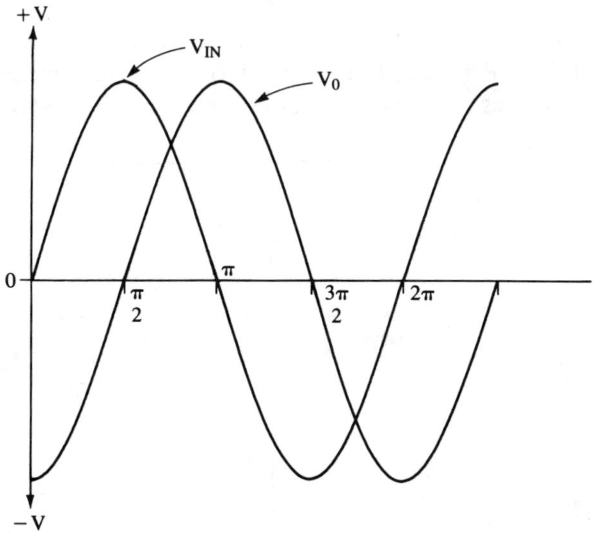

(b)

Figure 16-21 Phase shifter: A) Circuit, B) Input/output waveforms.

sponse of this circuit is

$$A_v = \frac{1 - 2\pi fR1C1}{1 + 2\pi fR1C1} \tag{16.59}$$

and the amount of phase shift is

$$\Delta\phi = -2\tan^{-1}(2\pi fR1C1) \tag{16.60}$$

STATE-VARIABLE ANALOG FILTERS

The state-variable filter is a variation on the multiple-feedback path design using three operational amplifiers. While the circuit is more complex than the other circuits presented in this chapter, it is also more versatile. The state-variable filter is capable of simultaneously providing bandpass, low-pass, and high-pass responses. Figure 16-22 shows the block diagram of the state-variable filter. The constituent parts include a summing amplifier, two integrators, and a damping network. The integrators are inverting Miller integrators. The summing amplifier is also a simple op-amp summer, and the damping network is a simple resistor voltage divider.

The damping factor (a) is the same for all three outputs and is defined as the reciprocol of Q:

$$a = \frac{1}{Q} \tag{16.61}$$

The state-variable filter shown in Figure 16-23 is normalized for 1 kHz, C = 0.0159 μF, and R = 10 kΩ. Designing for other frequencies is a matter of scaling these values to the new values required for the new frequency. As

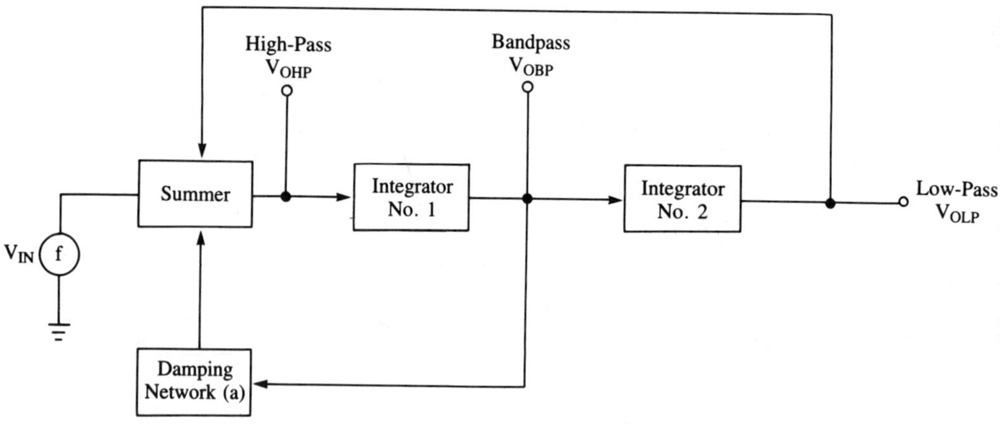

Figure 16-22 State-variable filter block diagram.

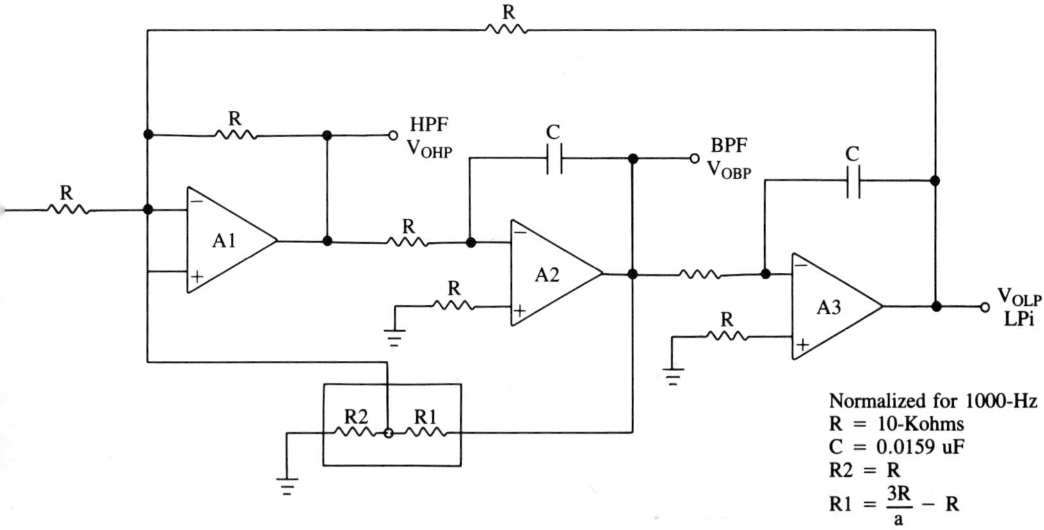

Figure 16-23 Typical state-variable filter circuit.

in the previous examples, change either R or C, but not both. The value of components for the damping network are found from

$$R1 = \frac{3R}{a} - R \quad (16.62)$$

and

$$R2 = R \quad (16.63)$$

VOLTAGE-TUNABLE FILTERS

A voltage-tunable filter is one in which the −3-dB response point is a function of an input control voltage. The analog multiplier and divider can be used to make a simple voltage-tunable filter in either low-pass (Figure 16-24A), high-pass (Figure 16-24B), or bandpass (Figure 16-24C) versions.

The low-pass filter is shown in Figure 16-24A. This circuit consists of a Miller integrator with an analog multiplier in a second negative feedback path through R2. The X-input of the multiplier is the output of the integrator, while the Y-input is the control voltage (V1) that tunes the filter. The output voltage of this circuit is:

$$V_0 = \frac{-V_{in}}{1 + \dfrac{2\pi fR1C1}{V1}} \quad (16.64)$$

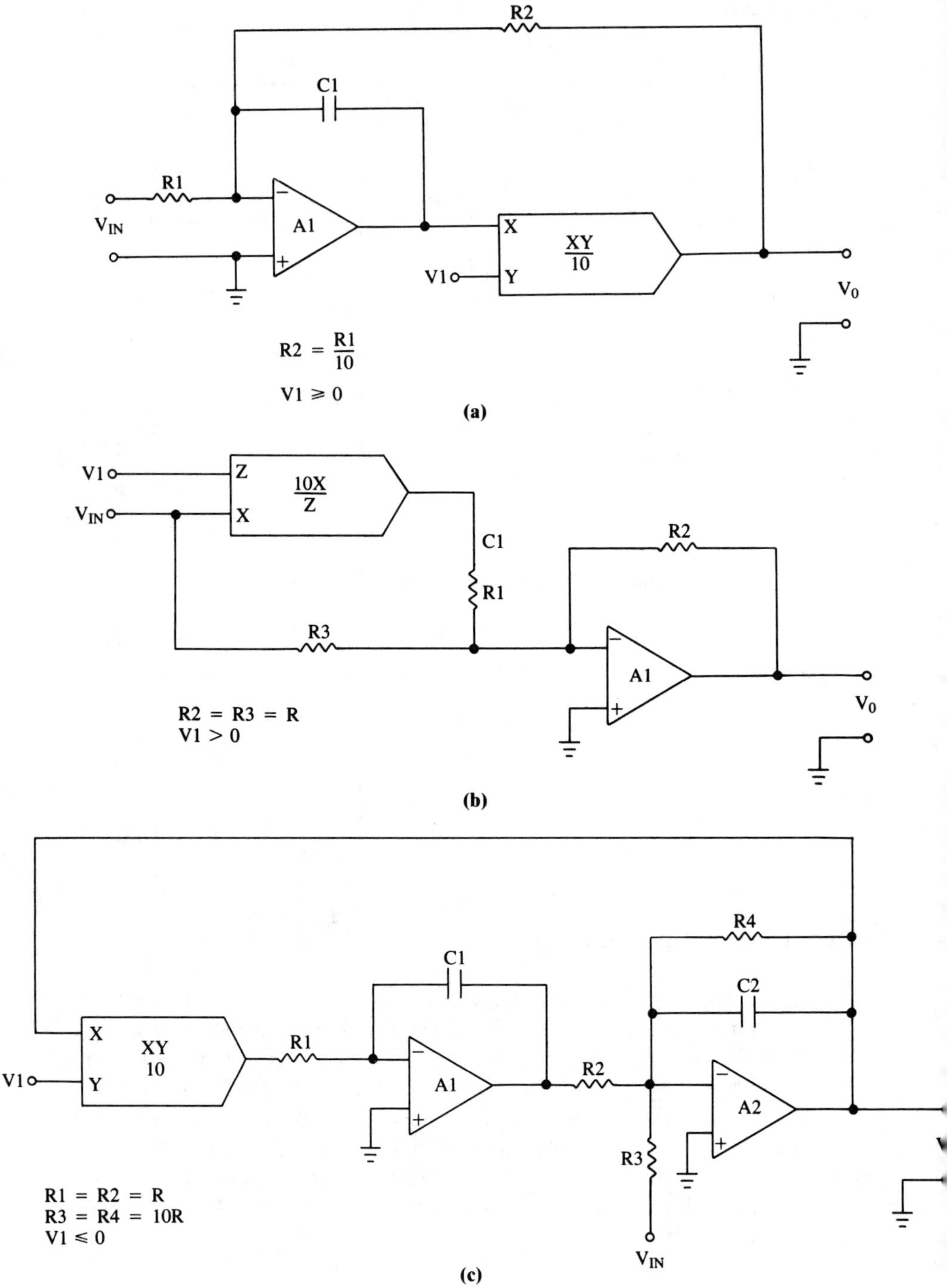

Figure 16-24 Filters based on analog multipliers: A) Low-pass, B) High-pass, C) Bandpass.

The high-pass filter circuit is shown in Figure 16-24B. This circuit consists of an analog divider driving an RC differentiator (R1C1). The output of the differentiator is summed with the input signal. For R2 = R3 = R, C1 = C, and R1 < R, the following transfer equation obtains:

$$V_0 = -V_{in}\left[1 + \frac{2\pi fRC}{V1}\right] \tag{16.65}$$

The bandpass circuit of Figure 16-24C is based on the analog multiplier and a pair of integrator circuits. This circuit produces a gain of -1 and a Q as follows:

$$Q = [-10V1]^{1/2} \tag{16.66}$$

The center frequency is

$$f_c = \frac{1}{2\pi RC}\left[\frac{-V1}{10}\right]^{1/2} \tag{16.67}$$

(assuming a control voltage $V1 <= 0$). The bandwidth is

$$BW = \frac{1}{20\pi RC} \tag{16.68}$$

All three of these multiplier- or divider-based circuits are dependent upon the properties of the device used. The multiplier gain, error, linearity, and response time determine the properties of the filter and the tuning rate.

17

DATA CONVERSION TECHNIQUES: A/D AND D/A

Many of the control, instrumentation, logging, and data reduction chores performed by small microcomputers involve data values taken from real-world sensors. Signals that are proportional to some physical parameter or another are said to be "analogs" of that parameter and are used extensively in electronic instrumentation and data-collection systems.

For example, let's suppose that a resistive Wheatstone bridge strain gage is used to measure human blood pressure. The output of the transducer will be a voltage that is proportional to the pressure applied to the strain gage diaphragm. Typically, in 0- to 10-V systems, the voltage analog is 10 mV per millimeter of mercury pressure (10 mV/mm Hg) or 10 mV/torr in modern units, which are not commonly used in clinical medicine. The output will, therefore, be 1,200 mV, or 1.2 V, when the patient's blood pressure is 120 mm Hg. But this fact is of little interest to the computer because it cannot interpret voltage levels. A computer will want to see a binary (base-2) number that represents the blood pressure, not a voltage analog.

The data converter is a device that converts data between the binary and analog worlds. An analog-to-digital converter (ADC or A/D), for example, could look at the 1.2-V signal and then output a representative binary word that the computer can accept. Similarly, a digital-to-analog converter (DAC) produces either a current or voltage output that is proportional to some binary word applied to its inputs. A DAC can be used in a computer-based instrument system to drive an oscilloscope or strip chart (i.e., paper) recorder so that the user can see the shape of the waveform that produced the data.

Returning to our medical example, suppose we wanted to make a microprocessor-based blood pressure monitor that would display the diastolic (i.e., lowest) and systolic (i.e., highest) values on the patient's blood pressure waveform and the waveform itself. We would create a system in which the amplified transducer output would be applied to the analog-to-digital converter to make a binary word that can be interpreted by the computer. The computer would then apply a bubble sort or some similar routine to find the peak maximum and minimum values and then display them on either LED readouts or a video monitor. The computer could also sequentially output the binary values to a DAC, which would reconstruct the waveshape. While it is true that the analog signal could be directly displayed, there are sometimes good reasons why the digitized (and then reconstituted) version might be used (e.g., when the data have to be transmitted over great distances that would be difficult for the analog waveform but not for the digital, or when the doctor wants to compare a previous waveform with the real-time waveform, which is sometimes done by a coronary care unit or anesthesiology doctors who wish to see the effect of some drug or treatment on the waveform).

WHAT ARE DATA CONVERTERS?

Analog circuits and digital instruments occupy mutually exclusive realms. In the analog world, a signal may vary between upper and lower limits and may assume any value within the range. Analog signals are continuous between limits. But the signals in digital circuits may assume only one of two discrete voltage levels (i.e., one each for the two binary digits, 0 and 1). A data converter is a circuit or device that examines a signal from one of these realms and then converts it to a proportional signal from the other.

A DAC, for example, converts a digital (i.e., binary) "word" consisting of a certain number of bits into a voltage or current that represents the binary number value of the digital word. An 8-bit DAC may produce an output signal of 0 V when the binary word applied to its digital inputs is 00000000_2, and, say, 2.56 V when the digital inputs see a word of 11111111_2. For binary words applied to the inputs, then, a proportional output voltage is created.

In the case of an ADC, an analog voltage or current produces a proportional binary word output. If an 8-bit ADC has a 0- to 2.56-V input signal range, then 0-V input could produce an output word of 00000000_2, while the +2.56-V level seen at the input would produce an output word of 11111111_2.

Data converters are used primarily to interface transducers (most of which produce analog output signals) to digital instruments or computer inputs, and to interface digital instrument outputs to analog-world devices such as meter movements, chart recorders, and motors.

In this chapter we will consider common DAC circuits and the following ADC circuits: servo (also called the binary counter or ramp ADC), successive approximation, parallel converter, voltage-to-frequency converter, and integrating types.

DAC CIRCUITS

Figure 17-1 shows a *binary weighted resistance ladder* and operational amplifier used as a binary DAC. The resistors in the ladder are said to be *binary weighted* because their values are related to each other by powers of 2. If the lowest-value resistor is given the value R, then the next in the sequence will have a value of 2R, followed by 4R, 8R, 16R, all the way up to the *n*th resistor (last one in the chain), which has a value of $(2^{n-1})R$.

The switches B1 through B_n represent the input bits of the digital word. Although shown in the figure as mechanical switches, they would be transistor switches in actual practice. The switches are used to connect the input resistors to either ground or voltage source E, to represent binary

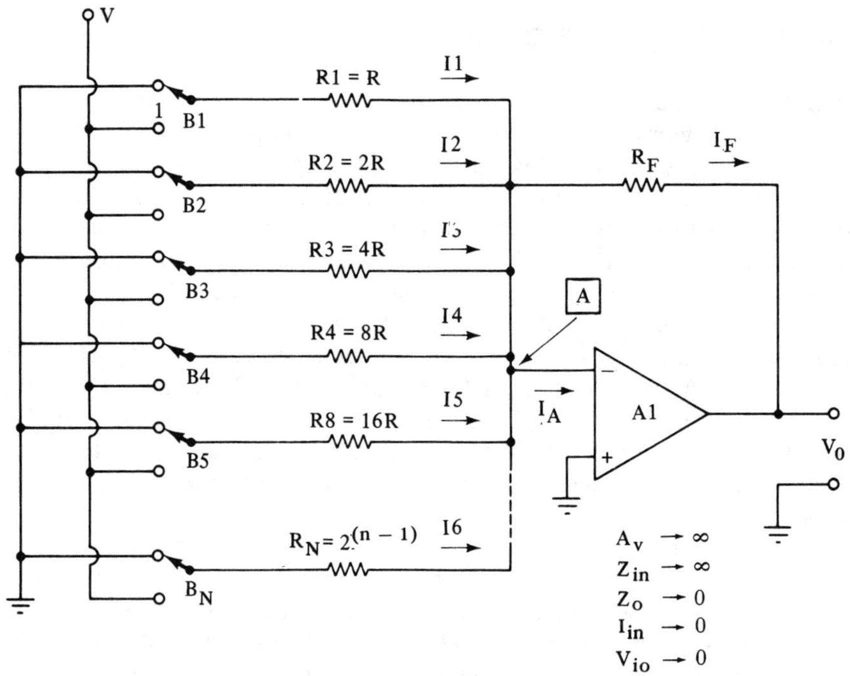

Figure 17-1 Binary weighted resistor ladder digital-to-analog converter (DAC) circuit.

states 0 and 1, respectively. Switches B1 through B_n create currents I1 through I_n, respectively, when they are set to the 1 position.

We know from Ohm's law that each current Il through I_n is equal to the quotient of E and the value of the associated resistor; that is,

$$I1 = \frac{E}{R1} = \frac{E}{R}$$

$$I2 = \frac{E}{R2} = \frac{E}{2R}$$

$$I3 = \frac{E}{R3} = \frac{E}{4R}$$

$$\vdots \quad \vdots \quad \vdots \quad \vdots \qquad \vdots$$

$$I_n = \frac{E}{R_n} = \frac{E}{(2_{n-1})R}$$

The total current into the junction (point A in Figure 17-1) is expressed by the summation of current Il through I_n:

$$I_A = \sum_{i=1}^{N} \frac{a_i E}{2^{i-1} R} \tag{17-1}$$

where

I_A = current into the junction (point A) in amperes (A)

E = reference potential in volts (V)

R = resistance of Rl in ohms (Ω)

a_i = either 1 or 0, depending on whether the input bit is 1 or 0

N = number of bits (i.e., the number of switches)

From operational amplifier theory we know that

$$I_A = -I_F \tag{17-2}$$

and

$$E_0 = I_F R_R \tag{17-3}$$

So, by substituting Equation (17.2) into Equation (17.3), we obtain

$$E_0 = -I_A R_F \tag{17-4}$$

and substituting Equation (17-1) into Equation (17-4) yields

$$E_0 = -R_F \sum_{i=1}^{N} \frac{a_i E}{2^{i-1} R} \tag{17-5}$$

Since E and R are constants, we usually write Equation (17.5) in the form

$$E_0 = \frac{-ER_F}{R} \sum \frac{a_i}{2^{i-1}R} \tag{17-6}$$

Example

A 4-bit (i.e., n = 4) DAC using a binary weighted resistor ladder has a reference source of 10 V DC and $R_f = R$. Find the output voltage E_0 for the input word 1011_2. (For input 1011, $a_1 = 1$, $a_2 = 0$, $a_3 = 1$, and $a_4 = 1$.)

Solution:

$$E_0 = \frac{-ER_F}{R} \frac{a_i}{2^{i-1}R}$$

$$= \frac{-10R}{R}\left[\frac{1}{2^{1-1}} + \frac{0}{2^{2-1}} + \frac{1}{2^{3-1}} + \frac{1}{2^{4-1}}\right]$$

$$= -10\text{ V}\left[\frac{1}{2^0} + \frac{1}{2^2} + \frac{1}{2^3}\right]$$

$$= -10\text{ V}\left[\frac{1}{1} + \frac{1}{4} + \frac{1}{8}\right]$$

$$= -10\text{ V}\left[\frac{1}{1.375}\right] = -7.27$$

Although not revealed by the idealized equations, the binary weighted resistance ladder suffers from a serious drawback in actual practice. The values of the input resistors tend to become very large and very small at the ends of the range as the bit length of the input word becomes longer. If R is set to 10 kΩ (a popular value), then R8 will be 1.28 MΩ. If we assume a reference potential E of 10.00 V DC, then I8 will be only 7.8 μA. Most common non-premium-grade operational amplifiers will not be able to resolve signals that low from the inherent noise. As a result, the bit length of the binary weighted ladder is severely limited. Few of these types of converters are found with more than 6- or 8-bit word lengths.

In commercial DACs all the resistors have a value of either R or 2R (Figure 17-2). The gain of the amplifier is unity, so E_0 can be expressed as

$$E_0 = E \sum_{i=1}^{N} \frac{a_i}{2_i} \tag{17-7}$$

(provided that $R_L >> R$ so that the voltage-divider effect between the ladder and R_L can safely be neglected).

Example

A 4-bit DAC using the R-2R technique has a 5.00-V DC reference potential. Calculate E_0 for the input word 1011_2.

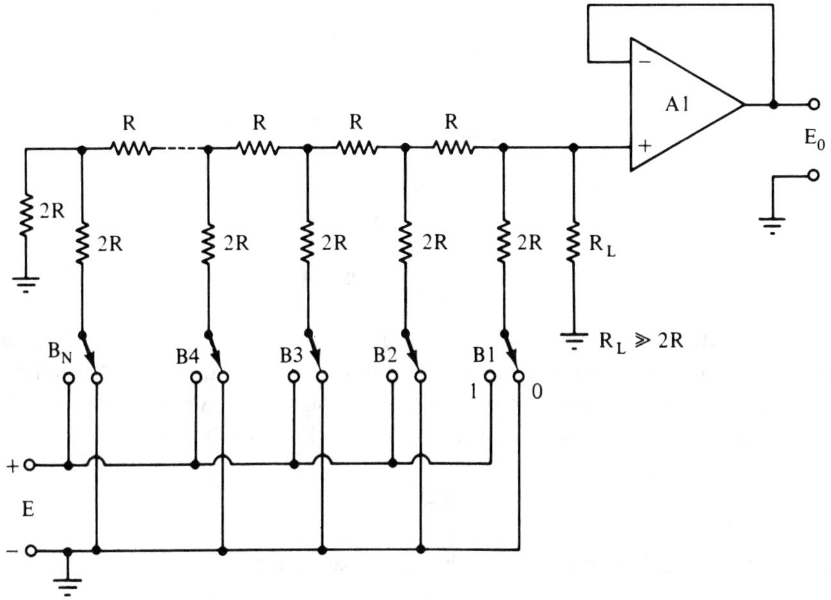

Figure 17-2 R-2R resistor ladder DAC circuit.

Solution:

$$E_0 = E\frac{a_i}{2^i} \tag{17-7}$$

$$= 5\text{ V}\left[\frac{1}{2^1} + \frac{0}{2^2} + \frac{1}{2^3} + \frac{1}{2^4}\right]$$

$$= 5\text{ V}\left[\frac{1}{2} + 0 + \frac{1}{8} + 0 + \frac{1}{16}\right]$$

$$= 5\text{ V}(0.688) = 3.44$$

The full-scale output voltage for any DAC using the R-2R resistor ladder is given by

$$E_{fs} = \frac{E(2^{n-1})}{2^n} \tag{17-8}$$

where

E_{fs} = full-scale output potnetial in volts (V)

E = reference potential in volts (V)

n = bit length of the digital input word

Example

Find the full-scale output potential for an 8-bit DAC with a reference potential of 10 V DC.

Solution

$$E_{fs} = \frac{E(2^{n-1})}{2^n} \qquad (17\text{-}8)$$

$$= \frac{10\ V\ (2^{8-1})}{2^8}$$

$$= \frac{10\ V\ (2^7)}{2^8}$$

$$= \frac{10\ V\ (255)}{256} = 9.96\ V$$

The output of a DAC cannot change in a continuous manner, because the input is a digital word (i.e., it can exist only in certain discrete states). Each successive binary number changes the output an amount equal to the change created by the least significant bit (LSB), which is expressed by

$$\Delta E_0 = \frac{E}{2_n} \qquad (17\text{-}9)$$

So, for the DAC in the example above, E_0 would be

$$\Delta E_0 = \frac{10\ V}{2^8}$$

$$= \frac{10\ V}{256} = 40\ mV$$

ΔE_0 is often called the 1-LSB value of E_0 and is the smallest change in output voltage that can occur. It is interesting that if we let 0 V represent 00000000_2 in our 8-bit system, the maximum value of E_0 at 11111111_2 will be 1 LSB less than E (confirmed by the result of the example).

There are numerous commercial DACs on the market in IC, function module block, and equipment form. The reader should consult manufacturer's catalogs for appropriate types in any given application.

SERVO A/D CONVERTER CIRCUITS

The *servo* ADC circuit (also called the *binary counter* or *ramp* ADC circuit) uses a binary counter to drive the digital inputs of a DAC. A voltage comparator keeps the clock gate to the counter open as long as E_0 does not equal E_{in}.

An example of such a circuit is the 8-bit ADC in Figure 17-3A, while the relationship of E_0 and E_{in} relative to time is shown in Figure 17-3B.

Two things happen when a start pulse is received by the control logic circuits: the binary counter is reset to 00000000_2, and the gate is opened to allow clock pulses into the counter. This will permit the counter to begin

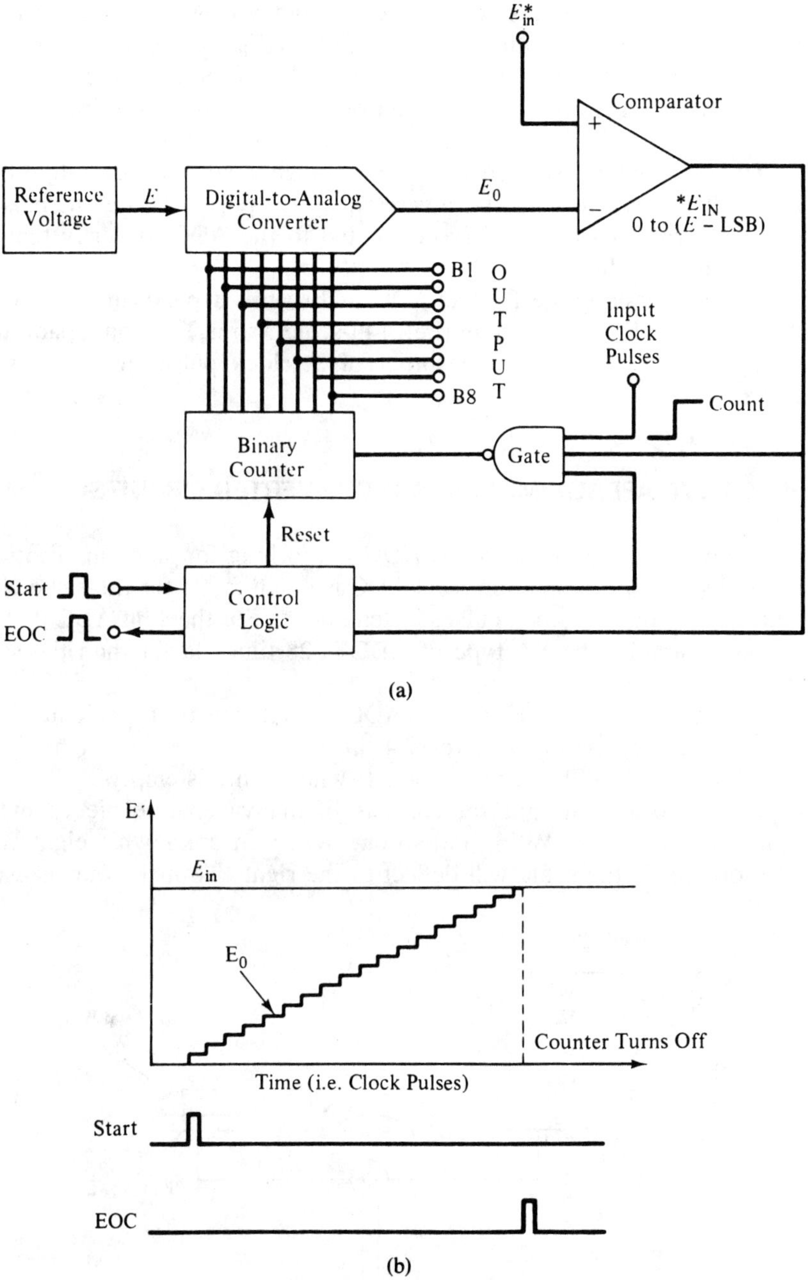

Figure 17-3 A) Binary ramp A/D converter circuit; B) Timing diagram.

incrementing, thereby causing the DAC output voltage E_0 to begin rising (Figure 17-3b). E_0 will continue to rise until $E_0 = E_{in}$. When this condition is met, the output of the comparator drops LOW, turning off the gate. The binary number appearing on the counter output at this time is proportional to E_{in}.

The control logic section senses the change in comparator output level and uses it to issue an end-of-conversion (EOC) pulse. This EOC pulse is used by instruments of circuitry connected to the ADC to verify that the output data are valid.

The conversion time T_c of an ADC such as this depends upon the value of E_{in}, so when E_{in} is maximum (i.e., full-scale), so is T_c. Conversion time for this type of ADC is on the order of 2^n clock pulses for a full-scale conversion.

SUCCESSIVE APPROXIMATION A/D CONVERTER CIRCUITS

The conversion time of the servo ADC is too long for some applications. The successive approximation (SA) ADC is much faster for the same clock speed; it takes (n + 1) clock pulses instead of 2^n. For the 8-bit ADC that has been our example, the SA type of ADC is 28 times faster than the servo ADC.

The basic concept of the SA ADC circuit can be represented by a platform balance, such as Figure 17-4, in which a full-scale weight W will deflect the pointer all the way to the left when pan 2 is empty.

Our calibrated weight set consists of many separate pieces, which weigh W/2, W/4, W/8, W/16, and so on. When an unknown weight W_1 is placed on pan 2, the scale will deflect to the right. To make our measure-

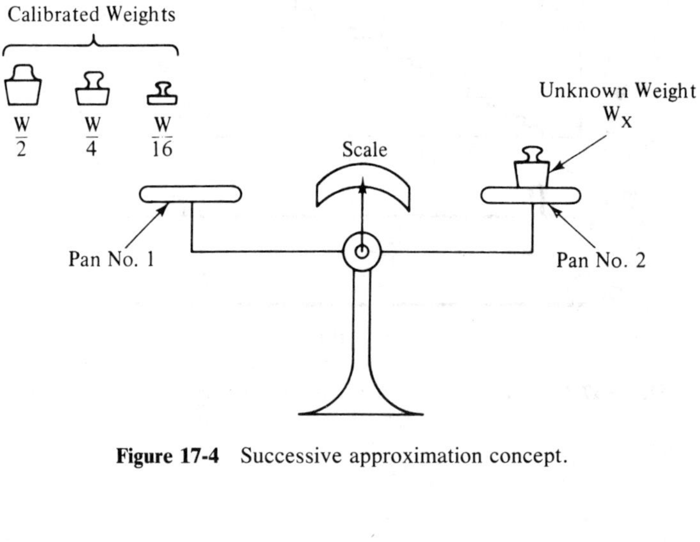

Figure 17-4 Successive approximation concept.

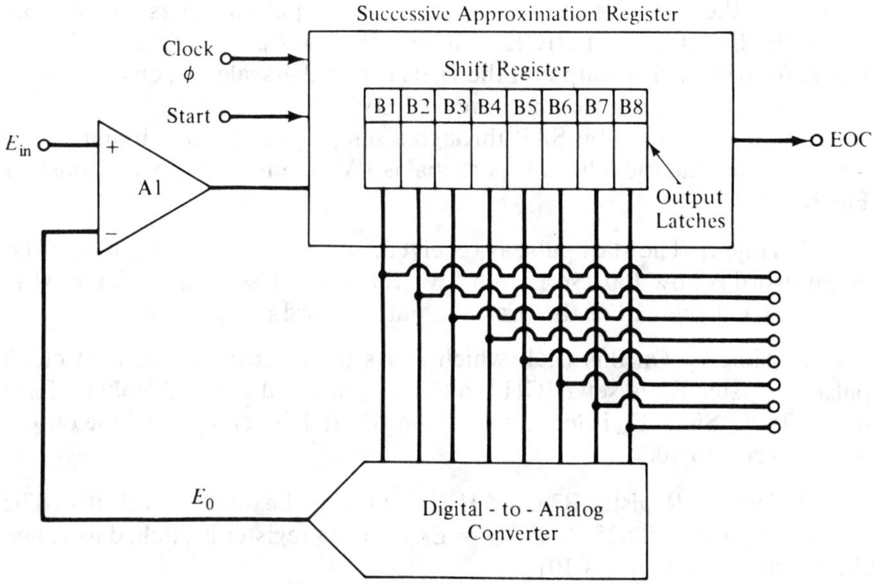

Figure 17-5 Successive approximation A/D converter.

ment, we start with W/2 and place it on pan 1. Three conditions are now possible:

$$\frac{W}{2} = W_x \text{ (scale is at zero)}$$

$$\frac{W}{2} > W_x \text{ (scale is to the left of zero)}$$

$$\frac{W}{2} < W_x \text{ (scale is to the right of zero)}$$

If $W/2 = W_x$, the measurement is finished, and no additional trials are necessary. But if W/2 is less than W_x, we must add more weights in succession (W/4, then W/8, etc.) until we find a combination equal to W_x.

If, on the other hand, W/2 is greater than W_x we must *remove* the W/2 weight and in the second trial start again with W/4. This procedure will continue until a combination equal to W_x is found.

In the SA ADC circuit we do not use a scale, but a shift register, as in Figure 17-5. A successive approximation register (SAR) contains the control logic, a shift register, and a set of output latches, one for each register section. The outputs of the latches drive a DAC.

A start pulse sets the first bit of the shift register HIGH, so the DAC will see the word 10000000_2 and therefore produce an output voltage equal to one-half of the full-scale output voltage. If the input voltage is greater than $1/2E_{fs}$, the B1 latch is set HIGH. On the next clock pulse, register B2 is set HIGH for trial 2. The output of the DAC is now 3/4-scale. If, on any trial, it is found that $E_{in} < E_0$, that bit is reset LOW.

Let us follow a 3-bit SAR through a sample conversion. In our example, let us say that the full-scale potential is 1 V, and E_{in} is 0.625 V. Consider Figure 17-6.

1. Time t_1: The start pulse is received, so register B1 goes HIGH. The output word is now 100_2, so $E_0 = 0.5$ V. Since E is less than E_{in}, latch B1 is set to 1, so at the end of the trial, the output word remains 100_2.

2. Time t_2: On this trial (which starts upon receiving the next clock pulse), register B2 is set HIGH, so the output word is 110_2. Voltage E_0 is now 0.75 V. Since E_{in} is less than E_0, the B2 latch is set to 0, and the output word reverts to 100_2.

3. Time t_3: Register B3 is set HIGH, making the output word 101_2. The value of E_0 is now 0.625 V, so $E_{in} = E_0$. The B3 register is latched to 1, and the output word remains 101_2.

4. Time t_4: Overflow occurs, telling the control logic to issue an EOC pulse. In some cases the overflow pulse *is* the EOC pulse.

Note that in the example we had a 3-bit SAR, so by our (n + 1) rule, we required four clock pulses to complete the conversion. The SA type of ADC was once regarded as difficult to design because of the logic required. But

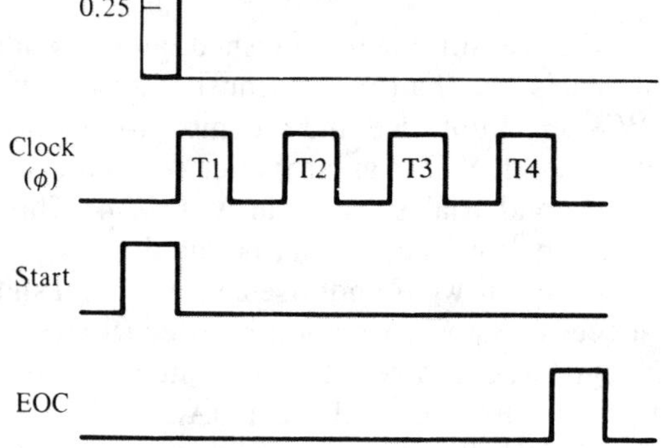

Figure 17-6 Typical operation of an SA A/D converter.

today IC and function blocks are available that use this technique, so the design job is reduced considerably. The SA technique can be implemented in software under computer control using only an external DAC and comparator. All register functions are handled in the software.

PARALLEL ("FLASH") A/D CONVERTERS

The parallel ADC circuit (Figure 17-7) is probably the fastest type of ADC known. In fact, some texts call it the "flash" converter in testimony to its

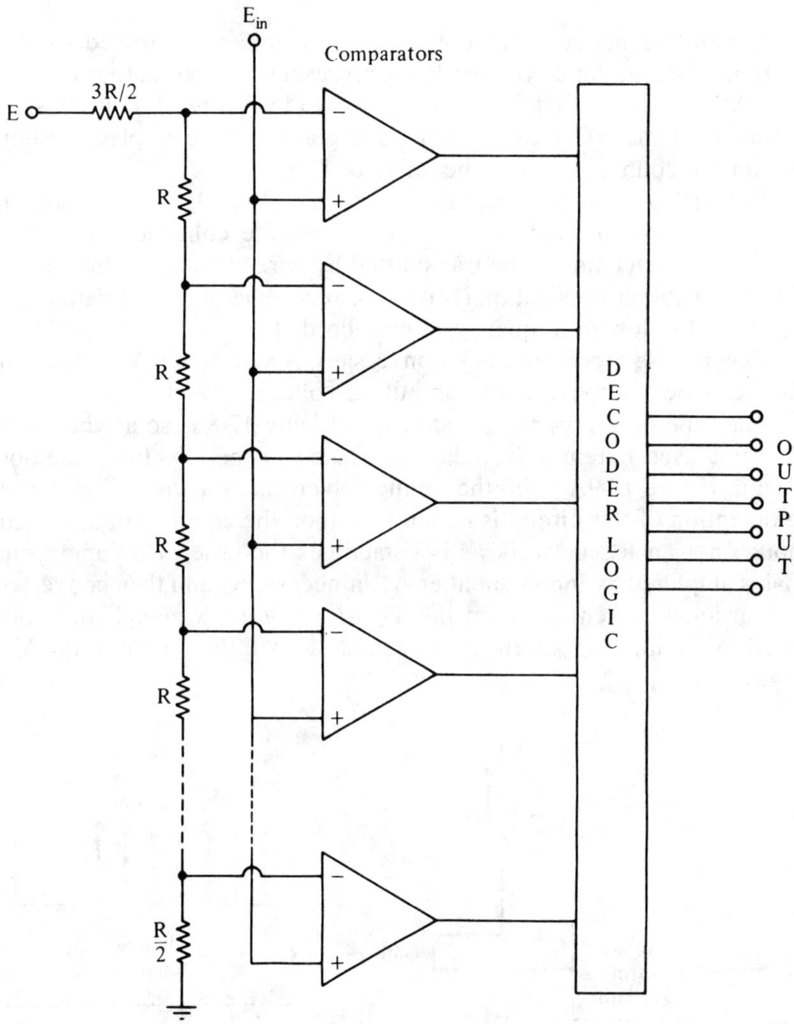

Figure 17-7 Flash or parallel A/D converter.

speed. It consists of a bank of (2^{n-1}) voltage comparators biased by reference potential E through a resistor network that keeps the individual comparators 1-LSB apart. Since the input voltage is applied to all the comparators simultaneously, the speed of conversion is essentially the slewing speed of the *slowest* comparator in the bank and the decoder propagation time (if logic is used). The decoder converts the output code to binary code, or possibly BCD in some cases.

VOLTAGE-TO-FREQUENCY A/D CONVERTERS

A voltage-to-frequency (V/F) converter is a voltage-controlled oscillator (VCO) in which an input voltage E_{in} is represented by an output frequency F. An ADC using the V/F converter is shown in Figure 17-8. It consists of no more than the VCO and a frequency counter. The display, or output state, on the counter gives us the value of E_{in}.

Voltage-to-frequency converters are used mainly where economics dictate serial transmission of data from a remote collection point to the instrument. Such data can be transmitted by wire or radio communications channels. Another application is the tape recording of analog data that are themselves too low in frequency to be recorded.

The inverse procedure, F/V conversion, is a form of DAC, in which an input frequency is converted to an output voltage.

The type of V/F converter shown in Figure 17-8 is somewhat archaic and is not used extensively today. A somewhat better circuit method is shown in Figure 17-9A, with the timing waveforms shown in Figure 17-9B. The operation of this circuit is dependent upon the charging of a capacitor, although not an RC network as is sometimes the case. The input voltage signal is amplified by input amplifier A1, if necessary, and then converted to a proportional current value in the *V-to-I converter stage*. If the voltage applied to the input, V_{in}, remains constant, so will the output of the V-to-I converter (I).

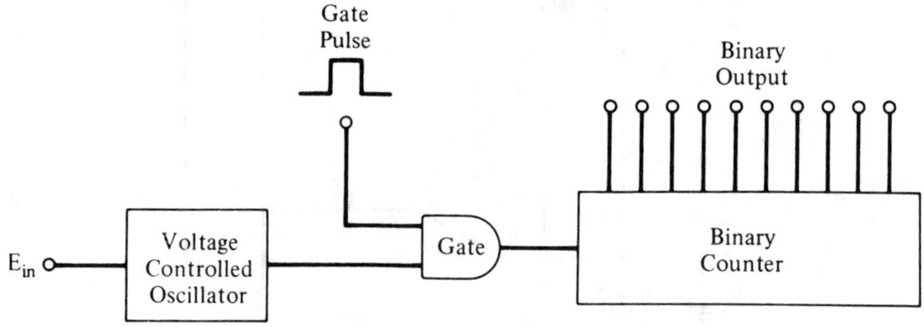

Figure 17-8 Gated counter V/F converter.

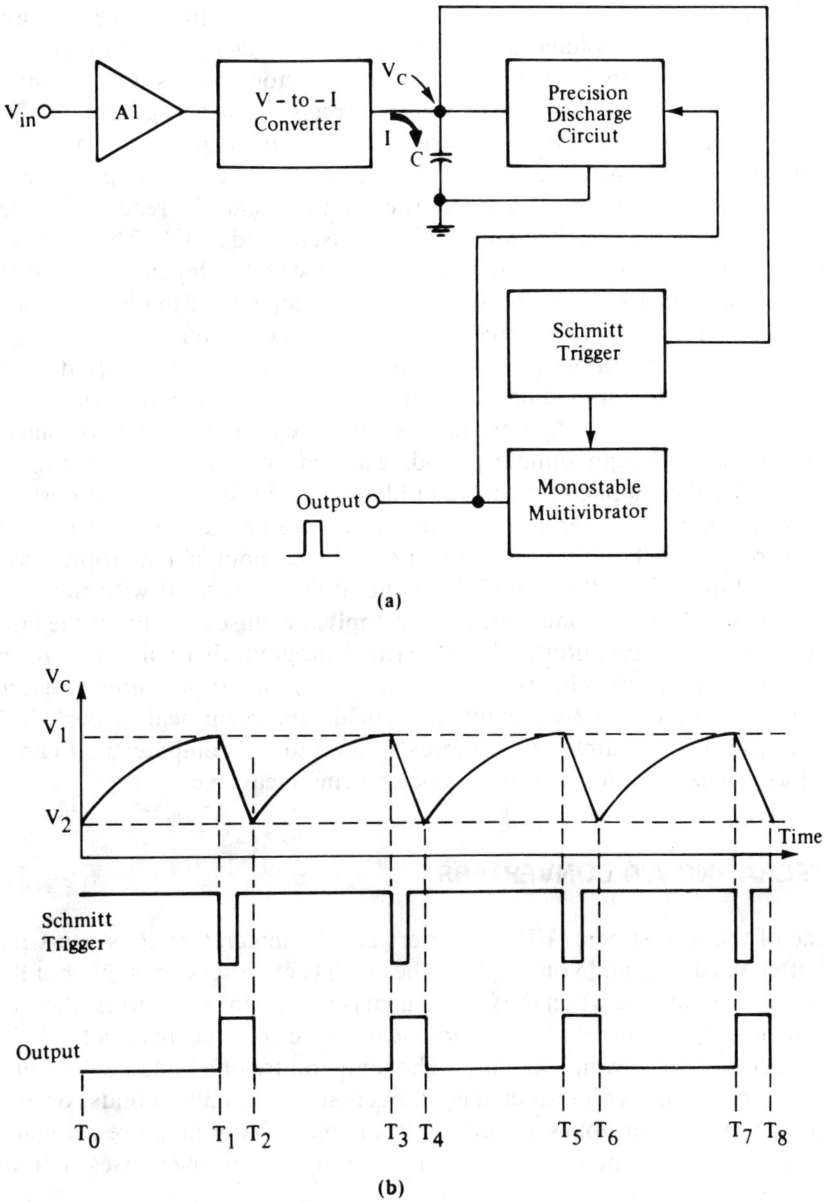

Figure 17-9 Analog V/F converter: A) Circuit, B) Timing.

The current from the V-to-I converter is used to charge the timing capacitor (C). The voltage appearing across this capacitor will vary with time as the capacitor charges (see the waveform in Figure 17-9B). The precision discharge circuit is designed to discharge the capacitor to a certain level (V2) whenever the voltage across the capacitor reaches a predetermined value (V1). When the voltage across the capacitor reaches V2, a Schmitt trigger circuit is fired that turns on the precision discharge circuit. The precision discharge circuit, in its turn, will cause the capacitor to discharge rapidly but in a controlled manner to value V1. The output pulse snaps HIGH when the Schmitt trigger fires (i.e., at the instant V_c reaches V1) and drops LOW again when the value of V_c has discharged to V2. The result is a train of output pulses whose repetition rate is exactly dependent upon the capacitor charging current, which, in turn, is dependent upon the applied voltage. Hence we have a voltage-to-frequency converter.

There are several ways in which the V/F converter can be used with a microcomputer to input data. One method is to use a binary or decade counter to count the output frequency (or at least the number of pulses) during a known-length sample period. The binary or BCD outputs of the counters are then applied (in a manner like Figure 17-8) to the input port of a microcomputer. An alternative version of this same method is to feed the pulses from the V/F output to the timer or counter input of a microprocessor support chip such as the Z80-CTC or the 6522 that is used with the 6502-series devices. Our second method is to apply the pulses to 1 bit of the input port of the microcomputer and then write a program that will measure the time between pulses. This is the basis for the *frequency-counter* programs that some people use; frequency is, after all, the reciprocal of period. In either case, we will input some representation to the computer that can be used as a binary analog of the parameter being measured.

INTEGRATING A/D CONVERTERS

Some of the lowest-cost A/D converters are the integrating converters that are often used in digital voltmeters. These A/D converters are slow, but that is often an advantage when the input signal is noisy. In other cases, the slow speed of conversion is of little practical concern if the parameter being measured is also slow in changing. The temperature of a large pot of metal, for example, is not going to change in microseconds, milliseconds, or even seconds; minutes or hours are more reasonable units of measure. It makes little difference in such a case whether the design engineer uses a 50-ms converter instead of a 10-μs version.

There are several different designs for the integrating A/D converter, but perhaps the most common are the *single-slope integrator* and the *dual-slope integrator*; of these, the latter is considered the better for most applications.

An example of the single-slope integrator is shown in Figure 17-10. The principal components of this circuit are the input amplifier (A1), which

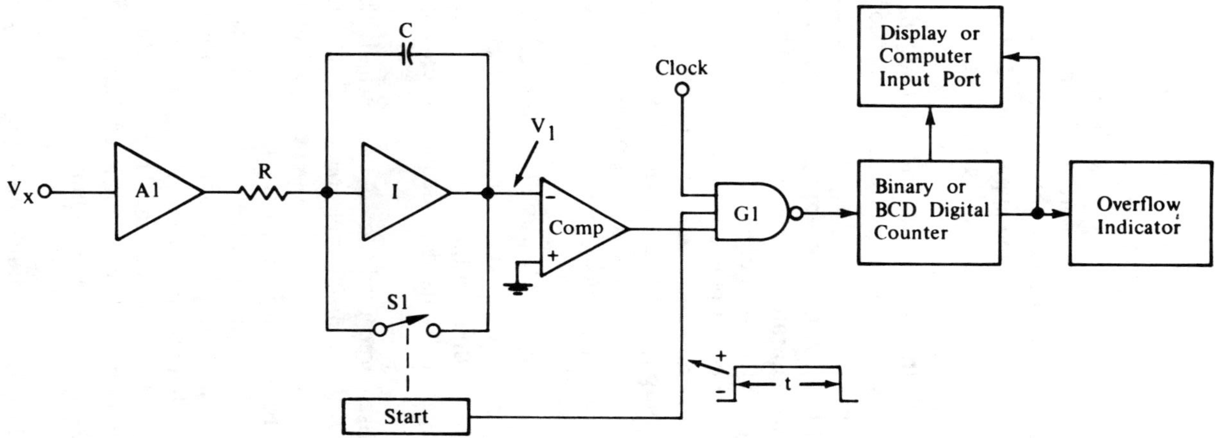

Figure 17-10 Single-slope integrator A/D.

is optional, a Miller integrator (I), a comparator (COMP), a main gate (G1), and a binary counter.

The input amplifier may or may not be used in any given single-slope integrator. It is optional and is merely used to scale the input signal to a level that optimizes the operation of the integrator circuit. Obviously, a too-small signal would take a long time to charge the integrator capacitor.

The Miller integrator uses an operational amplifier and a resistor-capacitor combination to produce an output voltage (V1) that is proportional to the integral (i.e., *time-average* of the applied input signal). For purposes of discussion, we will assume that A1 has a gain of unity, so the input to the integrator is the unknown input voltage V_x.

A comparator (COMP) is sometimes called an amplifier with too much gain. Indeed, we can make a comparator using an operational amplifier with an open feedback loop. This makes the amplifier gain 50,000 in cheap models to well over 1,000,000 in premium devices. Obviously, a few millivolts of input signal will saturate such an amplifier. This is the way a comparator operates. It produces an output level that indicates when the two input voltages are equal (i.e., output zero) or which of the two is highest. In the limited case shown here, one input of the comparator is grounded (at a 0-V potential), so the output will be zero when V1 is zero and HIGH when V1 is more than a few millivolts higher than ground. When the output of the comparator is HIGH, gate G1 is enabled and can pass pulses (provided that the Start pulse is also HIGH).

The counter will have either binary or BCD formats (the BCD format is used in digital voltmeter converters) and thus can be connected either to a display device (like the readout on the DVM) or to a microcomputer input port. An overflow indicator is used to denote when the input voltage is overrange.

When a start pulse is received (it may be internally generated), switch S1 is closed briefly in order to discharge capacitor C so that no accumulated charge will foul the results. Switch S1 is opened again almost immediately, thereby allowing C to begin charging at a rate that is determined by the input voltage V_x. As soon as the integrator output voltage is more than a few millivolts higher than ground, the output of the comparator will snap HIGH, thereby allowing pulses to flow into the counter. A third signal applied to G1 is the timing signal from the start section. This pulse stays on for a period t and prevents the integrator from continuing to charge until the counter overflows. The count at the end of the timing period is proportional to the applied voltage.

One major problem with the single-slope integrator is that it is sensitive to noise riding on the input signal. A noisy signal will usually produce an erroneous result.

The dual-slope integrator of Figure 17-11A solves the noise problem. The circuit is shown in Figure 17-11A and its timing diagram in Figure

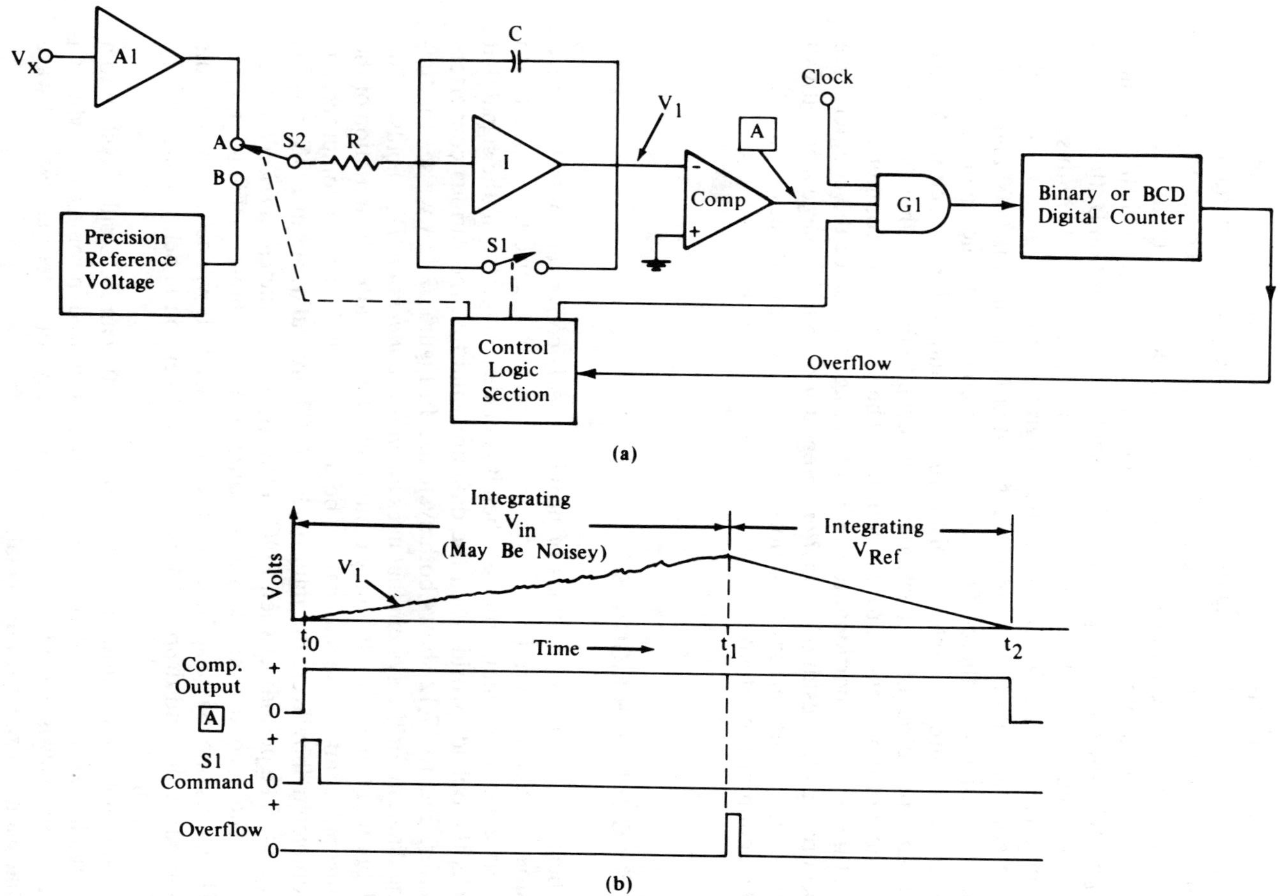

Figure 17-11 Dual-slope integrator A/D: A) Circuit, B) Timing.

17-11B. The principal difference between the two converters is that the dual-slope converter makes the conversion in two steps. During the first step, period t_0 to t_1 in Figure 17-11B, the integrator is charged from the unknown input signal V_x. The output voltage of the integrator continues to rise as long as the input switch S2 is connected to the input circuit. The binary/BCD counter is allowed to overflow (at time t_1), and this overflow signal tells the control logic circuits to switch S2 to position B, the output of a fixed precision reference voltage supply. The polarity of this voltage source is such that it will cause the integrator to discharge at a fixed rate. The counter state at the instant this switch is changed is 0000 and begins incrementing from there. The counter will continue to increment until the reference source completely discharges the integrator capacitor. At that instant, the output of the comparator will drop LOW again, thereby stopping the flow of clock pulses to the counter. The state of the counter output at that instant is proportional to the input voltage V_x. Since the counter state occurred as the result of a *constant* integrator output discharge slope, it will not contain noise errors. The noise errors are integrated out of the data by the action of the integrator.

A/D CONVERTER SIGNALS

There are two basic methods of providing A/D converter control systems. Most A/D converters have a *start* line, which will cause the converter to initiate the conversion process when it is made active. It is in the signal that tells the outside world when the data are valid that the various converters differ. Figure 17-12 shows both systems. In Figure 17-12A we see the timing diagram for the system that uses an *end-of-conversion* (EOC) pulse. The data output lines (B0–B7) may contain invalid data after the initiation of the *start* pulse, and these data cannot be used. When the conversion process is completed, however, the data on B0 to B7 are valid, so an EOC pulse is issued. The period between EOC pulses (t_1 to t_0) is the *conversion time*, T_c.

The second method uses a *status* signal as shown in Figure 17-12b. The status line may also be called the *busy* signal. It will be HIGH when the data are valid and drop LOW when the conversion is being made and the data are invalid.

Figure 17-13 shows a method for converting an EOC pulse system to a status line system. This conversion may be required in some cases when an A/D converter is being interfaced to an existing computer that uses software that wants to see a status signal.

The basis of this circuit is an R-S (reset-set) flip-flop that is made from a pair of cross-connected NAND gates. When the start pulse is received, the status line will drop LOW. The start pulse need only be momentarily

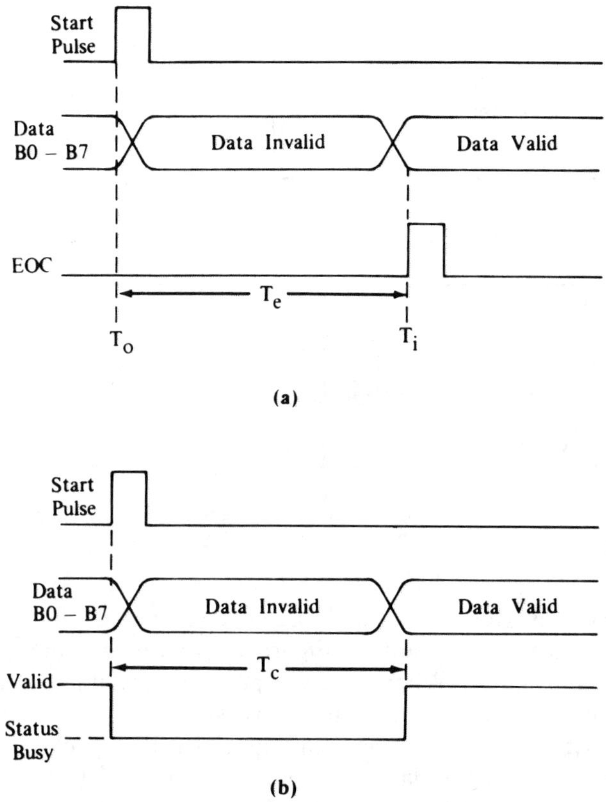

Figure 17-12 A/D timing: A) EOC pulse type, B) Status line type.

present, so the circuit is ideal for "catching" that temporary pulse. Similarly, when the EOC pulse is received, the status line goes HIGH again.

Another problem is that the A/D converter will be dormant between conversions. The device must be "tickled" by the start pulse, which is generated by the computer, before it can begin its work. But suppose we

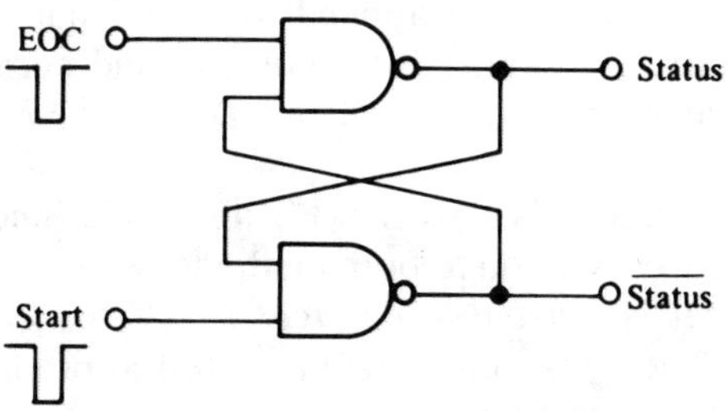

Figure 17-13 NOR R-S flip-flop circuit for EOC/START.

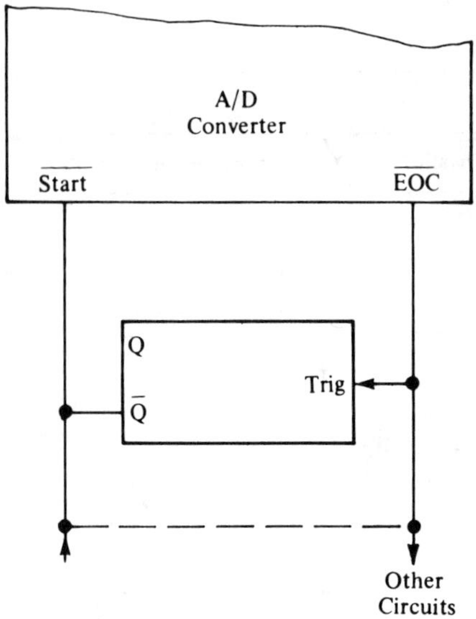

Figure 17-14 Asynchronous operation of A/D converter.

want the A/D converter to make continuous conversions, that is, to operate asynchronously. In that case (see Figure 17-14) we can connect the EOC output to the start input. When the EOC pulse occurs, it automatically tells the A/D converter to begin again the conversion process.

A problem that is sometimes experienced with asynchronous conversion, however, is that the data are valid for only one clock pulse. If that period is too fast, some data may be lost because the computer cannot input them fast enough. A solution to this problem (also shown in Figure 17-14) is to connect a monostable multivibrator (one-shot) stage between the EOC output and start input. The one-shot will insert a delay equal to its period between the two events.

INTERFACING DACS

DACs will convert binary words applied to their inputs to a proportional voltage or current at the output. Both voltage- and current-output DACs exist. There are two methods for interfacing DACs: *I/O-based* and *memory-mapped*.

The I/O-based method is shown in Figure 17-15. Since several aspects of DAC interfacing are common to both methods, we will show them only in this figure One facet is *current-to-voltage (I-to-V) conversion*; another is *low-pass filtering*. The I-to-V conversion is used to produce a voltage output (which is what is needed by oscilloscopes and most strip-chart paper recorders) from the output of the current type of DAC. In Figure 17-15, the

DAC output current (I_0) is converted to a voltage by passing it through a fixed-precision resistor R. The output voltage will be, according to Ohm's law;

$$V_0 = I_0R$$

The low-pass filter is used to smooth the output waveform. The DAC can produce only certain discrete output levels, so instead of a ramp function it would produce an "equivalent" staircase function. The low-pass filter will smooth the staircase to make it look more like a ramp.

The interfacing method shown in Figure 17-15 is used when there is an output port available that is latched (as are most); that is, the output port will contain the last valid data even after the computer has gone on to other chores. In that case, it is merely necessary to connect the output port bits to the DAC input bits on a one-for-one basis.

In a memory-mapped system, the DAC (or other peripheral; the method is not limited to data converters) is treated as a memory location and is assigned an address. Figure 17-16 shows the memory-mapped system. The OUT n signal is an output device-select pulse (see Chapter 4). For the case of a microcomputer that uses such a system, the elements that go into forming the OUT n signal are those that form a memory write operation. When the computer executes a write to the memory location defined in the OUT n signal, data on the data bus are transferred to the output of the 74100 dual quad-latch TTL IC. The outputs of the 74100 are used to drive the inputs of the DAC and are updated whenever the computer writes a new value to the memory location defined by the OUT n operation. Several IC devices other than the 74100 will do the same job.

The circuits shown thus far in this section are fine when used with 8-bit DACs connected to 8-bit computers. But how do we connect a DAC that

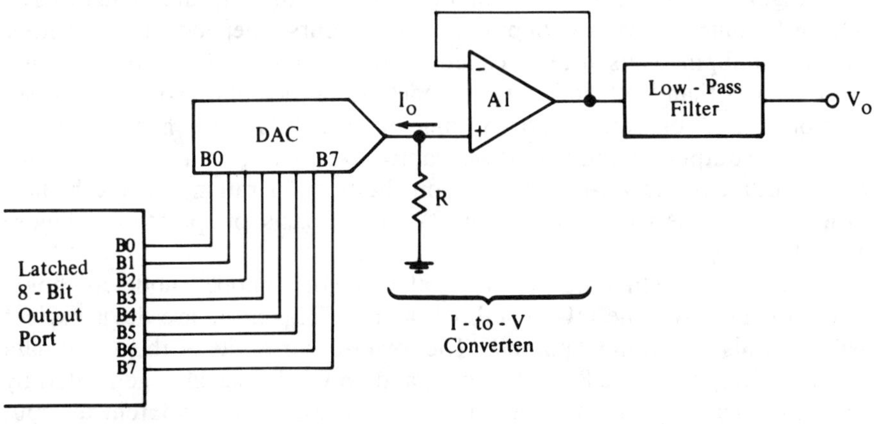

Figure 17-15 Typical current-output DAC (e.g., DAC-08) output circuit.

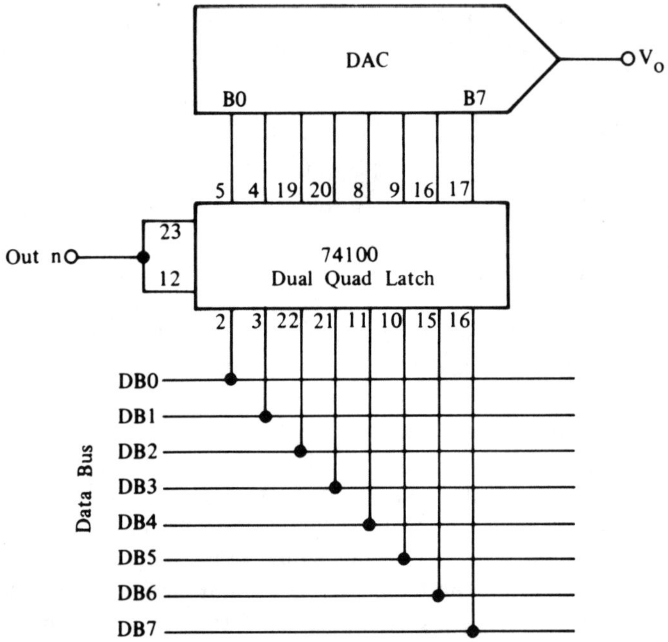

Figure 17-16 Interfacing DAC to data bus.

uses more than 8 bits? The 8-bit computer can easily handle greater than 8-bit input words because it can use double-precision programming techniques. Machines based on the Z80 microprocessor chip even have the ability to use 8-bit register pairs in single-instruction operations. We will, therefore, occasionally see the need for interfacing a larger-than-8-bit DAC to an 8-bit microcomputer.

Figure 17-17A shows the method for connecting the large-length DAC to an 8-bit output port or data bus (which, of course, depends upon whether memory mapping is used); it is called the double-buffered method. For any word greater than 8 bits in length, we can output the entire word using more than one output operation. For example, up to 16 bits can be handled by two successive output operations. If we wanted to output, say, a 12-bit word, we could output the lower-order 8 bits on the first operation and the higher-order 4 bits on the second operation. This is the basis for operation of Figure 17-17A.

Let's assume that the circuit is in the memory-mapped mode, as shown in Figure 17-17A. The OUT1, OUT2, and OUT3 signals are device-select pulses as discussed in Chapter 4. The lower-order 8 bits of the 12-bit data word are output on the 8-bit data bus, and an OUT1 signal is generated by the CPU. This signal will cause IC1, a 74100 eight-bit data latch, to input and hold the signal. Thus the lower-order 8 bits of the required 12 will be

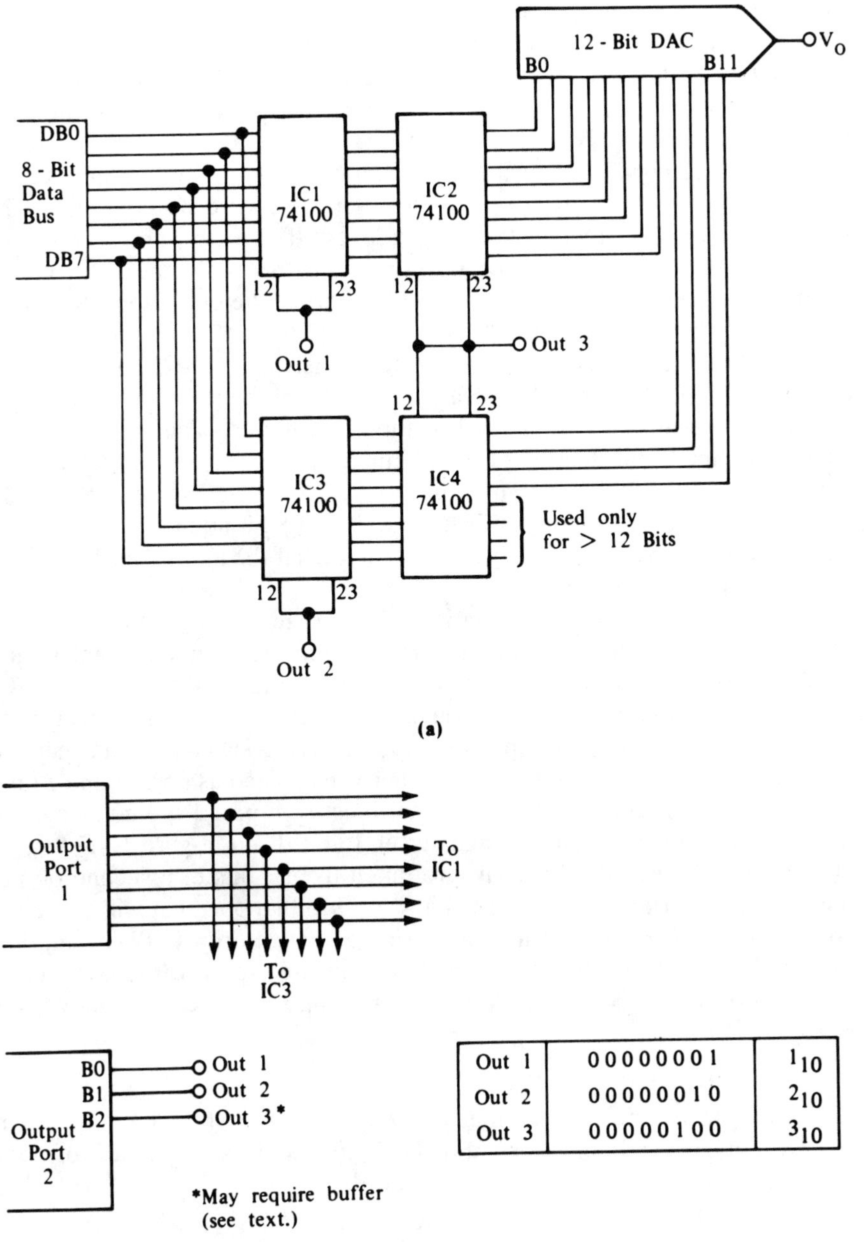

Figure 17-17 A) Circuit for interfacing longer DACs with 8-bit ports, B) Alternate scheme.

stored at the output of IC1 after the OUT1 signal disappears. On the next operation, the higher-order 4 bits of the 12-bit data word will appear on the lower 4 bits of the data bus, while simultaneously an OUT2 signal is generated. The effect of the OUT2 signal is to cause IC3 to input and hold the lower-order 4 bits; only 4 of the 8 bits are used on this operation. Hence after the OUT2 signal disappears, we will have the lower-order 8 bits stored in IC1 and the higher-order 4 bits of the 12-bit data word stored in IC3. The DAC is now ready to receive the entire 12 bits. If an attempt had been made to apply any of the data to the DAC prior to this time, the DAC would temporarily see an incorrect data word for part of the operation. Now that the entire 12 bits is available at the outputs of IC1 and IC3, we can crank the data into the 12-bit DAC-driver register consisting of IC2 and IC4. An OUT3 pulse will turn on both IC2 and IC4 and thereby transfer the data that are on the outputs of ICl and IC3 to the DAC inputs. The DAC will now have an entire 12-bit data word on its inputs.

The specific circuit shown in Figure 17-17A will accommodate up to 16 bits because each 74100 device is essentially an 8-bit latch (actually, it is a dual-quad latch, but it is effectively an 8-bit latch if the two strobe lines, pins 12 and 23, are tied together).

Figure 17-17B shows a variation of the basic circuit that allows interfacing with a pair of 8-bit output ports. If you are using a commercially available computer or intend to use one of the commercial "no frills" SBCs that are frequently sold as "controllers," it may be more cost-effective to use extra I/O ports for this application rather than design a memory-mapped add-on. Two output ports are needed, here designated as ports 1 and 2 (any designation could be used). The output lines from port 1 are connected to the input lines of the 74100s designated IC1 and IC3 in Figure 17-17A. The OUT1, OUT2, and OUT3 signals are taken from 3 bits of a second output port: we must write an appropriate data word to that port that will cause the correct bit to go HIGH for a short period and then drop LOW again. For example, the line for the OUTl signal is connected to bit B0 HIGH and all others LOW (i.e., 00000001). A typical program sequence will follow these steps:

1. Write 01 hex to port 2.
2. Jump to a timer subroutine that will provide a short delay (1 ms is usually a good selection, and the 74100 devices will react to much faster pulses).
3. Clear OUT2 by writing 00 hex to port 2.

In the initialization section of the main program, it will be necessary to ensure that port 2 is reset (i.e., 00000000) or there may be unwanted HIGH conditions on some of the lines at certain times following power-up. The binary words needed to create the three select signals are given in the inset table in Figure 17-17B.

Because of the limitation of output port drive current, it may be that bit B2 of port 2, which drives the OUT3 signal, will require a high-current buffer stage. This is because there are four TTL inputs connected to this line (i.e., pins 12 and 23 on IC2 and IC4), and many microcomputer output devices are limited to 3.6 mA, which will support only two TTL inputs. This extra buffer will not be used if the output port device used in some particular application will support four TTL lines.

INTERFACING ADCS

Perhaps the simplest method for interfacing an A/D converter to a microcomputer is shown in Figure 17-18; Figure 17-18A shows the method used when the operation is synchronized under program control, and Figure 17-18B shows the method used when the A/D converter operates asynchronously. In both cases, the eight data lines from the A/D converter are connected directly to the eight lines of an 8-bit parallel input port.

Figure 17-18A shows the circuit for the case when the A/D converter is under direct program control (i.e., the program issues the start pulse that begins the conversion process). The start line of the A/D converter is connected to 1 bit (B0 selected here) of output port 1. Any bit or port could be selected, and the unused bit remaining can be used for other applications.

The EOC (end of conversion) pulse is applied to 1 bit of a second input port (other than the data input port). A typical program sequence would be as follows:

1. Write 01 hex to port 2. This step causes B0 of port 2 to be HIGH.
2. Reset port 2 by writing 00 hex to port 2.
3. Loop until bit B0 of input port 1 is made HIGH.
4. Input data on port 1.

The method shown in Figure 17-18A is wasteful of one output port (i.e., the port used for the start pulse) and requires the program to be continuously dedicated to that task. The method of Figure 17-18B is asynchronous and will free up the computer somewhat, provided that the A/D converter uses a latched output stage and the conversion time is sufficiently long. The asynchronism is gained by the simple expedient of connecting the EOC and start lines together. The EOC pulse becomes the start pulse for the next conversion cycle. The computer will loop until it sees the EOC pulse on bit B0 of input port 2. Again, the assumption is made that the A/D converter has a latched output stage.

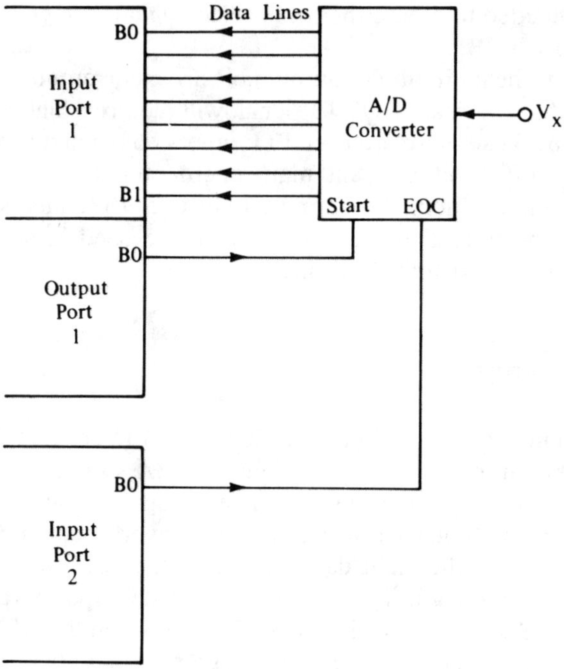

(a)

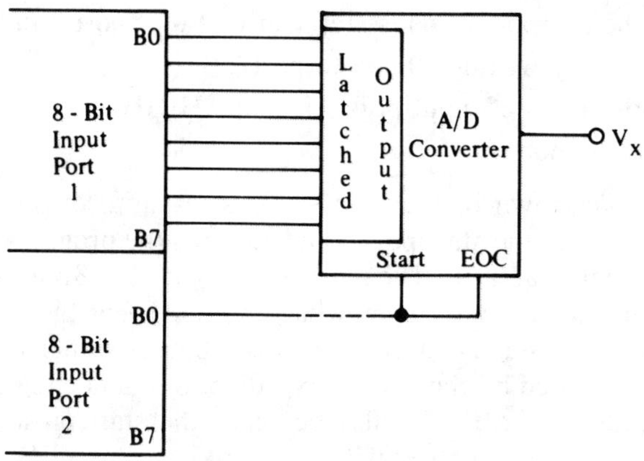

(b)

Figure 17-18 A) A/D converter interfacing; B) Alternate method.

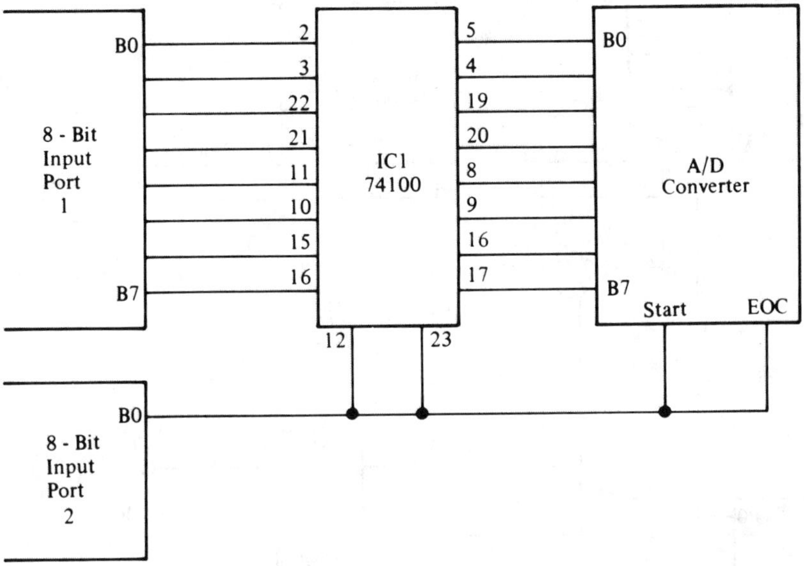

(c)

C) Use of 74100 to interface DAC.

We can add a latched output stage to an A/D converter that lacks such capability by using the circuit of Figure 17-18C. The data latch is a 74100 dual-quad latch (or some similar chip). The two halves of the latch are each activated by a separate strobe terminal (pins 12 and 23), which are here wired together in order to accommodate the 8-bit word length. When the EOC/start pulse is generated, indicating that the data on the output of the A/D converter are valid, the data will be transferred from the inputs of the 74100 to the respective outputs.

We can both gain freedom from keeping the computer tied up looking for the A/D converter data and free up one input port by adding a few components to the basic circuit (see Figure 17-19). The 8-bit input port is driven by IC2, a tristate 8-bit buffer, and tristate inverter G2, which is connected to only 1 bit of the port (bit B7). The reason for the tristate components is to permit them to be floated at high impedance when not in use. Otherwise, any LOW on the specific line affected would automatically affect the other devices connected to the line. For example, in the case of Figure 17-19, a LOW on the output of G2 would short to ground the B7 output of IC2.

The transient nature of the EOC pulse requires the computer in our previous examples to loop while searching for the HIGH EOC pulse; this is very wasteful of CPU time. Although the use of the interrupt capability of

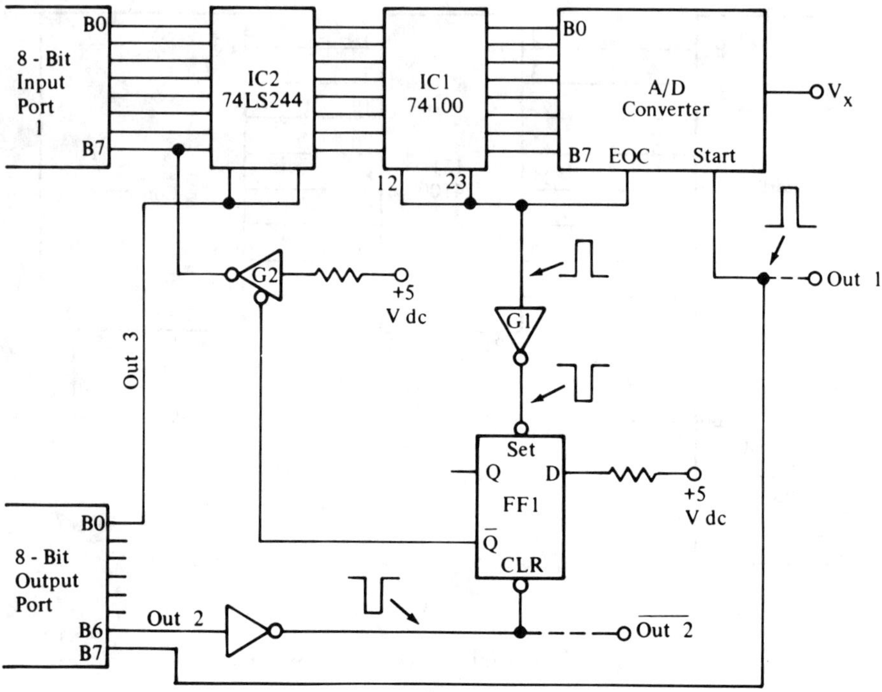

Figure 17-19 A/D converter interface.

the CPU is a better selection, this method allows interfacing with an existing computer I/O port without the necessity of gaining access to the interrupt terminals. Flip-flop FF1 is used to store the EOC pulse, thereby making it essentially a status signal.

The control signals for the A/D converter circuit of Figure 17-19 are designated OUT1, OUT2 (and its inverse @OUT2), and OUT3. These signals are generated by making specific bits of the output port HIGH for a short period of time. A typical operation consists of making the appropriate bit HIGH for a specified period of time and then resetting it LOW. Such a program that will, for example, generate OUT1 will have to do the following:

1. Write 80 hex (i.e., 10000000_2) to the output port.
2. Jump to a timer subroutine, if necessary, to generate a delay during which B7 is HIGH. This step sets the duration of the OUT1 pulse.
3. Write 00 hex to the output port, thereby resetting it and canceling the OUT1 signal.

The A/D converter must have either latched output lines or incorporate IC1 in order to store the output data temporarily. The purpose of IC2 is to provide a tristate buffer between the TTL outputs of IC1 and the input lines of the I/O port. Otherwise, we would not be allowed to bus the output of inverter G2 to B7 of the I/O port.

The first step in the program sequence will be to generate the OUT2 signal in order to ensure that flip-flop FF1 is cleared. This step will be performed when the computer is first turned on or, alternatively, when the A/D converter program is first invoked. When the NOT-Q output of FF1 is LOW (i.e., the FF is in the *set* state), inverter G2 is turned on and a LOW is applied to bit B7 of the input port.

When the program calls for the A/D converter to begin a conversion cycle, it will generate the OUT1 signal. This pulse is connected to the start input of the A/D converter and so will initiate the conversion process. The computer is then free to perform other chores while the conversion is taking place. In most applications the computer program will have to inspect B7 of the input port every few milliseconds (this depends upon the conversion time of the A/D) to see if a LOW is present. When the EOC pulse is generated by the A/D converter, it will cause the A/D output data to be stored in IC1, and the pulse is also inverted (in G1) and used to set FF1. The NOT-Q output of FF1 thereby goes LOW and turns on G2. Since the input of inverter G2 is HIGH, its output will be LOW. When the computer program returns to check B7, it will jump to the A/D input subroutine when it sees B7 LOW. An appropriate program will then issue an OUT2 signal to set FF1, followed by an OUT3 signal that turns on IC2 and thereby gates the A/D converter data input to the input port. The sequence will be as follows:

1. Ensure FF1 is reset by generating an OUT2 signal during program initialization when the computer is first turned on or when the A/D program is first invoked. Alternatively, a system power-on reset pulse could be used by adding an OR gate to the circuit.
2. Generate a start pulse by generating signal OUT1.
3. Periodically check bit B7 of the input port for a LOW.
4. When the LOW is sensed at B7, generate an OUT2 pulse to reset FF1.
5. Generate an OUT3 signal to gate the contents of IC1 onto the input port lines. Hold this signal until the data are input.
6. Read the input port data and store it at the memory location required.
7. Reset signal OUT3.

A similar system can also be used for memory-mapped A/D converter applications. In those cases, however, we might also want to consider a simpler system that makes use of genuine device-select pulses (see Chapter 4). Figure 17-20 shows such a method in which OUT1 and IN1 signals are used to control the A/D converter.

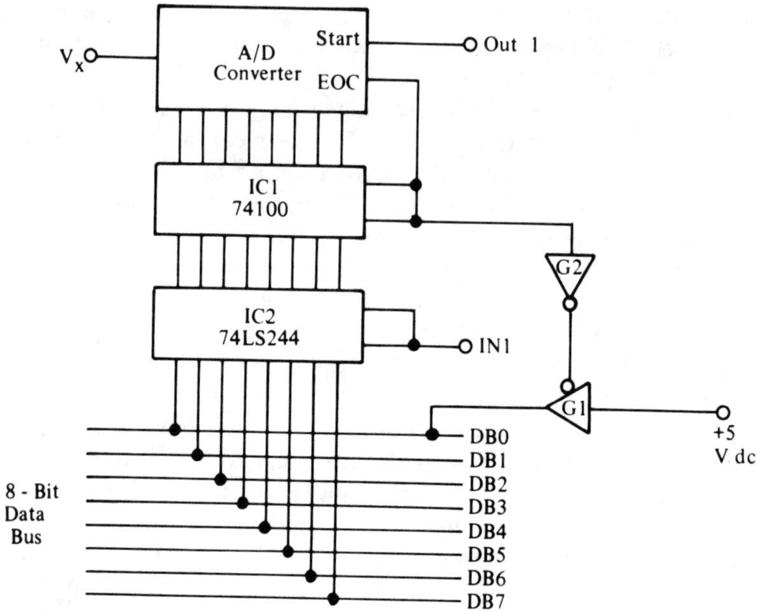

(a)

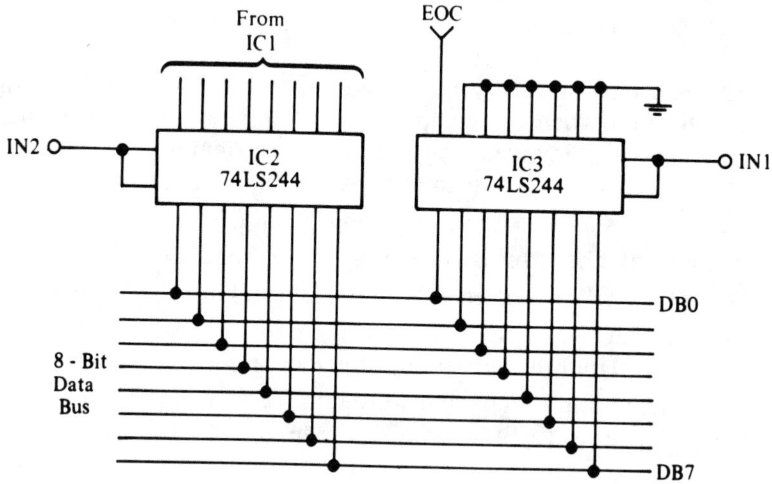

(b)

Figure 17-20 Latched tristate A/D interfacing.

In a memory-mapped system, the A/D converter is equipped with its own dedicated *input port*, which in Figure 17-20A consists of the 74100 data latch (IC1) and the 74LS244 tristate 8-bit buffer (IC2). The output lines of IC2 are connected to the data bus lines. Since the lines from IC2 are tristate, they will not load down the data bus lines when IC2 is in the inactive state. But when an IN2 signal is generated, it will gate the contents of the 74100 latch onto the data bus.

The conversion sequence is initiated by generating the OUT1 device-select pulse, which is applied to the start input of the A/D converter. When the conversion process is completed, the EOC pulse will be generated. This pulse has two effects; one is to latch the A/D data, while the other is to turn on G1 (the EOC pulse is inverted by G2 before being applied to the enable terminal on G1). When G1 is enabled, a HIGH is placed on bit DB0 of the data bus. This signal tells the program that the A/D converter data are ready. The program will generate an IN1 device-select signal, thereby gating the data from the 74100 outputs onto the data bus. Figure 17-20B shows a method for using one section of a second 74LS244 to form gate G1. Although this illustration shows the unused inputs of the 74LS244 as grounded, they could be used for other applications if the need arose.

The A/D converter interfacing techniques presented thus far have limited applications. The method of Figure 17-20, for example, requires the program to occasionally check for the EOC signal, and such an arrangement is at best clumsy. A somewhat better approach that is open to those who can access the interrupt line(s) of the computer is shown in Figure 17-21. In this circuit, a J-K flip-flop (FF1) is used to send an interrupt signal to the CPU.

Again, we require either an A/D converter that has latched outputs or a 74100 arrangement as in the previous cases. Also, we are once again using an 8-bit tristate buffer to control entry of the A/D output data onto the data bus. This buffer is actuated by the IN1 device-select signal. In the circuit of Figure 17-21 the A/D converter is connected in the asynchronous mode, so the EOC pulse becomes the start pulse for the next conversion cycle. The A/D converter, therefore, will continuously convert the input signal. Every time the data are ready anew, the EOC pulse will cause the latest value to be input to the data latch. This same EOC pulse also clears FF1, thereby making the Q output LOW. The Q output of FF1 is connected to the active-LOW interrupt line to the CPU, so this action will cause the CPU to be interrupted.

The CPU will not respond to the interrupt immediately but will wait until the execution of the current instruction is completed. In most microprocessor CPUs, the interrupt line is examined during the last clock pulse of the execution cycle. If the interrupt line is active (i.e., LOW in this case), the address of the next instruction to be executed in the normal program sequence is stored on an external stack somewhere in memory and the

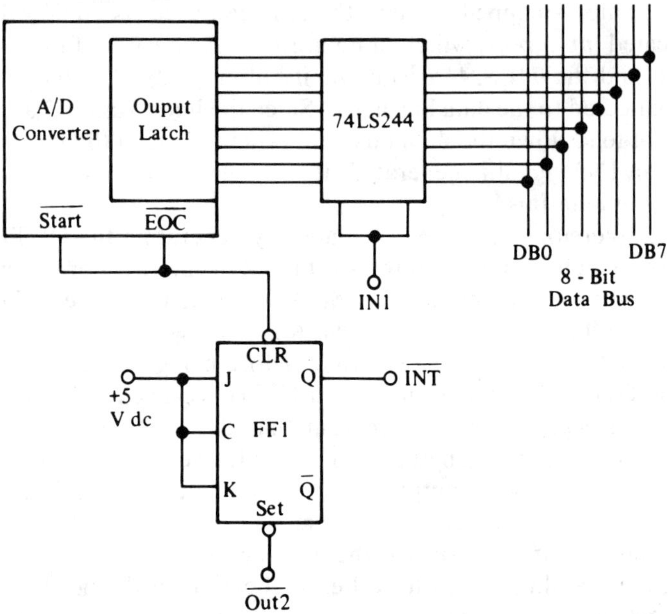

Figure 17-21 Interfacing A/D to data bus.

program control jumps to an interrupt service subroutine. In this case, the service program will be the A/D data input routine. Such a subroutine must accomplish the following:

1. Generate an @OUT2 signal to reset the interrupt flip-flop (FF1).
2. Generate an IN1 signal to gate the A/D data onto the data bus.
3. Store the A/D data at some appropriate point in memory.

The principal advantage of this arrangement is that the computer can be used for other chores while awaiting the A/D converter data. Except for a few really high speed A/D converters, most A/D converters have conversion times that are very long compared with the clock period of the computer. Some A/D converters will perform an 8-bit conversion in 5 to 10 μs, but there are many that have 50- to 100-ms conversion times. Of course, we could always specify the speedy versions, but that would be wasteful of both money and CPU time.

The conversion time should be related to the maximum frequency component in the analog input signal that is considered significant. For example, the human electrocardiogram (ECG) waveform has a 0.05- to 100-

Hz bandwidth. According to a well-known criterion, we must sample this signal at a rate that is equal to at least twice the highest frequency, so a 200 sample/second rate is sufficient; many experts would select a rate five times faster than the minimum required, so we would sample at 1000/s. This rate means that the A/D converter should be capable of 1-ms conversion time.

We can get an idea of the scale of time involved by comparing the rate of operations with the time that the CPU would idle uselessly while awaiting the EOC signal. In one model, the CPU can execute 400,000 operations/second, so the CPU could perform 400 operations during the 1 ms required for the A/D converter to do its job. Most applications where an A/D converter is needed will require some signal processing or further data message other than a simple storage operation. We could use that 1-ms "lost time" to perform some of these operations.

An example is the evoked potentials computer used in medical and physiological studies to examine the component of the EEG brain waveform that is due to some specific stimulus. This type of computer will either sum or average successive input data in a coherent manner, thereby processing out the randomness and leaving only the desired data. For the 30-Hz EEG waveform, we could sample at 200/s (5-ms conversion time) and then use the lost time to either sum or average previous data. The A/D converter will interrupt the signal processing any time that new data are available.

Many computers allow more than one device to interrupt the CPU but have only a single general interrupt line or a single maskable interrupt and one nonmaskable interrupt. The Z80 chip allows eight different devices to be used in a *vectored interrupt* mode. In a vectored interrupt on the Z80, all eight devices will drive the @INT line but will also place onto the data bus an 8-bit RST n code that tells the program counter where to jump to find the interrupt service routine; the term n directs the CPU to a specific memory location. An RST 3, for example, causes an automatic jump to location 00 18 hex. According to the Z80 instructions, we must place the hex code DF (11011111_2) on the data bus during the time when the @INT line is active. In Figure 17-22 we see the previous circuit modified to add this capability.

The @INT* signal in this circuit is the same as the @INT signal in Figure 17-21. Since more than one device may interrupt the computer, we use the @INT* signal to drive one input of a 7430 NAND gate (G2). We invert the output of G2 in order to form the @INT signal that is actually connected to the computer interrupt line.

The INTAWK (interrupt acknowledge) signal turns on IC2, a 74LS244 8-bit tristate buffer. The input lines of IC2 are connected either HIGH or LOW as required for the specific n code, in this case 11011111. When the INTAWK signal is generated by the CPU, the data word 11011111 is applied to the data bus and tells the CPU to jump to the beginning of the service subroutine.

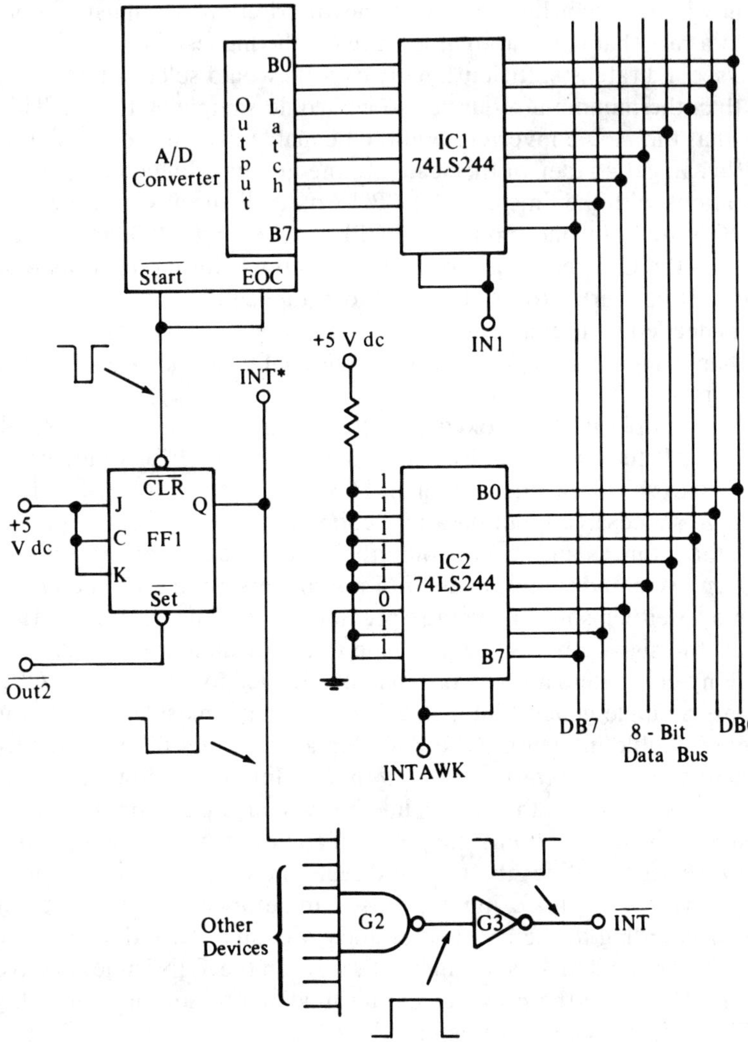

Figure 17-22 A/D interfacing to data bus.

MISCELLANEOUS TOPICS

Figure 17-23A shows a simple circuit that can be used to implement a software A/D converter. Both ramp and successive approximation methods can be accommodated by this arrangement. The elements of the circuit are a DAC and a voltage comparator. The comparator output will be LOW whenever the input voltage is higher than the DAC output voltage (V_0). When

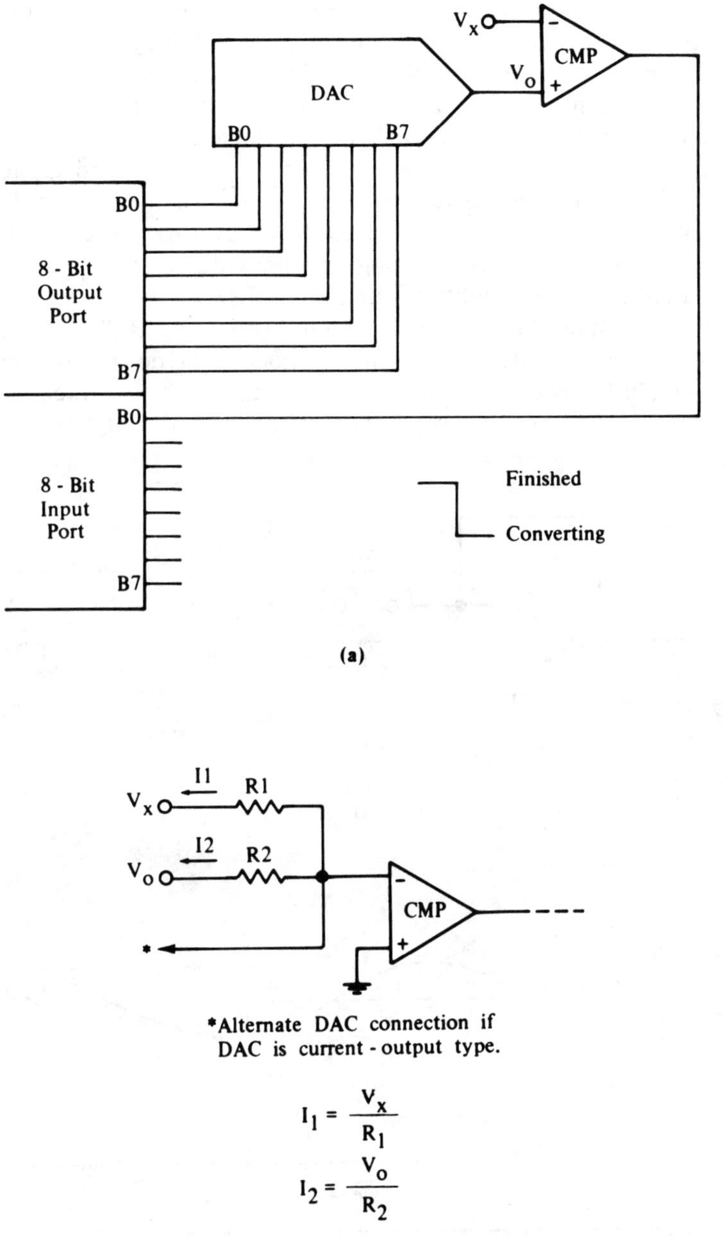

Figure 17-23 DAC used for simple A/D converter routine: A) Circuit, B) Current-mode comparator.

$V_x = V_0$, the comparator output snaps HIGH. The output of the comparator is applied to 1 bit of an input port. The program will vary the data applied to the DAC inputs according to the specific method and algorithm selected (either binary ramp or SA) and then examine the comparator output to determine whether or not the DAC output voltage matches the unknown. When a match is found, the computer accepts the binary word applied to the DAC via the output port as representative of the input voltage value.

Software A/D conversion reduces the hardware overhead of the computer system at the expense of using up more CPU time. The designer will have to determine the validity of the trade-off.

A variation on the theme is shown in Figure 17-23B. In this circuit we are using the comparator in the current mode, a tactic that will make the operation a little faster for many forms of comparator. It is the comparator settling time that limits many of these circuits, and that time is less for

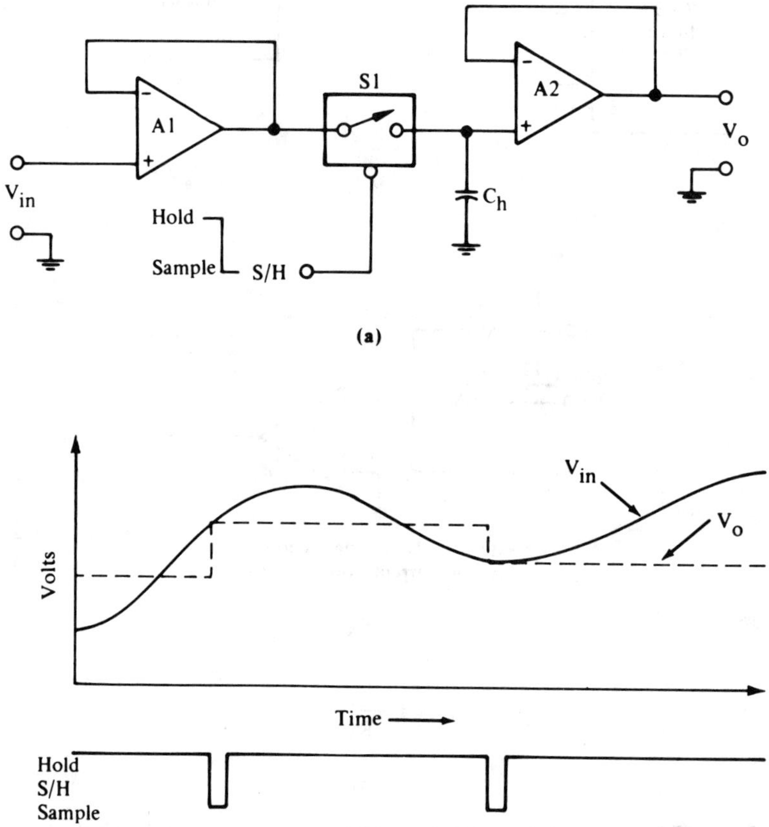

(b)

Figure 17-24 A) S&H circuit; B) Timing and operation.

current-mode operation. In current-mode operations, the two voltages are applied to the comparator inverting input through resistors that convert the voltages to currents. The noninverting input of the comparator is grounded, so output transition occurs when I1 = I2. We can reduce the component count by one resistor if a current-output DAC (1408, DAC-08, etc.) is used instead of a voltage-output model.

Our last topic involves the sample and hold (S&H) circuit of Figure 17-24A. There are times when we want to sample the signal prior to conversion. In some cases, for example, the A/D output will contain substantial error terms if the input analog signal varies during the conversion period. The S&H circuit will hold that signal constant for the brief period required to make the conversion. In other cases, we may wish to convert simultaneous instantaneous values (i.e., values that are related to each other in time) but use only one converter to do the job. While this requirement may be met by using parallel A/D converters that operate from the same start pulse, such use is uneconomical even in this era of low-cost A/D converters. The answer is to use a separate S&H circuit for each signal and drive them from the same sample line. The A/D converter can then make the conversions in sequence while retaining the simultaneity of the data.

The elements of the S&H are an input amplifier (A1), an electronic switch (S1), a hold capacitor (C_h), and an output buffer amplifier (A2). Both amplifiers are unity gain noninverting types and act to isolate the switch and the hold capacitor from the outside world. The tightest restrictions are applied to A2, which must have an extremely high input impedance in order to avoid discharging capacitor C_h while the signal is supposedly being held; amplifier A2 is usually specified as a BiFET or BiMOS type. Figure 17-24B shows the sampling pulse signal and the relationship to the input and output voltages.

18

ANALOG OUTPUT DISPLAY DEVICES

Analog displays of waveforms are particularly useful for discerning truth and beauty from these signals, so many instruments include such displays. The most usual form of display is an *amplitude versus time* waveform tracing. Two principal forms of display are used: *oscilloscopes* and *paper recorders*. In this chapter we will take a detailed look at each form.

OSCILLOSCOPES

The *cathode ray oscilloscope* (CRO), also called either *oscilloscope* or simply *'scope*, is an electronic instrument that uses a beam of electrons to "paint" a light picture on a phosphorous viewing screen, much like a TV set. If vertical (Y) and horizontal (X) directions each represent a different signal, then the pattern traced on the CRO viewing screen represents the relationship between the two signals. This principle is used in medical instruments such as the *vectorcardioscope*. Although differences exist between clinical medical, research laboratory, and medical instrument service shop oscilloscopes, the similarities are so great as to make it best to examine all forms in the same chapter.

Briefly restated, an oscilloscope is an instrument which uses a cathode ray tube (CRT), a device similar to a television picture tube, to display a voltage waveform. The waveform might be amplitude versus time, or it might be a complex thing that compares two amplitudes. The CRT is designed such that a beam of electrons from an "electron gun" will strike the phosphor-coated viewing screen (Figure 18-1), leaving a spot of light wherever the beam strikes. The beam can be deflected either magnetically or electrostatically, but for service, workbench, and laboratory type oscilloscopes electrostatic deflection is used. Television receivers, video/analog monitors (video monitors and some medical patient monitor 'scopes) use a magnetic deflection CRT in which a "yoke" electromagnet around the electron path in the neck of the CRT forms the controlling magnetic field. The frequency response and nonlinearity of magnetic deflection make it unsuitable for oscilloscope use at frequencies above about 1000 Hz.

Electrostatic deflection CRTs use a pair of *deflection plates* (as in Figure 18-1) to deflect the electron beam in the vertical and horizontal directions. Figure 18-2 shows how the beam on the screen of the CRT moves

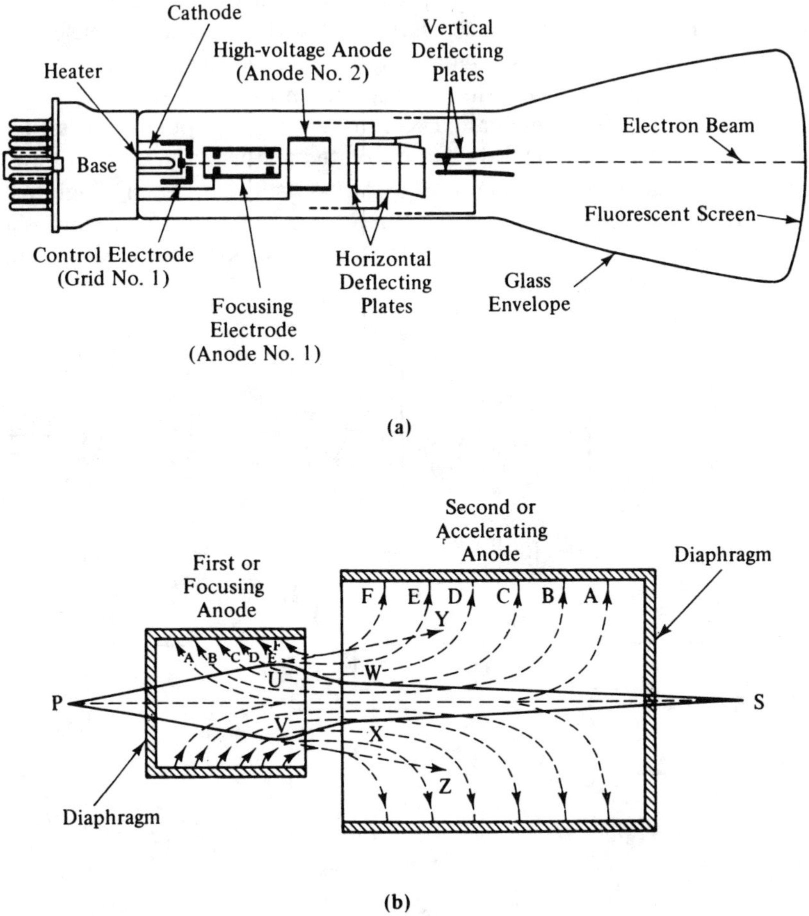

(a)

(b)

Figure 18-1 Cathode ray tube internal construction.

under different situations. In Figure 18-2A the beam is centered, which fact indicates that the voltages across the deflection plates are balanced (or zero). The beam is travelling straight down the neck of the CRT and striking the screen in the center. In "Figure 18-2B, however, the vertical plates remain balanced, while the horizontal plates are unbalanced such that the left side is now more positive (or less negative) than the right side; thus the beam (and light spot) is yanked to the left. Similarly, in Figure 18-2C we see the situation where the horizontal plates are balanced or zero, and the vertical plates are unbalanced. In this case the top vertical plate is more positive (or less negative) than the bottom plate. In Figures 18-2D and 18-2E we see situations where both sets of plates are unbalanced, so the beam is deflected up and to the right in Figure 18-2D, and down and to the right in Figure 18-2E.

If a sawtooth waveform is applied between the horizontal deflection plates, then the beam will sweep left to right, forming a timebase (Figure 18-3). If a time-varying input signal is applied across the vertical plates, then the electron beam will trace the waveshape on the CRT screen.

Similarly, instruments called X-Y oscilloscopes apply input signals to both horizontal and vertical deflection plates to form a vector pattern that allows us to discern the relationship between them. Such diagrams are called *Lissajous patterns* (see Figure 18-4A). Color-TV repair shops, cable-TV operators, and TV broadcast studios use X-Y 'scopes, as do medical instrument manufacturers.

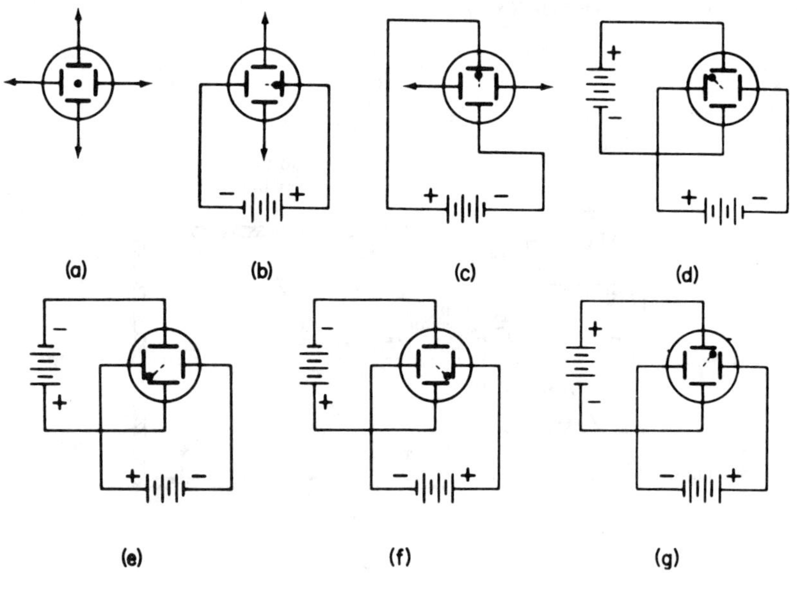

Figure 18-2 Response of light beam to different deflection plate situations.

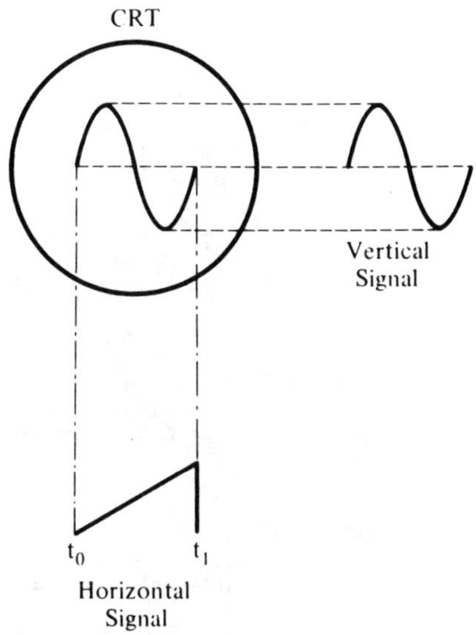

Figure 18-3 Using sawtooth on X-axis to make a time base.

You can measure audio frequencies with the Lissajous pattern. The ratio of the number of loops along the vertical and horizontal sides of the pattern tell you the relationship between frequencies applied to vertical and horizontal inputs (Figure 18-4B).

SOME OSCILLOSCOPE EXAMPLES

Figure 18-5 shows a low-cost oscilloscope intended for service shop applications. Like all modern oscilloscopes, this one is a triggered sweep model. This means that the beam does not start to sweep left-to-right across the CRT screen until a signal is present in the amplifier. By controlling the sweep triggering point we can control what part of the waveform is examined. Note that the CRT screen on the left side of the instrument has a graticule grid inscribed on the face. The divisions (usually 1 cm each) are used to measure voltages and times, which we will discuss further later.

Older model oscilloscopes were not triggered sweep. In those models, the horizontal deflection of the beam was asynchronous with respect to the input signal. The horizontal sweep controls were calibrated roughly in units of frequency, i.e., the number of times per second the beam swept left to right. Such instruments are of limited usefulness today. Most of them have a low-frequency vertical amplifier bandwidth, so they therefore cannot display many of the signals that are of interest.

The CRO in Figure 18-5 is a dual-trace model. These instruments were once rare and expensive and only available in the highest-priced model

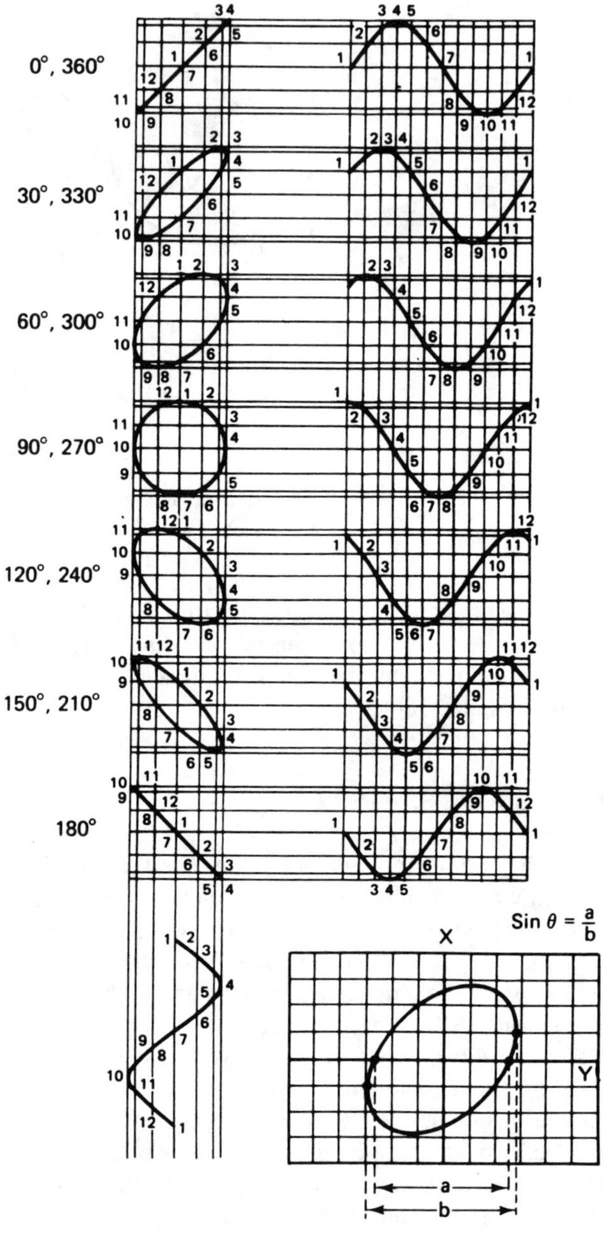

Figure 18-4 A) 1 : 1 Lissajous figure

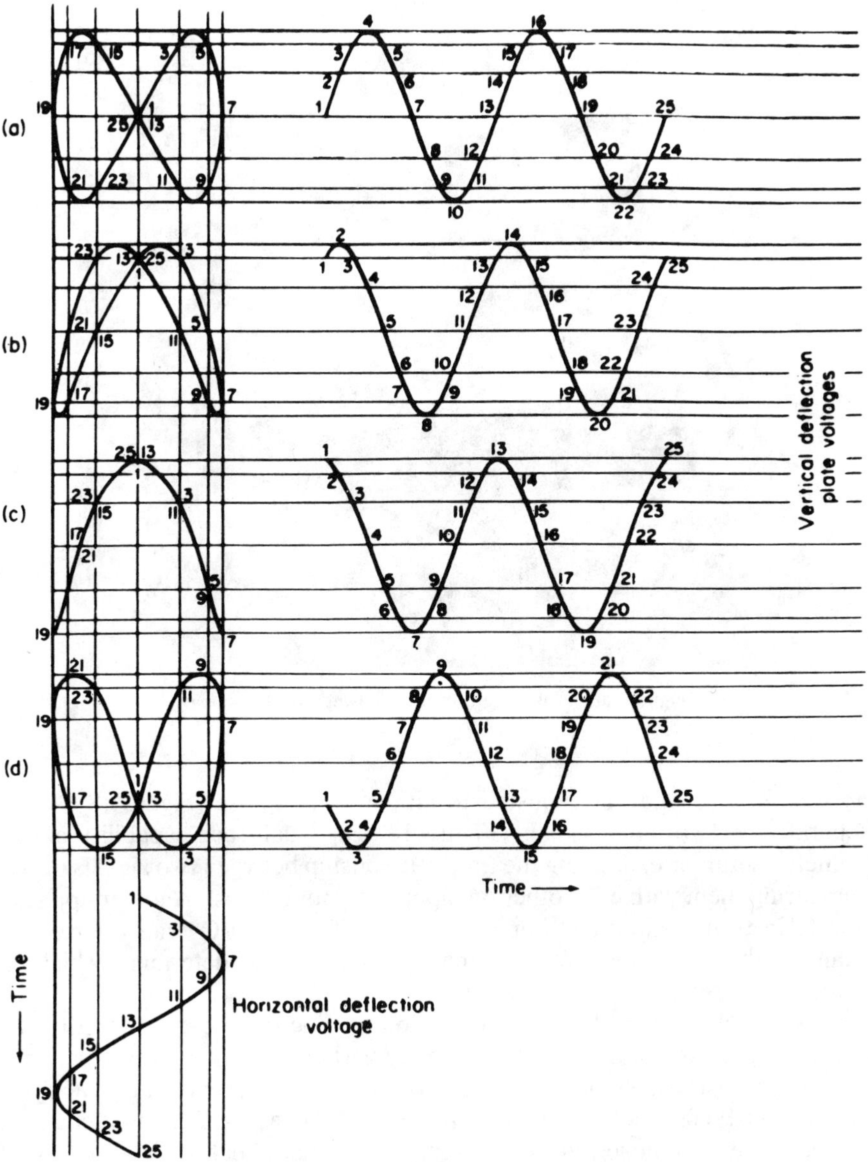

Figure 18-4 (continued) B) 2 : 1 Lissajous figures

lines. But today manufacturers can offer dual-trace models in their most modest lines, so the instruments are now quite common. There are two separate vertical inputs on these CROs, so two separate signals can be viewed simultaneously against the same time base (Figure 18-6). The two

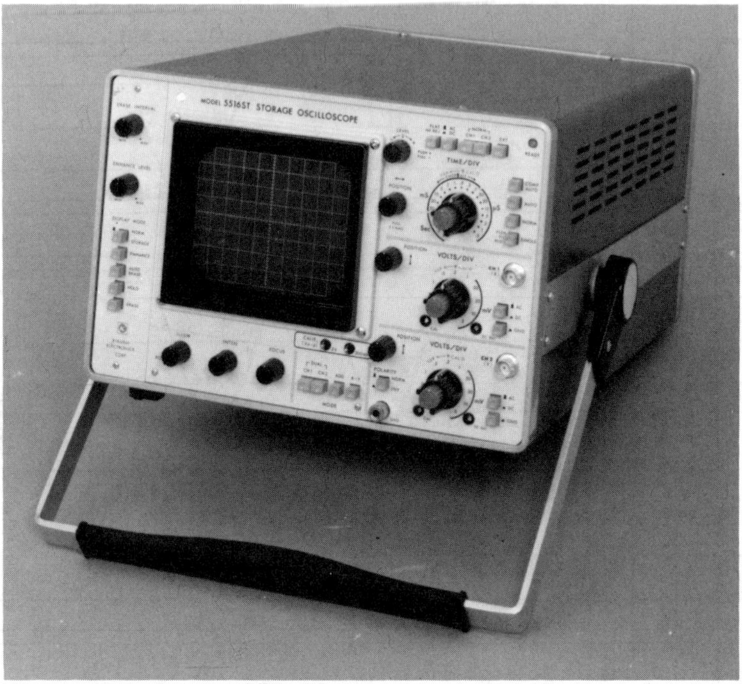

Figure 18-5 Dual-beam triggered sweep oscilloscope.

position controls can be adjusted to either superimpose (Figure 18-6A) or separate the two input signals (Figure 18-6B). Dual-trace capability is extremely useful for examining the time relationship between two signals or for comparing them with each other on a point-by-point basis. Another application is in troubleshooting. For example, dual-trace CROs can be used to examine the input and output signals of a circuit to determine which is malfunctioning.

Vertical bandwidth is a specification of much interest in selecting an oscilloscope. At one time, high-frequency vertical amplifiers were available only in the costliest engineering laboratory grade instruments. For example, in the mid-1960s one expected to pay \$4000 for a 35-MHz, two-channel oscilloscope (and those were 4000 "then-year-dollars" before inflation). But today we can buy 35-MHz 'scopes for less than \$1000 in inflated "now-dollars," and 100-MHz models are only about \$2300. As a rule of thumb that is valid for most cases, select a model with as high a vertical bandwidth as you can afford. It is difficult to go wrong with too much vertical bandwidth, but too little can cost you capability.

Now we will briefly describe typical controls found on a modern oscilloscope and give you some insight as to how they function. Keep in mind

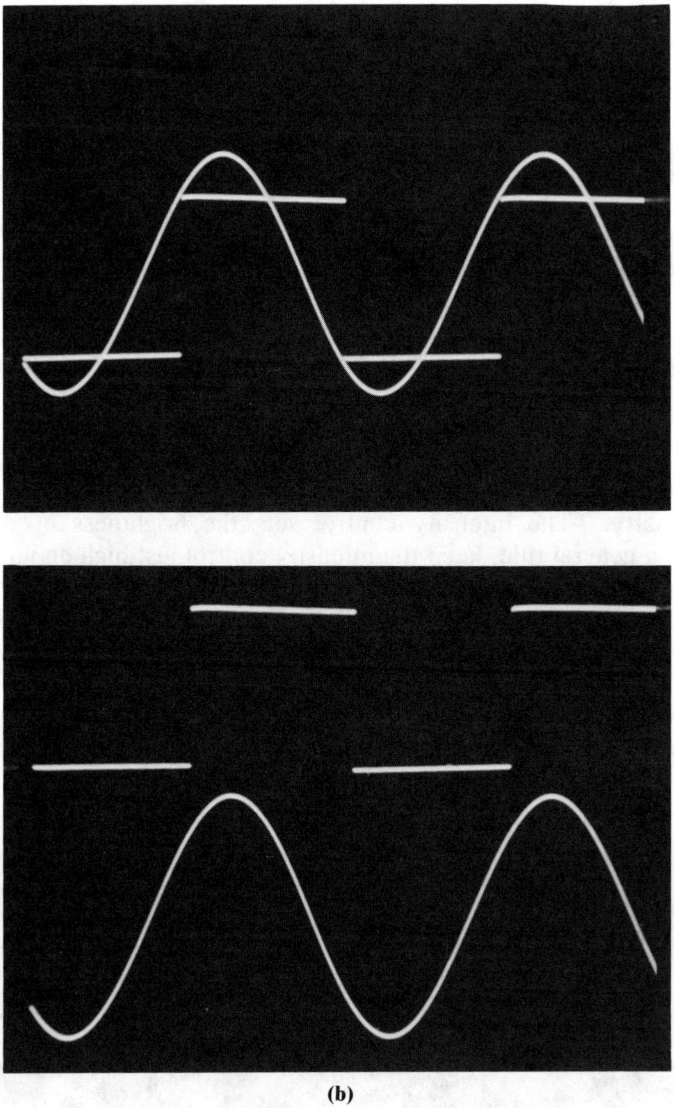

(b)

Figure 18-6 Operation of the position control: A) Signals superimposed, B) Signals separated.

that not all 'scopes have all of the controls, especially some of those in the specialized triggered sweep system that the selected model offers. Also, on some models the same controls may carry slightly different names on the labels but function pretty much the same. We will divide the controls into the following groups: *control, vertical*, *horizontal*, and *triggering*.

Control Group

The control group (some of which may appear on the rear of the 'scope), shown in Figure 18-7, includes the *on-off (power)*, *intensity*, *focus*, *astigmatism*, *illumination*, *trace rotation*, *beam finder*, and *internal calibration signals*. These controls operate as described in the following paragraphs.

On-off/power. The power control turns the instrument on and off; it is the AC mains power switch (or battery switch in portable models). In some 'scopes, the on/off switch is part of another control (usually intensity), similar to the on/off switch on a radio being ganged to the volume control. In other 'scopes, perhaps on most modern models, the on/off switch is a separate entity. Although on older models it may have been a toggle or rotary switch, on most 'scopes today it is a push button switch.

Intensity. The intensity control sets the brightness of the CRT beam. As a general rule, keep the intensity control just high enough to see the entire waveform comfortably. If the same waveform is expected to remain on the CRT screen for a long time (or there is no waveform), then keep the intensity low in order to prevent burning of the CRT phosphor.

Focus. This control adjusts the size of the electron beam spot on the CRT screen. It is sometimes interactive with the astigmatism control.

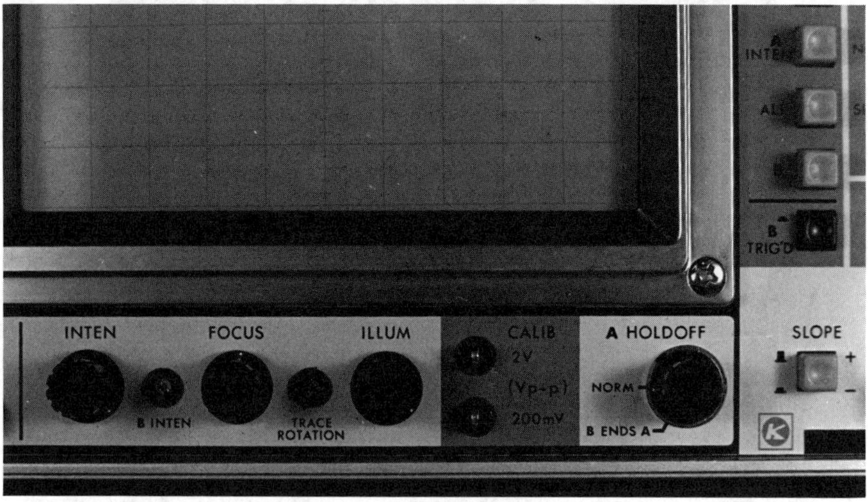

Figure 18-7 Control group panel.

Astigmatism. This control adjusts the "roundness" of the CRT spot, and is often interactive with the focus control. A good way to adjust the astigmatism control is to set it for a uniform line thickness when the CRT is swept horizontally (but no signal is present in the vertical channel). On many 'scopes, the astigmatism control is a screwdriver adjustment available on either the front or rear panel, while the focus control is knob-adjustable on the front panel.

Illumination. The illumination control adjusts a lamp that lights up the graticule lines inscribed on the CRT screen. At the lowest settings of this control no light appears on the graticule. Care must be used with this control when photographing the CRT screen, as the graticule lighting can overexpose the ASA 3000 film typically used in 'scope cameras.

Trace Rotation. Also called *trace align* on some models, this screwdriver adjustment compensates for the effects of local magnetic fields on the CRT trace. This control is adjusted so that the CRT beam is horizontal with respect to the graticule.

Beam finder. This control disconnects the horizontal and vertical inputs so that the beam collapses to a spot of higher intensity than the original trace. It helps the operator locate the beam, which may be off the screen at times.

Calibration. The calibration control provides a standard signal to calibrate the vertical amplifier controls. Typically, such signals are 400 or 1000 Hz, either 1 V or 2 V peak-to-peak. In some cases it is a sine wave, in others a square wave.

Vertical Group

The vertical group controls the vertical position of the beam, amplification factor, and input selection of the oscilloscope. Controls in this group (shown in Figure 18-8) include: *input, input selector, position, step attenuator, vernier* (variable attenuator), *ground, 5X magnification, channel-2 polarity,* and *vertical mode.* Also shown in Figure 18-8 is an internal trigger selector (INT TRIG), which is technically part of the trigger control group (its purpose is to select the channel that supplies the triggering signal).

Input connector. The input connector is the point at which a signal is applied from the external world to the oscilloscope vertical amplifier. In the model shown in Figure 18-8 the input connector is a coaxial BNC-style chassis-mounted connector. This form of connector is the standard on modern oscilloscopes. On older 'scopes, the input connector might be an SO-239

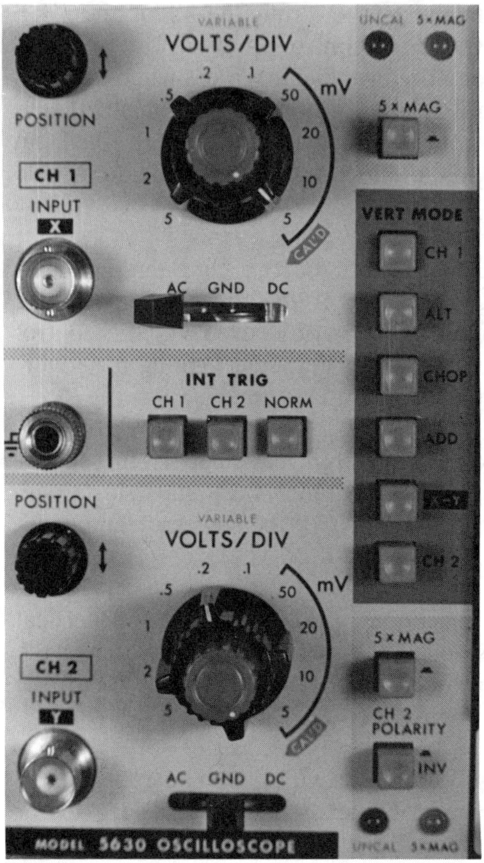

Figure 18-8 Vertical group control panel.

female UHF coaxial connector or even a pair of five-way binding posts spaced on 0.75-inch centers. Because most modern probes and 'scope accessories are now BNC-equipped, owners of older models usually opt to buy either an SO239-to-BNC or Post-to-BNC adapter (as appropriate).

Input selector. This switch is marked "AC-GND-DC" and is used to select the coupling of the input connector to the input of the vertical amplifier. The DC setting means that the connector is DC (direct current) coupled to the amplifier input; the AC setting means that a blocking capacitor is in series with the connector center conductor (hence only AC signals will pass—DC is blocked); in the GND setting the input of the vertical amplifier is shorted to ground and the input connector center conductor is open-circuited (GND does NOT ground the input connector, which would short-circuit the signal source!).

Position. The position control moves the electron beam up and down on the CRT face. On a dual-trace 'scope the two vertical position controls are normally used to prevent overlapping of the two traces (and resultant confusion). Otherwise, the control can be used to precisely position the trace over the graticule markings for amplitude measurements. It is common practice, for example, to set the AC-GND-DC input selector to GND and then use the position control to set the straight-line trace over either the bottommost or center graticule line, which then becomes the zero-volts reference point.

Step attenuator. The sensitivity of the oscilloscope amplifier is a measure of its gain and is expressed in terms of the voltage required to deflect the CRT beam a specified amount, i.e. volts/division (or volts/cm). The step attenuator (see close-up in Figure 18-9) is a resistor/capacitor voltage divider that allows the instrument to accommodate higher potentials that would otherwise over-deflect the CRT beam. Each position of the step attenuator is calibrated in volts or millivolts per division (V/Div or mV/Div). The actual peak-to-peak voltage measure of the input signal is made by noting how many CRT screen divisions the signal occupies and then multiplying that figure by the sensitivity factor in V/Div. For example, suppose a sine-wave signal occupies 5.6 divisions peak-to-peak when the step attenuator is set to 0.2 V/Div. The peak-to-peak voltage is (5.6 Div) × (0.2 V/Div) = 1.12 V (i.e., 1,120 mV).

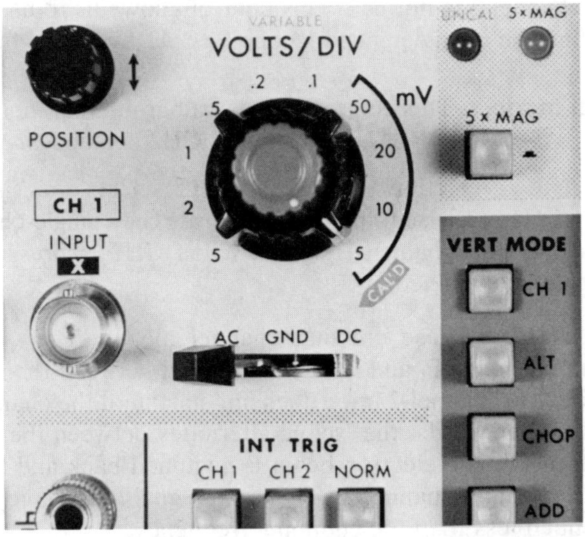

Figure 18-9 Vertical attenuator.

Vernier attenuator. This variable control is concentric to (in the center of) the step attenuator, and it allows continously variable adjustment of the sensitivity factor (hence also the trace vertical size). The calibration of the step attenuator is valid only when the vernier attenuator is in the CAL'D (also called CAL) position, which on most 'scopes is detented for easy location. When the vernier control is not in the CAL'D position, a red UNCAL lamp on the front panel warns the operator that the step attenuator settings are not to be trusted.

Ground. This ground jack is connected to the chassis ground at the input of the vertical amplifiers. It can be used to provide a proper "star" ground in order to eliminate (or prevent) ground loop errors.

5X mag. The 5X magnification control increases the sensitivity factor by five times, which means that all of the V/Div and mV/Div calibrations must be divided by five. For example, when the "VOLTS/DIV" knob is in the 50 mV/Div position and the "5X MAG" button is pressed, a 5X MAG light turns on to warn the operator, and the sensitivity increases fivefold (i.e., to 10 mV/Div in the example case). This feature is especially useful when dealing with low-level signals that are ordinarily below the threshhold of the normal settings, so it effectively doubles the number of available sensitivity factors.

Channel-2 polarity. The polarity control inverts the channel-2 vertical signal when pressed. If left unoperated, the polarity of the signal on the screen from channel-2 is normal. This control allows us to have a pseudo-differential input on a single-ended 'scope (see ADD control below).

Vertical mode. This control forms a subgroup that includes the following: CH1, ALT, CHOP, ADD, X-Y, and CH2 submodes.

CH1, CH3. Selection of CH1 or CH2 selects the single-channel mode. When CH1 is pressed, the 'scope operates as a single-channel model and displays only the channel-1 signal. When CH2 is pressed, only the channel-2 signal is examined.

ALT, CHOP. These are dual-channel modes. There is only one electron beam in the CRT, and it must be shared between the channels. In the ALT mode, the channel-2 trace does not start until the channel-1 sweep is finished. In other words, the 'scope alternates between the two signals. In the CHOP mode, the electron beam is switched back and forth rapidly between channel 1 and channel 2. The input signal must have a frequency that is very much less than the chopping frequency.

ADD. With ADD, the signals are combined into one, with the resultant amplitude being the algebraic sum of the two channels (CHl + CH2). If the CH2 polarity control is pressed, then the inputs become pseudo-differential and the summation is CH1 − CH2.

X-Y. In this mode the internal oscilloscope time base is disconnected and the instrument becomes a vectorscope. Channel 1 becomes a horizontal (X) input, while channel 2 is the vertical (Y) input. In this mode, the oscilloscope can be used for modulation measurements, color-TV Lissajous patterns, and so forth.

Horizontal Group

The horizontal control group (Figure 18-10A) determines the horizontal deflection and sweep characteristics. These controls consist of *sweep time, sweep vernier, horizontal position, 10X mag,* and *sweep mode.*

Sweep time. This is the main horizontal timing control, and it is used to determine the amount of time required per division to sweep the beam across the CRT face left-to-right. The calibration of this control is in units of time/division (s/Div, mS/Div, or μs/Div). The period of a signal can be determined from this control and the number of divisions occupied by one cycle of the signal. For example, if exactly one cycle of a sine wave occupies 6.2 divisions of the horizontal graticule, and the switch setting is 2 ms/Div, then the period of the signal is (6.2 Div) × (2 ms/Div) = 12.4 ms. Because frequency is the reciprocal of period, we can calculate the frequency as F = 1/T = 1/0.0124 = 80.65 Hz.

Sweep vernier. This is a continuously variable time control that allows us to interpolate between step time settings. The step time settings are accurate only when the vernier is in the CAL'D position. The vernier is ganged to the step attenuator and is thus concentric to and in the center of that control.

Horizontal position. The horizontal (or "fine") position control moves the trace left and right on the CRT screen. Like the equivalent vertical control, the horizontal position control is used to place key features right over graticule points for purposes of precision measurement.

10X mag. The 10X magnification control speeds up the sweep 10 times. For example, if the time/div sweep control is set to 10 ms/Div, then the 10X mag control would force it to become 1 ms/Div.

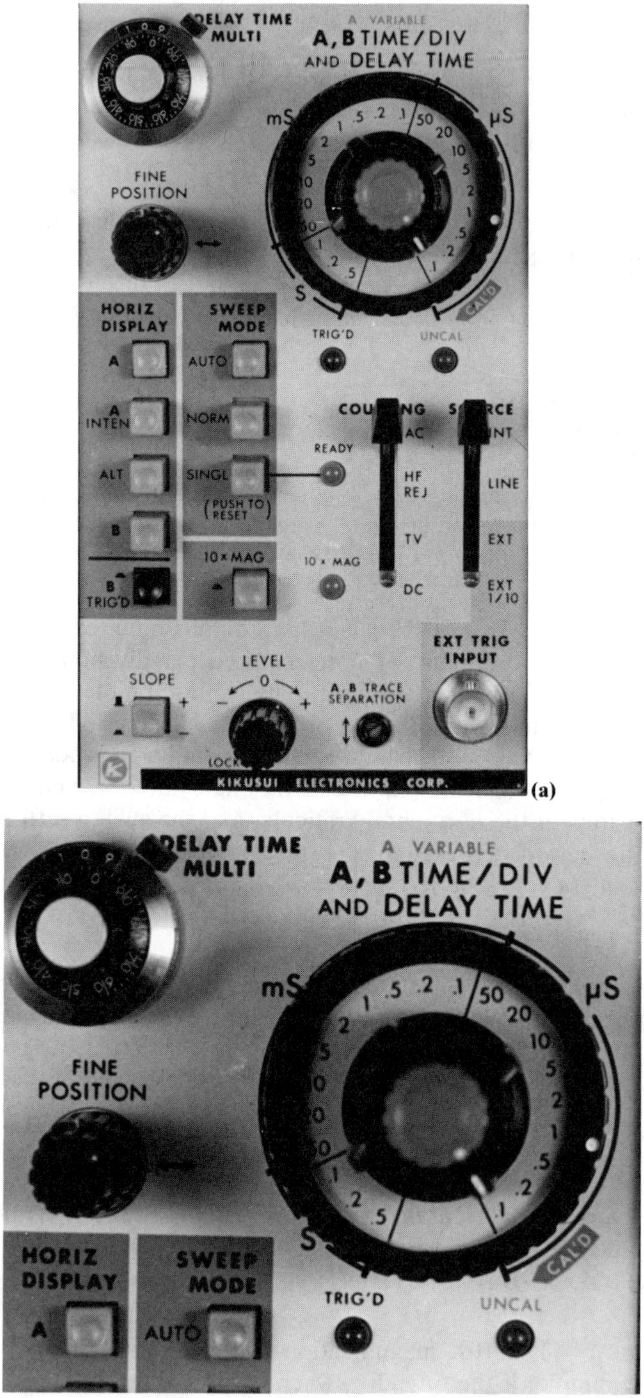

Figure 18-10 A) Horizontal/trigger control group, B) Horizontal sweep speed control.

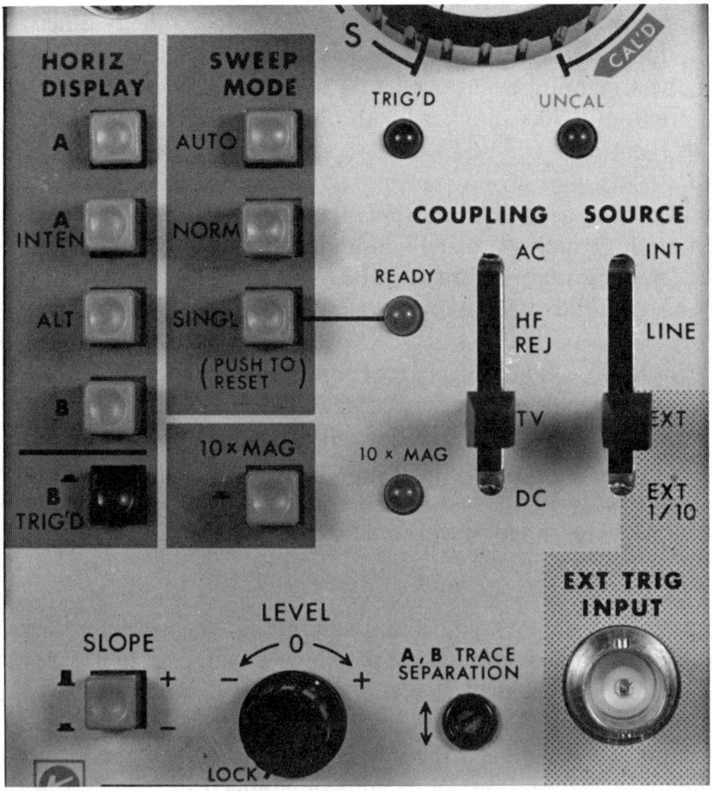

(c)

Sweep mode. The sweep mode control selects automatic (AUTO), normal (NORM), and single sweep (SINGLE) submodes. In the AUTO mode, the sweep will periodically retrigger even if no signal is present in the vertical amplifier. The NORM mode requires a vertical signal to begin sweeping the CRT, and the screen will remain blank otherwise. In the SINGLE mode, the CRT beam will sweep only once. Two means of operation are noted. If there is a periodic signal present, pressing the button in the AUTO position will force one sweep to take place. If the NORM mode is selected, then the SINGLE button will reset the circuit, which will sweep only after a valid input signal is received.

Trigger Group

The triggered sweep oscilloscope is considerably more useful than the old untriggered forms. The triggered sweep 'scope will not allow the CRT beam to sweep across the screen unless a signal is in the vertical amplifier to trigger the sweep generator. Some models also allow a delay time triggered sweep function. That is, the sweep will not actually begin until some preset time after the triggering event occurs. The trigger group of controls is also shown in Figure 18-10A along with the horizontal controls. The controls include: *trigger level* (*LEVEL*), *slope, source, coupling, external trigger input,* and a *horizontal display* selector. In addition, some models (such as in Figure 18-10A) also have a *time delay vernier* control. Keep in mind that the sweep mode, which we discussed under the horizontal controls, is also part of the trigger group (Figure 18-10B), as is the CH1-CH2-NORM switch shown along with the vertical group controls.

Trigger level. The trigger level control determines the minimum amplitude vertical signal required to trigger the horizontal sweep. Figures 18-11A and 18-11B show the effect of this control. Both traces were taken moments apart with the same signal input to the same oscilloscope. The only difference in these displays was the setting of the LEVEL control (see Figure 18-11A). The range of the control runs from negative, through zero, to positive voltage values.

Slope. This control determines whether the trigger occurs on a negative-going or a positive-going edge of the input waveform. In Figures 18-11A and 18-11B the slope control is set to the "+" position, so the triggering occurred on the positive-going edge of the sine wave. In Figure 18-11C, on the other hand, we see exactly the same signal with the level control set as it was in Figure 18-11B and the slope control changed to "−." Note that the triggering now occurs on the negative-going slope of the waveform.

Source. The source control selects the source of the signal applied to the triggering circuits. The selections are INT, LINE, EXT, and EXT/10. The INT is the internal selection, which means that the source is selected by the CH1/CH2/NORM switch in the vertical section. For example, with the source control in INT and the other switch in CH1 position, the signal in the CH1 vertical amplifier will cause triggering. The LINE selection means that the 60-Hz AC line will cause triggering, a feature useful in some measurements. The EXT means that the signal applied to the external trigger input (EXT TRIG INPUT) will trigger the sweep circuits. Again, some useful measurements are possible with this feature. The EXT/10 is the same as EXT, but a 10 : 1 attenuator is in place.

Coupling. The coupling control allows us to tailor the triggering, and it has selections of: AC, HF REJ, TV, and DC. The AC and DC are self-explanatory and are similar to the same markings on the vertical selector switch. The HF REJ selection uses a low-pass filter at the input of the trigger circuit that rejects high frequencies. This system will, for example,

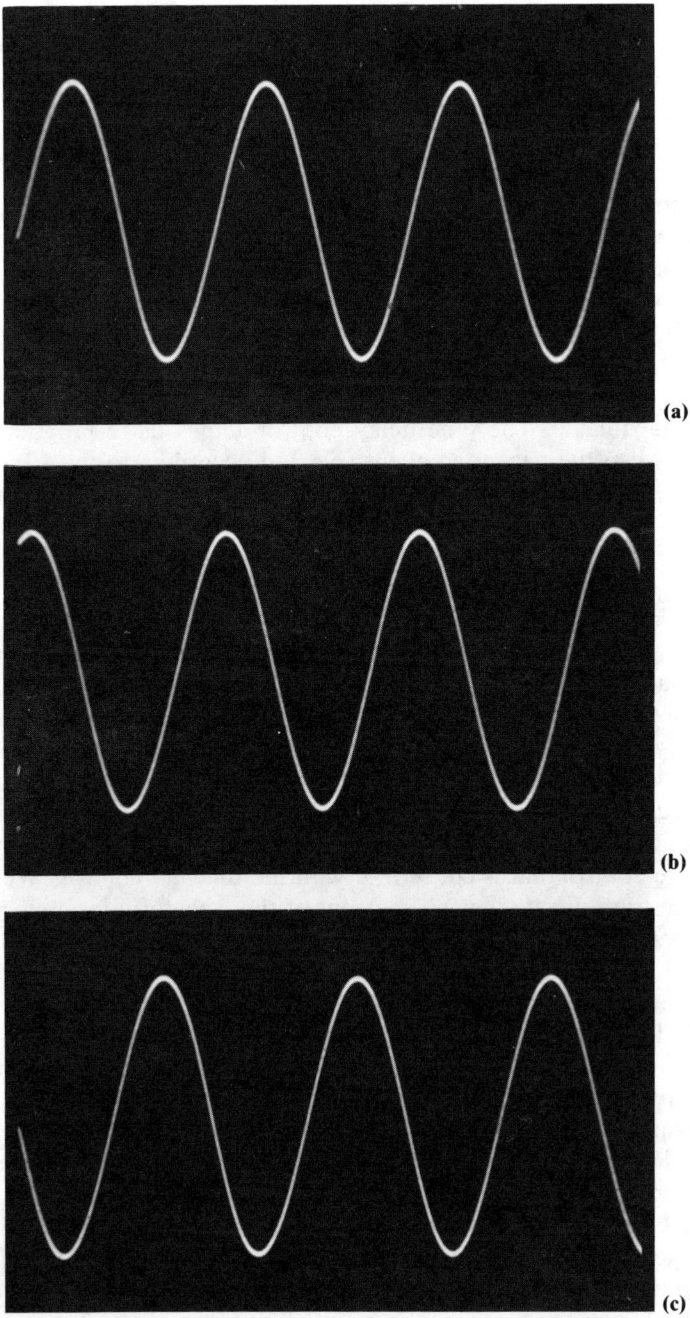

Figure 18-11 Effect of various trigger control settings: A) Level = 0, slope = (+), B) Level > 0, slope = (+), C) Level = 0, slope = (−).

allow us to trigger on the modulation of a modulated RF carrier while ignoring the RF signal. Some 'scopes also have a LF REF, which is similar except that the filter is a high-pass filter. The TV selection allows us to sync to the horizontal and vertical frequencies used in television sweep systems. Some models have separate TV VER and TV HOR selections.

External trigger input. The external trigger selection is an input connector that provides the trigger circuit with an external signal for special-purpose syncing and triggering. Some 'scopes also have a TRIG GATE function on this connector in certain switch selections (or a separate TRIG GATE output). That feature produces a short-duration pulse for synchronizing external circuits to the sweep system.

Time delay. The time delay control allows us to program a short delay between the triggering event selected according to the LEVEL and SLOPE controls and the actual onset of the sweep. Using this control we can view small segments of the waveform, while using the main signal as the trigger event.

Horizontal display. The HORIZ DISPLAY is a switch bank (Figure 18-10B) that allows certain submodes. Not all 'scopes have this feature, even though it is very useful. Figure 18-12 shows the operation of certain features of this selector. When button "A" is pressed, the 'scope operates as any triggered sweep 'scope operates. But in the A INTEN mode, we see a trace such as Figure 18-12A. Note the segment of the waveform that is intensified. The position of this intensified segment is a function of the time delay control, while the length of the intensified portion is a function of the delay time control that is concentric with the time/div control. We can use this mode to designate a small segment of the waveform for a closer look.

When the "B" switch is pressed, that portion of the waveform is displayed, as in Figure 18-12B. A slightly different function is shown in Figure 18-12C, which is the trace that results when the ALT button is pressed. In this case, we see both the main waveform and the time-delayed "close-up" portion. A screwdriver "A-B separation" control allows us to either separate or superimpose these waveforms.

Figure 18-13A shows the most basic form of input probe for oscilloscopes. Here we see a length of shielded cable, usually coaxial cable, with a BNC (banana plugs or PL-259 on older instruments) on one end and a pair of alligator clips on the other end. This method works well for signals with frequencies from DC up to a certain point, and for many readers this probe set is all that is required. But there is a problem that must be recognized. The cable has capacitance on the order of 20 pF per foot. The input impedance of a typical oscilloscope is a 1-MΩ resistance shunted with a 20-pF capacitance. If the cable is 3 feet long, then it has a capacitance of 3×20

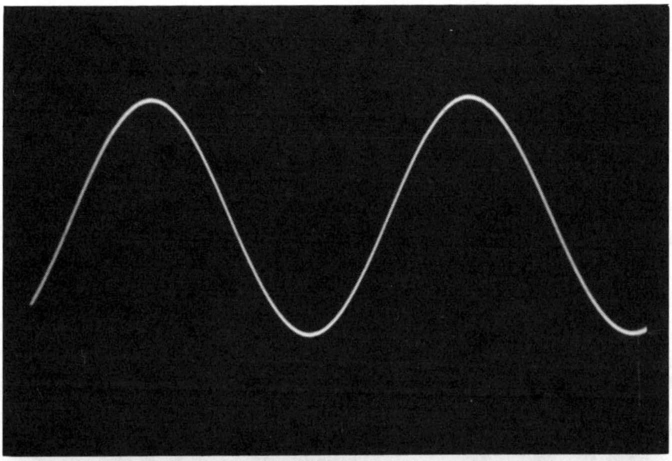
(a)

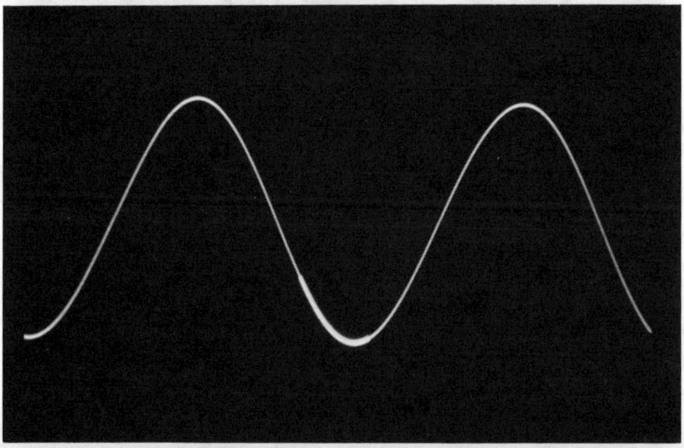
(b)

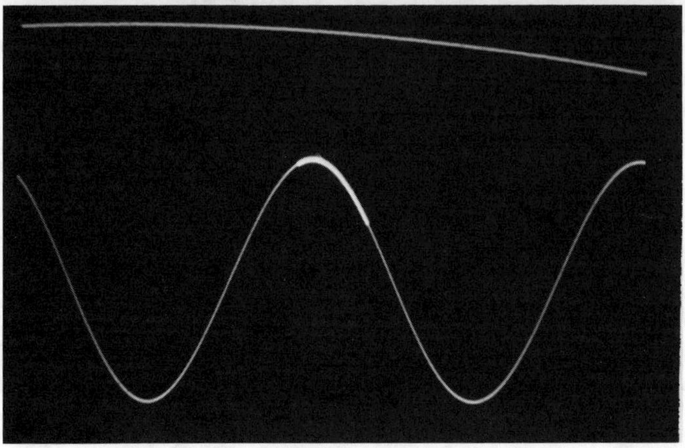
(c)

Figure 18-12 A) Nonintensified input signal, B) A-intensified, C) Expanded.

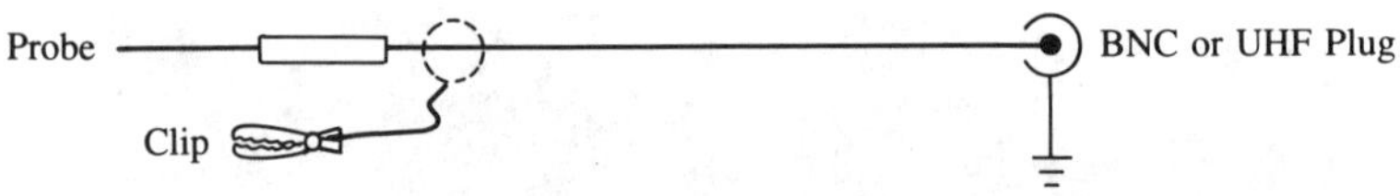

(a.1)

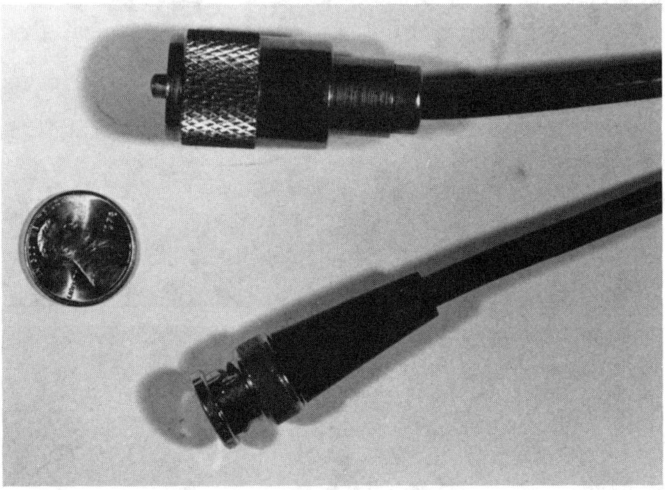

(a.2)

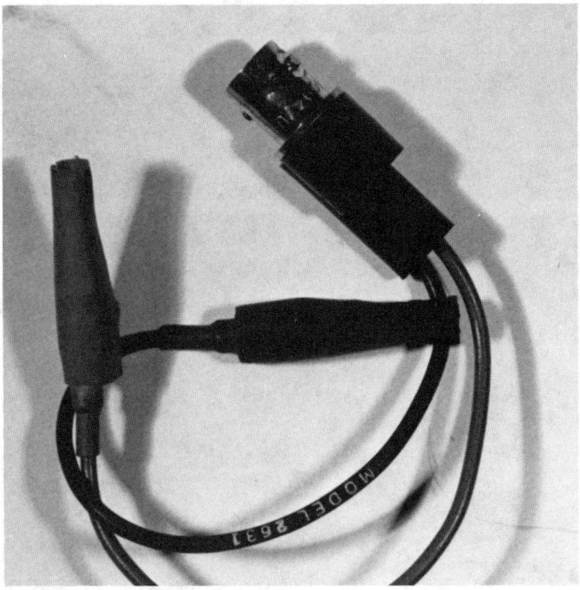

(a.3)

Figure 18-13 A) Standard 1 : 1 probe: A.1) Circuit, A.2) Connectors, A.3) Possible end-connector.

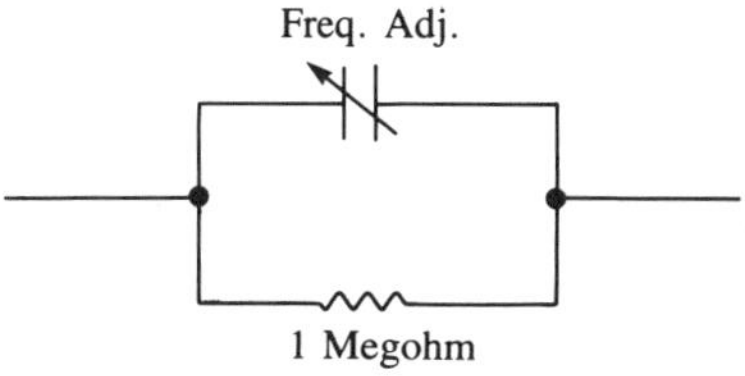

Circuit

(b.1)

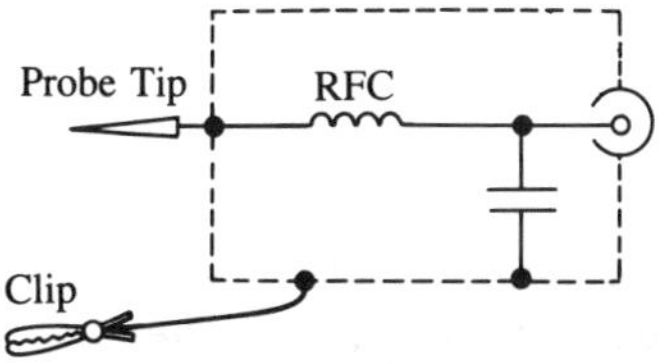

(c)

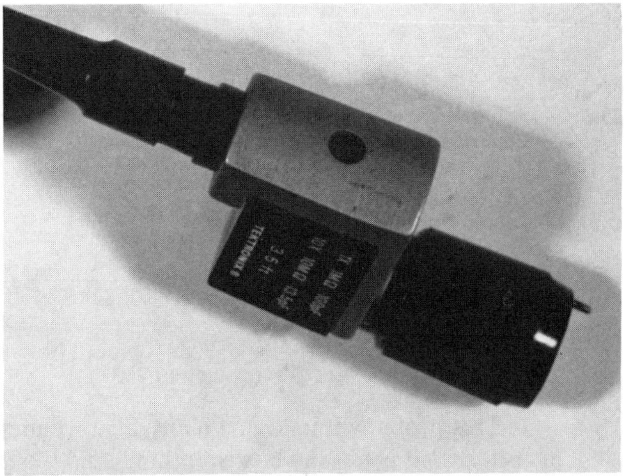

(b.2)

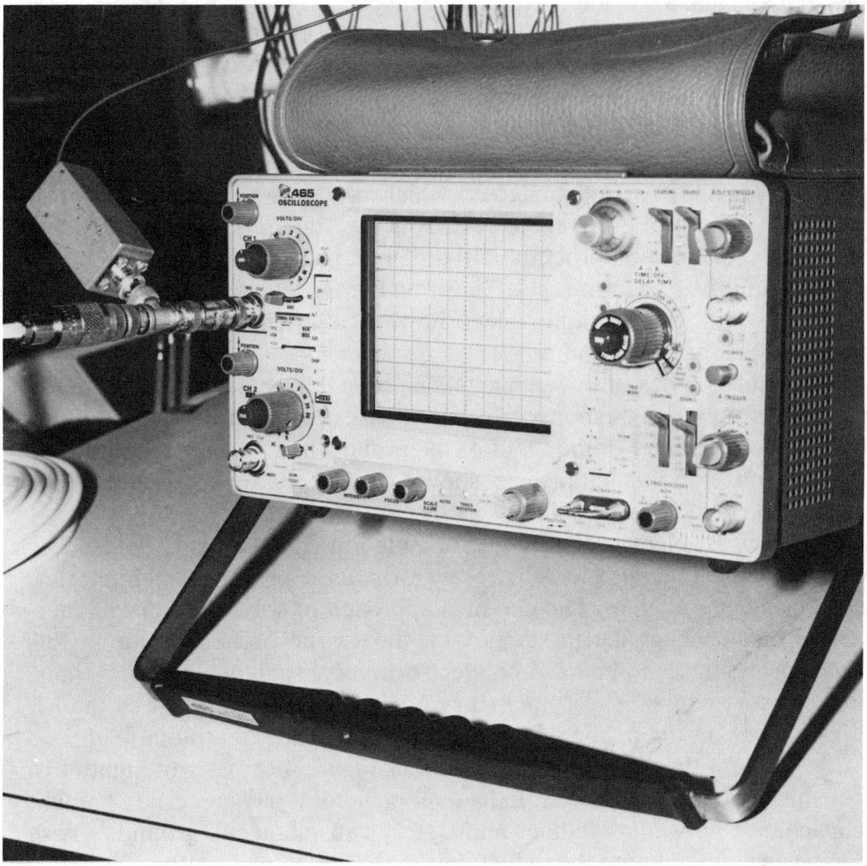

(d)

Figure 18-13 (continued) B) 10 : 1 probe, B.1) Circuit, B.2) Actual probe connector; C) RF probe; D) Special TDR RF probe.

pF, or 60 pF, which when added to the natural input capacitance results in 80 pF shunting 1 MΩ. The RC network thus created has a low-pass filter characteristic that rolls off −6 dB/octave above a −3 dB frequency of

$$F = \frac{1}{2\pi RC}$$

$$F = \frac{1}{(2)(3.14)(10^6\ \Omega)(8 \times 10^{-11}\ F)} \tag{18-1}$$

$$F = \frac{1}{0.0005} = 1{,}990\ Hz$$

This probe will load down any high-frequency circuit that it is used to measure, so it is not the best solution. And the fundamental frequency need not be anywhere near the cutoff frequency for there to be problems. Nonsinusoidal signals are made up of a collection of sine waves consisting of a fundamental plus harmonics. Thus a 100-Hz fast rise-time square wave is made of a 100-Hz sine wave plus even harmonics up to the zillionth or so. The low-pass filter effects of the probe in Figure 18-13A will roll off the higher harmonics and round off the shoulders of the square wave.

The answer to the frequency response problem is to use a low-capacitance probe, two examples of which are shown in Figures 18-13B and 18-13C. The probe in Figure 18-13B is the standard 10 : 1 ratio probe. The output signal of this probe is 1/10 the input signal. If the resistors used are precision types, then the scale factor on the 'scope vertical attenuator is multiplied by 10. For example, when the vertical attenuator is set to 0.5 V/cm, the actual scale factor is 5 V/cm.

In all three types of low-capacitance probe the capacitor is adjusted to flatten the frequency response. In most cases a fast rise-time 1000-Hz square wave is applied to the input of the probe when it is connected to the 'scope. Adjust the capacitance to show as square a square wave on the screen of the 'scope as possible.

Another problem is the matter of isolation from external fields. The classic problem is taking a look at a waveform in the presence of an interfering electromagnetic field. The classical approach to solving this problem is insertion of an RF choke in series with the 'scope probe. Figure 18-13C shows a probe that can be used on electrosurgery machine and radio transmitter measurements. The 1-millihenry (mH) RF choke suppresses the RF that is present on the probe when it is in the presence of a radio field.

A problem that exists on the probe in Figure 18-13C is the matter of self-resonance. All RF chokes, indeed all inductors, have a certain amount of capacitance between windings and a stray capacitance to ground. These capacitances interact with the inductance of the coil to make either (or both!) series or parallel resonances . . . and that spells trouble in some cases.

A different kind of oscilloscope input device is shown in Figure 18-13D. Certain RF and computer measurements require special adapter devices to make the oscilloscope work. This particular adapter is a device used in time domain reflectometry, a method for "doping out" coaxial cable transmission lines such as those used to interconnect the elements of the receiver and antenna system of the cardiac telemetry system in a hospital.

MAKING MEASUREMENTS ON THE 'SCOPE

The standard oscilloscope display (Figure 18-14) is calibrated in two axes: vertical and horizontal. The horizontal axis is calibrated in units of time, typically time per division (most modern 'scopes use the centimeter as the division). The vertical axis is the amplitude (usually voltage) of the applied signal. The graticule shown in Figure 18-14 has two forms of divisions on the screen. The major divisions are each 1 cm, while the minor divisions are 0.2 cm each. These minor divisions are inscribed only on the center vertical and horizontal axes. In the example of Figure 18-14, the vertical displacement is approximately 4.4 divisions and the horizontal is 3.25 divisions.

Medical oscilloscopes used in ECG and similar applications are often calibrated in units of distance and time. For example, the standard external lead ECG is usually calibrated at 25 millimeters per second (mm/s). Other instruments will be calibrated at 50 mm/s or 100 mm/s. Some multipurpose instruments are available with all three calibrations. The graticule on many

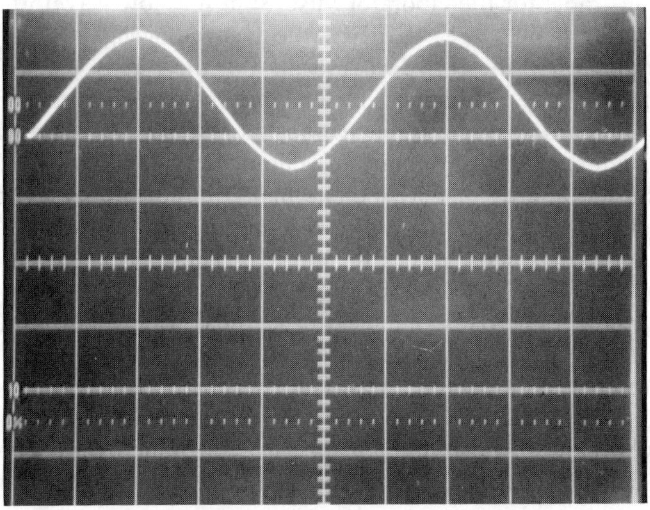

Figure 18-14 Oscilloscope grid with a signal.

medical 'scopes is inscribed with millimeter marks, and the major divisions represent 5 mm each. This calibration makes it easy to measure ECG feature parameters.

A newer form of oscilloscope now on the market is intended for use with computer graphics displays. These 'scopes are often calibrated in the units being measured, distance, or some other physical parameter. The computer will inscribe the CRT screen with a calibrating scale or will directly label the features as to calibration.

BIOMEDICAL AND LIFE SCIENCES OSCILLOSCOPES

In the sections to follow we will take a closer look at oscilloscopes designed specifically for medical, biological, and other low-frequency applications. There are three types of 'scope typically found in the medical world.

First, there are the straight analog "bouncing ball" models. The nickname "bouncing ball" comes from the fact that the beam of light seems to bounce as the ECG waveform parades across the 'scope face. These 'scopes are medical variants of the straight oscilloscope discussed earlier, so they will be discussed only briefly.

While the medical models tend to have magnetic deflection of the CRT electron beam (via a CRT neck yoke), they are otherwise very similar. The phosphor on the CRT screen is also a bit different. The regular scope will probably use a P1 or P4 phosphor. These phosphor mixtures are relatively fast, so they fade immediately after the electron beam passes a point. This feature is desirable for the laboratory user, but for slow waveforms such as the ECG, arterial pressure, or EEG, the fast-fading light beam does not allow examination of the entire waveform. For these reasons, a P7 or similar "long-persistence" phosphor is usually selected by the 'scope designer.

The standard long-persistence phosphors have an interesting property. The light emitted is a mixture of yellow, green, and violet-blue colors. The standard unfiltered display looks violet-colored to most people (especially under flourescent lighting). But if a filter is used over the CRT screen, then either green or yellow colors are emitted. These colors are a bit dimmer (especially the yellow) than the unfiltered versions.

The second type of 'scope used in medical applications is the analog storage 'scope. These instruments use trickery inside the CRT to store the signal on the face of the CRT for a long period of time. These instruments are used in older vectorcardiogram 'scopes and other applications. The modern approach, however, is to use a digital storage oscilloscope. These are the so-called "nonfade" 'scopes seen in medicine (of which more will be said later).

Analog Storage Oscilloscopes

On a regular oscilloscope the input waveform must be periodic for the display to remain stable on the screen. Further, each repeated cycle must be identical to the previous cycle. But in medical oscilloscopes, the waveform might be periodic, aperiodic, or episodic. In addition, in cases where there is a periodic rhythm (as in an ECG), the successive waveforms might not be identical (and the differences may be clinically significant). In all of these cases a single event or feature might pass too rapidly to be viewed properly on a bouncing ball display unless the 'scope is equipped with a storage feature. There are two forms of storage 'scope: analog and digital. In this section we will take a look at the analog form.

The analog storage 'scope uses a special CRT; it is shown in block diagram form in Figures 18-15A and 18-15B. A storage 'scope retains the image on the screen for a period of time before it eventually fades out. In the class of instruments using special CRT designs, there are several subclasses that operate on slightly different principles. Figure 18-16 shows three different types of storage CRT.

The two types of CRT shown in Figures 18-16A and 18-16B depend for operation on a phosphor screen in which individual particles of phosphor are insulated from each other. In the case of Figure 18-16A the phosphors lie in the same plane to form target dots, while the other type of CRT (Figure 18-16B) uses layers of scattered particles of increasing weights. A special flood gun in the CRT emits high-energy electrons that excite the phosphors. The phosphor particles struck by these electrons take on a charge of 150 to 200 V, but unenergized particles remain at 0 V. When electrons from the main electron gun have preenergized certain phosphors, forming an image on the CRT screen, these phosphors attract more flood gun electrons. Erasure of the screen is accomplished by grounding the phosphor screen, thus returning all phosphor particles to a 0-V potential.

The other type of special CRT is the wire-mesh variety shown in Figure 18-16C. The flood gun charges one mesh so that no further electrons can pass through, although the write gun or main gun electrons will pass through if sufficiently energetic (a function of the high-accellerator-anode voltage).

A split-screen storage oscilloscope allows the operator to store waveforms in either the top or bottom half of the screen. If both halves are turned on, then the device will store waveforms in any portion of the screen. This feature allows subsequent waveforms (for example, a before and after condition) to be examined together even though separated in time.

Some storage 'scopes, which find wide application in medical electronics, have a feature called variable persistence, which allows the operator to vary the length of time that the image will remain on the screen.

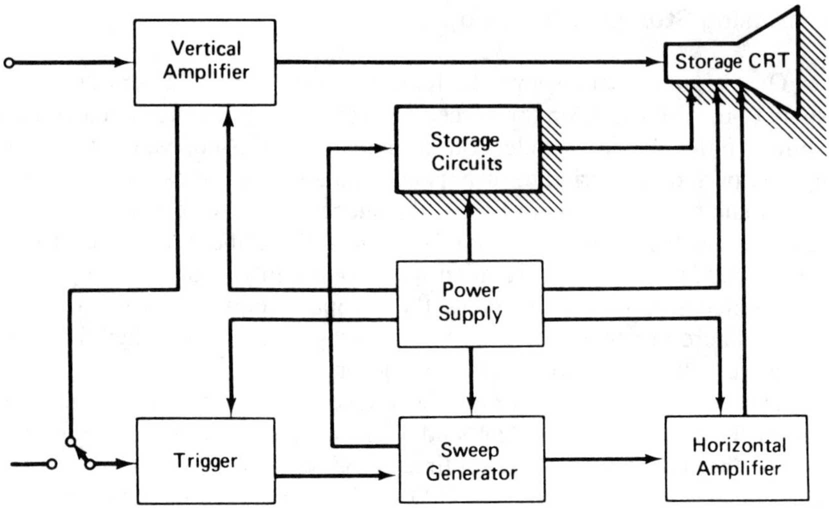

(a)

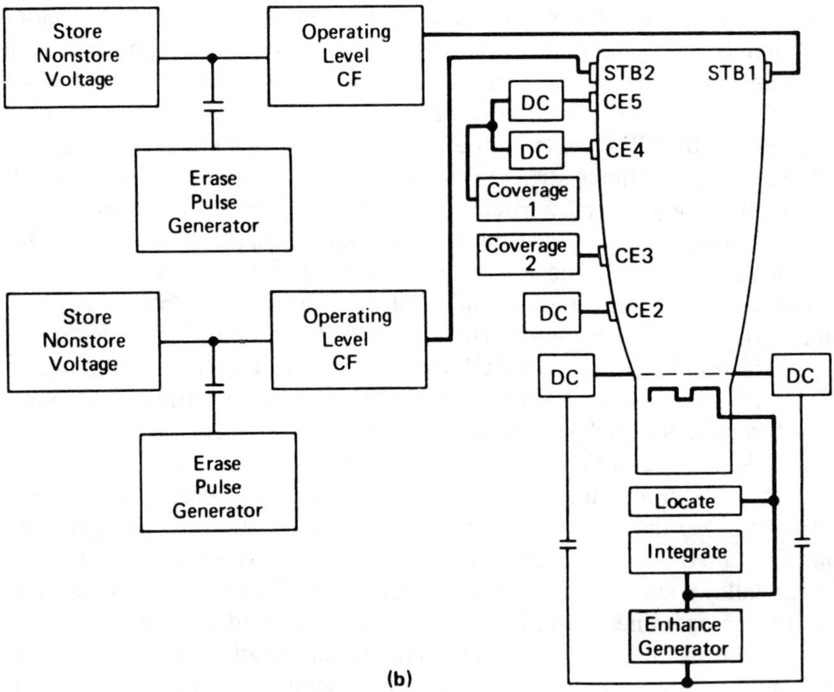

(b)

Figure 18-15 Function of a typical analog storage-type oscilloscope: A) Block diagram, B) Subdivision of storage circuits.

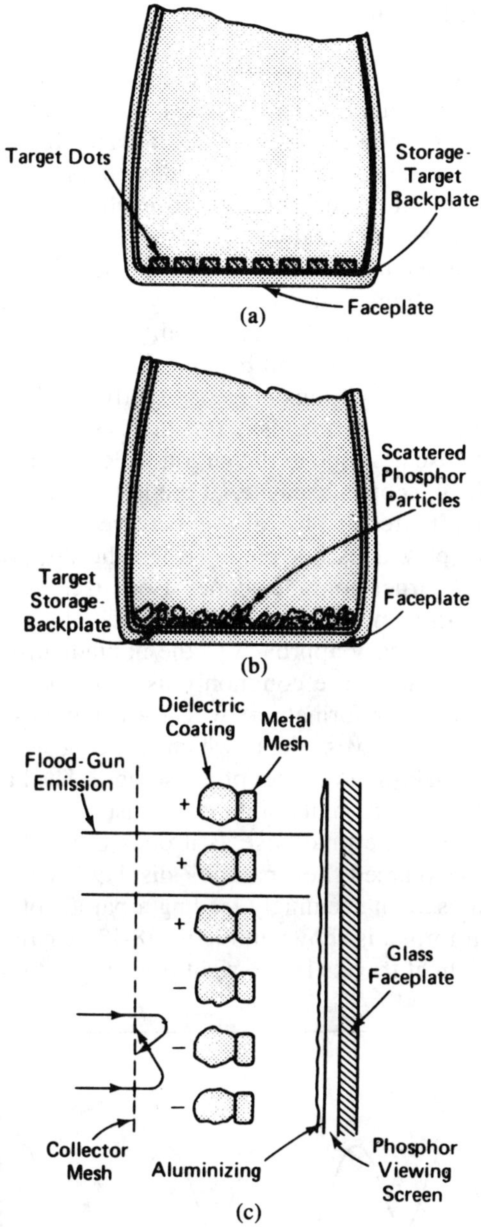

Figure 18-16 Three types of storage CRT: A) Target dot, B) Scattered phosphor, C) Wire mesh.

Nonfade Oscilloscopes

The traditional form of oscilloscope uses a beam of electrons to sweep the screen, writing the analog waveform as it is deflected. Even with long-persistence phosphors, however, the trace vanishes from the CRT shortly after it is written onto the screen. To the medical personnel using this sort of bouncing ball display, it is very difficult to evaluate waveform anomalies because the trace fades too rapidly.

The analog storage oscilloscope is a partial solution to the problem for some research applications, but it is generally not suited to monitoring and other clinical applications. A problem with the analog storage 'scope is that the trace tends to "bloom" out and become fuzzy a few minutes after it is taken . . . or if circuit conditions are not exactly right.

The solution to these problems is the digital storage oscilloscope, also called the non-fade oscilloscope in the jargon. CRT storage systems are not used in most medical 'scopes because, at the low frequencies involved, the digital types offer a better display at competitive prices. Also, the digital type of nonfade display does not bloom when the display is either old or erased. While most earlier nonfade scopes used discrete digital logic (as do many today), modern computer techniques (which are all but ubiquitous in medicine now) offer an easy approach to the nonfade display.

Two different formats are commonly used: *parade* or *erase bar* (see Figure 18-17). With parade format, the waveform marches across the screen from right to left. The newest data, which is being written in real time, appear in the upper right-hand corner of the screen. The light beam bounces up and down at a fixed horizontal point in response to the vertical waveform; it does not move along the time base as it does in regular analog 'scopes. The waveform is nicknamed the "parade" display because the oldest data marches across the screen seemingly leading a parade of waves.

The erase bar format is shown in Figure 18-18. On this type of nonfade 'scope the beam of light travels left to right (it is not stationary as on parade

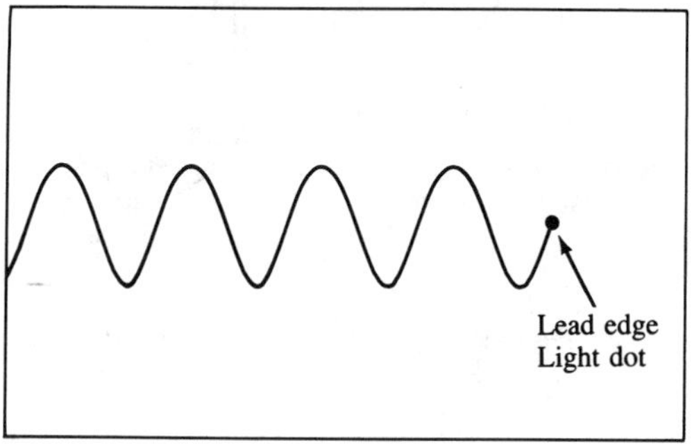

Figure 18-17 Parade-type nonfade display.

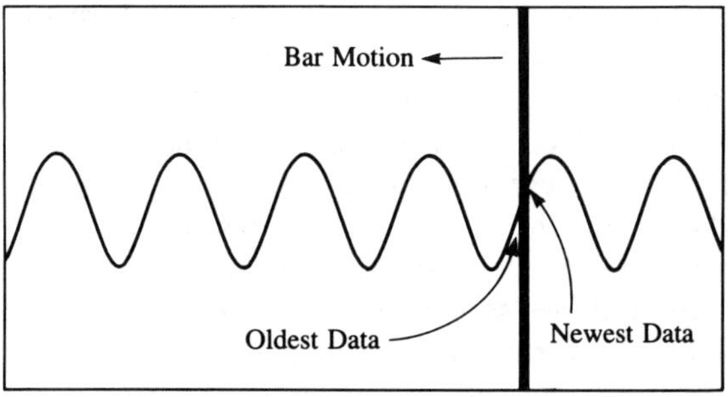

Figure 18-18 Erase-bar nonfade display.

models). There is an erase bar (i.e., a dark region) traveling ahead of the beam that obliterates the oldest data so that the new data can be written onto the CRT screen.

There are two forms of digital circuit for nonfade oscilloscopes. Figure 18-19 shows a complex form that uses a regular computer memory, as might be found in a personal computer. The signal is applied to the input amplifier and is scaled to match the dynamic range of the A/D converter, which is used to convert the analog voltage to an equivalent binary word. For example, in a unipolar system, with a 0- to +10-V range, the binary word 00000000 might represent 0 V while 11111111 might represent +9.96 V (because of the offset, it is not possible to represent exactly +10 V if zero is to be represented properly).

The output of the A/D converter is stored first in a short-term "scratch pad" memory. The contents of the scratch pad memory is periodically dumped to the main display memory. The transfer from the scratch pad to the main memory is done at a rate faster than the eye can see (typically 40 to 80 times per second). The memory is scanned via a signal from the control

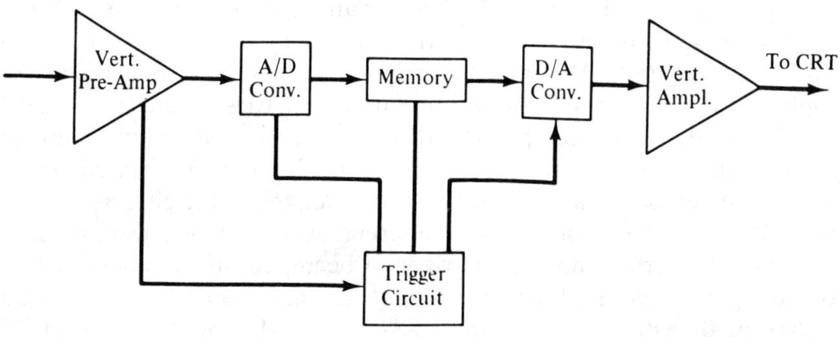

Figure 18-19 One form of computer-memory digital nonfade 'scope circuit.

logic section and output in sequence to a DAC that forms the signal that is applied to the oscilloscope vertical amplifier.

The horizontal sweep signal should be a sawtooth waveform. This signal is formed by a DAC that is driven by a binary counter. As the clock causes the counter to increment through its range, the output of the DAC ramps upwards. When the counter overflows, however, the DAC output snaps back to zero so the process can start anew.

The type of nonfade 'scope depicted in Figure 18-19 can store an immense amount of waveform data. In fact, it can also transfer that data to a computer or to a long-term magnetic medium such as a floppy disk, tape, or main disk. But in simpler instruments a different form of nonfade 'scope might be used.

The simple form of nonfade 'scope system is shown in Figure 18-20. The system is similar to Figure 18-19, but the main memory is replaced with a recirculating shift register. The data is updated by outputting the contents of the scratch pad memory to the input of the recirculating shift register periodically. Between updates, however, the output data is fed back and reentered into the shift register, preserving the image.

Computers have added a new dimension in the medical instruments industry, and medical oscilloscopes have fared well with these developments. Figure 18-21 shows a form of graphics display for a computer that can be used for medical applications (Hercules, CGA, EGA, VGA, and special-format graphics protocols are used extensively). The screen of the CRT is broken into a matrix of tiny square or rectangular zones called picture elements, or *pixels*. Each storage location in the digital memory represents one pixel on the CRT screen. The pixel can be lighted to an intensity defined by the protocol (6 to 32 shades of gray). It is lighted by the electron beam as it raster scans the CRT surface.

Figure 18-22 shows a typical medical monitor 'scope used in an intensive care unit. The pattern on the screen is the self-test pattern that comes on the screen either on operator initiation or at power-on. This type of monitor is used in a computer-based system that actually forms the image.

The monitor of Figure 18-22, like certain other displays, uses a "touch screen" method for the selector switches. Figure 18-23 shows how most such 'scopes operate. Positioned along the edges of the display are a series of infrared (IR) sources (light-emitting diodes, LEDs, operating in the IR region) and infrared detectors. The IR light is invisible to the naked eye, so it is not seen by the operator. The function labels are either painted onto the CRT by the computer (as in Figure 18-23) or affixed to the edge as in Figure 18-22. When the operator touches the screen over any label, his or her finger interrupts one vertical and one horizontal beam, causing a unique pattern. For example, suppose all detector outputs are at binary LOW when the IR reaches the detector. The X outputs X1-X2 are L-L, and the Y outputs Y1-Y2-Y3-Y4-Y5-Y6 are L-L-L-L-L-L. And then someone touches the SELF

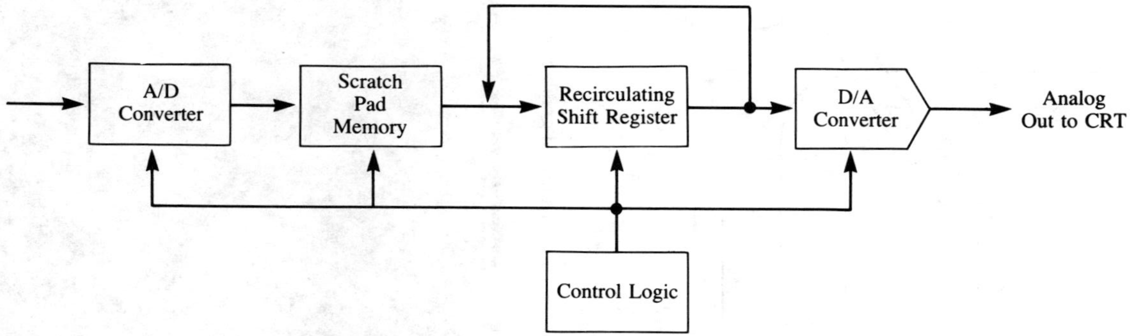

Figure 18-20 Recirculating shift register nonfade circuit.

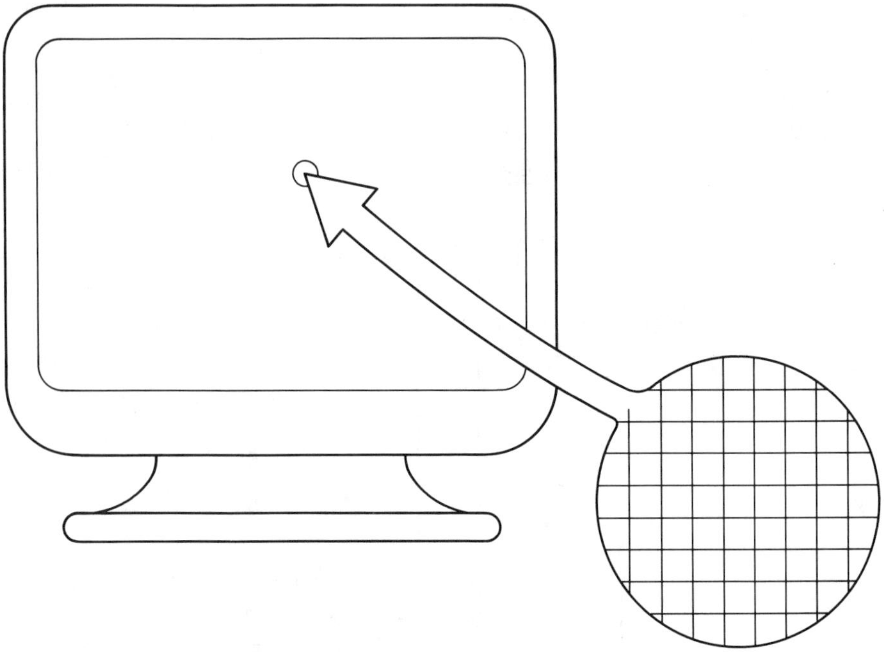

Figure 18-21 Modern digital graphics monitor.

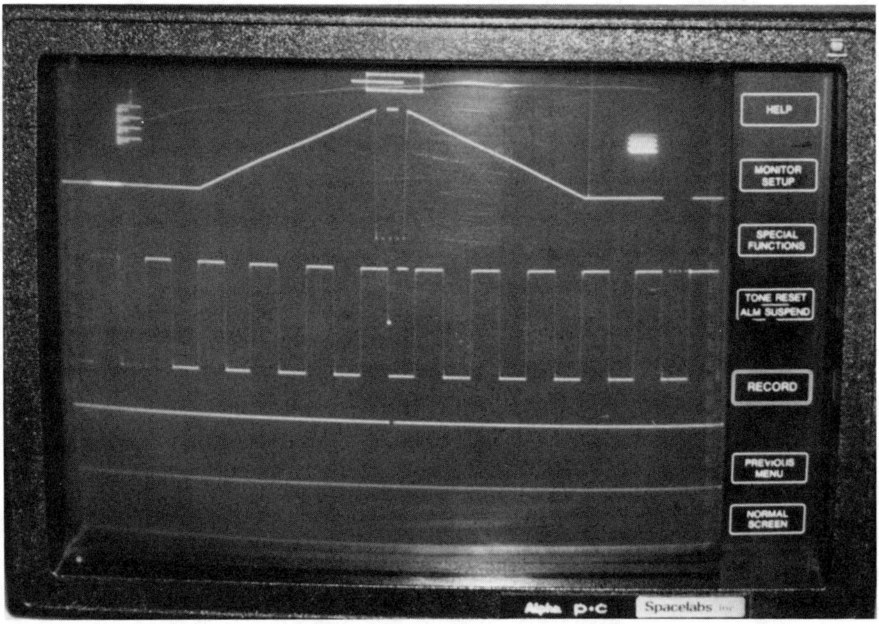

Figure 18-22 Nonfade medical oscilloscope.

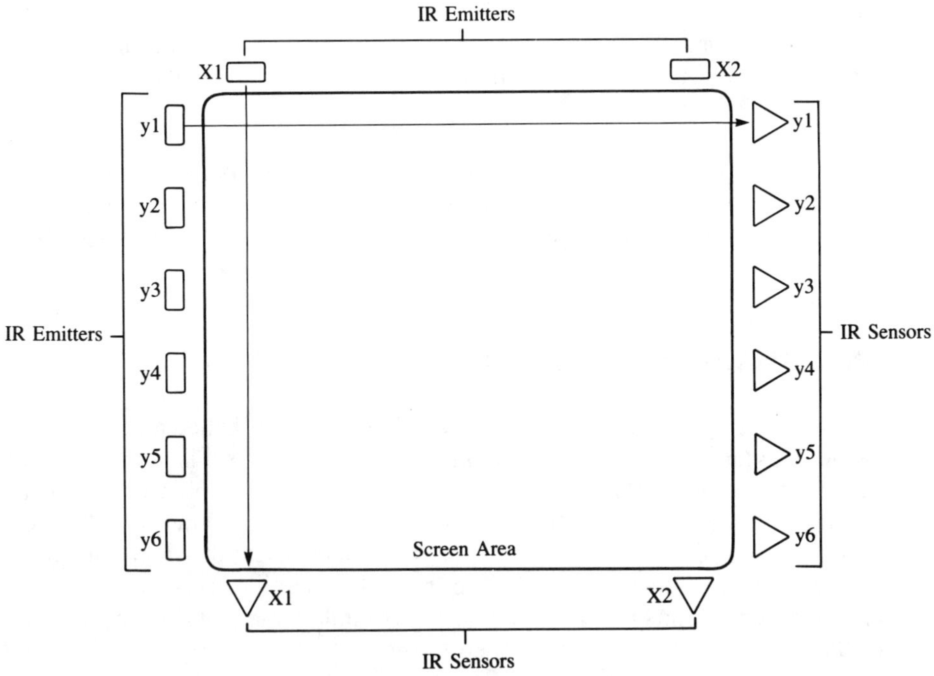

Figure 18-23 "Magic screen" touch control scheme.

TEST label. In this case, X1 and Y1 are both interrupted and their respective outputs go HIGH. Thus the patterns become H-L and H-L-L-L-L-L. The computer recognizes this as an operator command to branch to the SELF TEST software stored in program memory.

MECHANICAL ANALOG RECORDERS

The term "mechanical recorders" refers to a broad class of devices that make a permanent paper record of analog waveforms. To the analog data system, such as an ECG machine, the recorder serves the same function as a printer in a personal computer system. That is, it creates a permanent record ("hardcopy") of the data output produced by the instrument. As you will see later in this chapter, for some modern analog recorders (which are actually digitally based) the analogy to a computer printer is more than merely metaphorical.

Several different categories of analog recorders are used in medicine and other life sciences: strip chart, X-Y servorecorders, and plotters are among the most common. We also now have analog-digital dot matrix printing mechanisms for analog recorders. The strip chart recorder uses paper that is continuous on either a roll (like adding machine paper) or in a Z-fold pack (like personal computer printer paper). The X-Y servorecorder uses a single sheet of paper and an analog pen. The plotter is actually a sophisticated digitally controlled version of the analog X-Y recorder. Of these, the strip chart recorder is by far the most commonly used in medicine (especially on emergency equipment). For this reason we will begin our study with this category of recorder. But first, we must examine the principal means for writing in the strip chart recorder: the permanent magnet moving coil (PMMC) galvanometer mechanism.

PMMC GALVANOMETER MOVEMENTS

The heart of any standard strip chart recorder is a device called a permanent magnet moving coil (PMMC) galvanometer, or ''galvie'' as it is known in the trade (see Figure 18-24). The moving coil consists of a bobbin on jeweled bearings, with a lightweight coil of wire wound over it. When an electric current is applied to the coil, a weak magnetic field is set up around the coil. The signal being recorded is applied to the coil so that it causes a currrent to flow in the coil. Thus the magnetic field surrounding the coil is proportional to the amplitude of the signal applied to the coil.

The permanent magnet applies a strong magnetic field across the space occupied by the moving coil. The magnetic field of the moving coil therefore interacts with the field of the permanent magnet. Recall the standard rules for magnetic field interaction: opposite poles attract and like fields repel. The reaction of the coil will be to move when a current flows. The jewelled bearings allow rotational motion only. By attaching a pen or stylus (discussed later) to the coil's pivot point we produce an angular deflection proportional to the applied signal amplitude.

The pen tip is positioned over a strip of chart paper that is pulled under

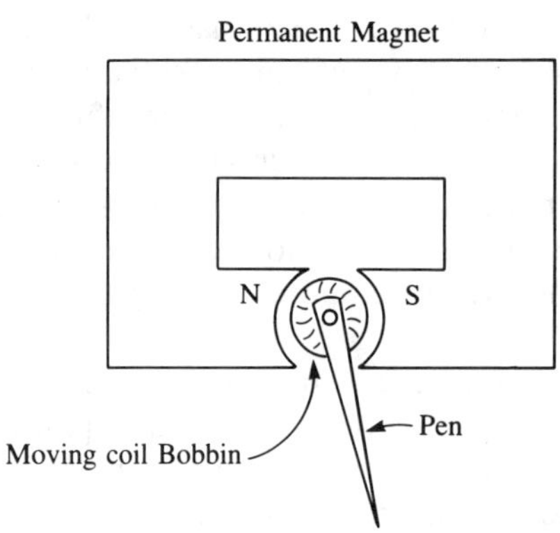

Figure 18-24 Permanent magnet moving coil galvanometer.

the pen tip at a constant speed, thereby establishing a time base. Deflection of the pen across the paper reproduces the waveshape of the signal applied to the coil.

A PMMC galvanometer with a short pen arc will write in a curvilinear manner (Figure 18-25A); the pen will scribe an arc at its tip instead of a straight line. In some applications this may be tolerable, and the user merely records on a paper that has a semicircular grid marked on it. In medical applications, however, it is limited to such jobs as recording the time-varying temperature of blood lockers, incubators, and other machines that must maintain an internal temperature constant. In these cases, the amount of deflection (e.g., a temperature) is more important than the shape of the wave. Thus this type of pen is almost useless for recording medical information and waveforms from a patient.

A solution to the problem of curvilinear recording is shown in Figure 18-25B. The PMMC mechanism is not connected directly to the pen, but through a mechanical linkage that translates the curvilinear motion of the PMMC coil bobbin to a rectilinear motion needed at the pen tip. This type of mechanism is used extensively in medical applications, although not as extensively as the knife edge method discussed below.

A pseudorectilinear technique is shown in Figure 18-25C. In this type of assembly the pen is very long compared with the arc that it must scribe (and also the chart paper width). The pen tip, therefore, travels in an arc that is very small compared with its radius, that is, the pen length. The trace, then, is nearly linear. Some curvature will exist, however, so this method is only used on some noncritical medical applications.

There are several writing methods used in strip chart recorders, but two are amenable to a special type of pseudorectilinear writing. Both of these writing systems use a knife edge (also called "writing edge" in some service manuals) to effectively linearize the trace. The method is shown in Figure 18-25D. The pen tip still travels in a curvilinear method (which allows the simplest form of PMMC pen mechanism), but the trace is very nearly linear because the knife edge is straight. This technique works in both thermal writing and direct pressure systems (as opposed to ink) because the pen, which is actually a heated stylus, can write anywhere along its own length—not just at the tip. By keeping a straight knife edge in a fixed position beneath the stylus, and allowing the stylus's point of contact with the heat- or pressure-sensitive paper to vary, we obtain a rectilinear recording. It is the straightness of the knife edge that creates the rectilinear operation.

The thermal "knife edge" recording system is used more widely than the direct contact system. The direct method uses specially treated paper similar to the "carbonless" multicopy form paper marketed by companies such as NCR for commercial purposes. The heat-sensitive paper is paraffin coated and turns black when heated. We will shortly discuss several writing methods used in medical equipment, including the method mentioned above.

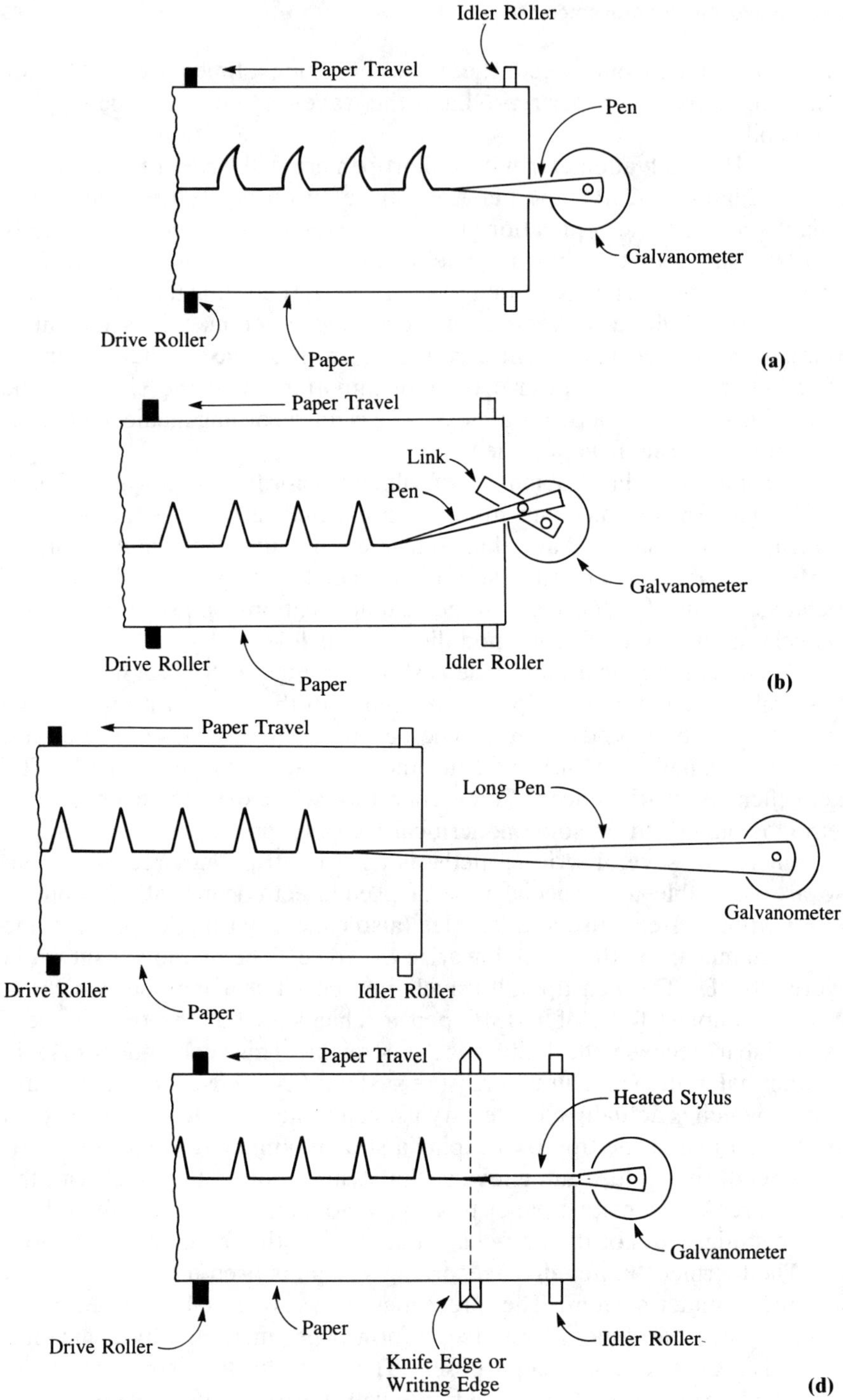

Figure 18-25 A) Curvilinear recorder, B) Mechanical linkage pseudo-rectilinear recorder, C) Long-pen pseudo-rectilinear recorder, D) Knife-edge writer.

PMMC WRITING SYSTEMS

There are several different writing systems used on PMMC recorders: direct pressure, thermal, ink pen, and optical.

Those mechanisms that use pens or styli, that is, direct pressure, thermal, or ink pen, tend to have relatively low "frequency response" due to the mechanical inertia of the massive pen or stylus. Most such recorders have an upper frequency limit of 100 to 200 Hz. Ink jet and optical recorders, on the other hand, have responses up to 1000 or 2000 Hz. Why is this important? All analog waveforms, such as ECG, EEG, and EMG, are made up of components of frequencies higher than the apparent "fundamental" frequency. For example, a 60-BPM ECG has a fundamental frequency of one beat per second, or 1 Hz. But an analysis of its components in a special electronic device called a Fourier spectrum analyzer will show that it contains frequency components in its spectrum of 0.05 to 100 Hz. Similarly, a phonocardiogram may have spectrum components up to 1500 Hz. Therefore, a stylus recorder can easily handle an ECG waveform but not a phonocardiogram waveform. In other words, the type of recorder used for any given medical purpose is very important.

The thermal recording system uses a specially treated paper that turns black when heated. The paper is paraffin coated. The writing instrument is an electrically heated stylus. Early models formed the stylus tip from a U-shaped electrical resistance element. More modern models use a resistance element inside of a cylindrical metal stylus. In both cases a low-voltage electrical current is passed through the element, heating it almost to incandescence. The heated stylus leaves a black mark on the treated paper wherever it touches.

Ink pen systems write using a hollow pen and an ink supply. In some machines, the ink is relatively lightweight (i.e., it has a low viscosity), and pressure is applied manually through an atomizer type of pump (also called a "squeeze ball" pump). Other machines, such as the Gould-Brush instruments and Hewlett-Packard Model 7402, use a thick, high-viscosity ink. The pressure is applied through a spring-driven piston (Figure 18-26A) inside the ink supply cartridge. An ink manifold (Figure 18-26B) distributes the ink to the several pens that might be used in the system. Pressure is applied to the manifold by a solenoid that is energized when the machine power is turned on.

A related type of writing system, used in some high-frequency response phonocardiogram machines, is the high-velocity ink jet. In that type of recorder, ink under high pressure is fed to a nozzle mounted on the PMMC galvanometer in place of the pen. The ink jet is directed at the moving paper and, when the system is properly adjusted, produces a line that is almost as fine as that produced by thermal and ink pen recorders. Only a small amount of trace fuzziness is apparent, and this is due to ink splattering.

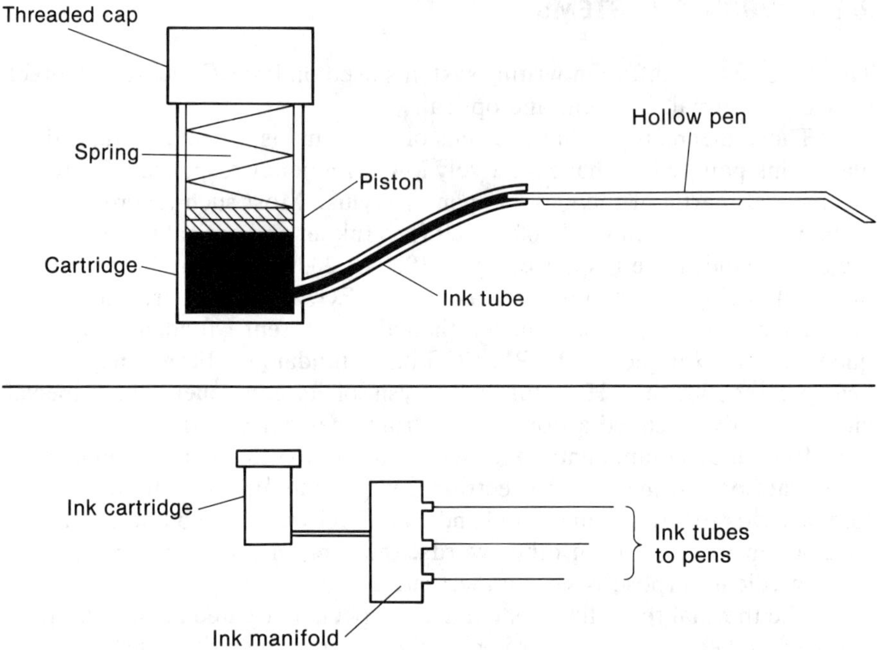

Figure 18-26 A) Spring-driven ink cartridge, B) Multi-manifold pen system.

The high-velocity ink jet recorder finds use wherever a high-frequency response is needed. The low mass of the ink jet nozzle, compared with the relatively bulky mass of the ink pen and thermal stylus, gives the machine a 1000- to 2000-Hz frequency response. These recorders are sometimes used for recording phonocardiograph (''heart sound'') waveforms.

An example of an optical PMMC recorder is shown in Figure 18-27A, while a cathode ray tube optical recorder is shown in Figure 18-27B. The PMMC optical writer uses a mirror mounted on the PMMC galvanometer to reflect the light beam from a collimated light source onto the photosensitive chart paper. In most such cases, the paper is at least six inches wide, so a greater span (or resolution) is possible. Additionally, on a multichannel recorder, the time relationships between two different traces can be more easily seen because the beams can be made to cross each other—a difficult trick to accomplish with pen or stylus recorders.

The paper is exposed to an ultraviolet light source that develops the trace as the paper comes out of the recorder. The trace will fade over a long period of time, or if it is exposed to a strong light (sunlight will cause damage). Therefore, for long-term storage a light-tight box is used. Alternatively, some machines allow the paper to be wet-developed like photographic printing paper, after which process the image is stable.

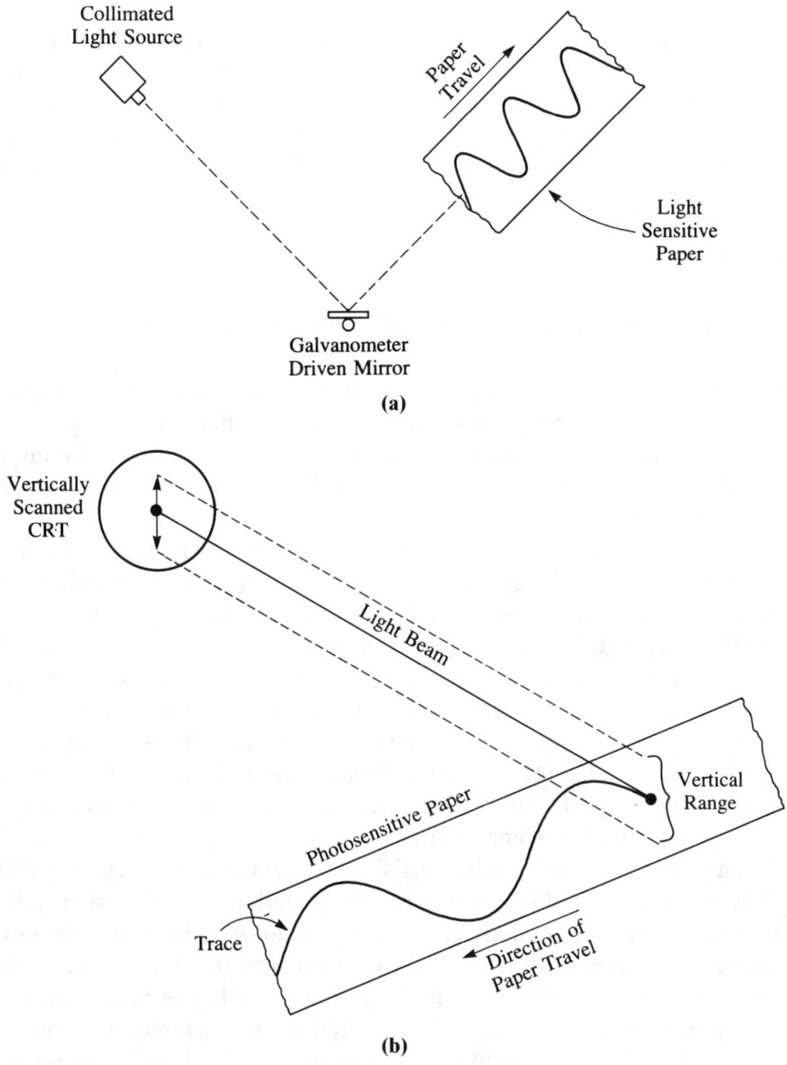

Figure 18-27 A) PMMC optical recorder, B) CRT camera optical recorder.

A few models of older recorders use a CRT as shown in Figure 18-27B. The old VR-6 and VR-12 recorders by Electronics for Medicine are still in use. This type of chart recorder is called a camera recorder.

There is no time-base sweep on the CRT screen. The beam sweeps back and forth along one axis in response to amplitude variations of the input signal (examine Figure 18-27B closely). A time base is provided by the constant speed at which the photosensitive paper is drawn in front of the CRT screen.

The frequency response of the PMMC optical recorder is better than that of any other PMMC system because of the low mass of the reflecting mirror. The CRT camera has an even higher frequency response because it is limited only by the writing time of the photosensitive paper. As a result, CRT cameras are often used for high-frequency medical signals such as phonocardiograph, electromyograph, and Bundle of His (and other intracardiac ECG) recordings.

RECORDING POTENTIOMETERS AND SERVORECORDERS

A potentiometer is a three-terminal device that acts as a variable resistor. Two ends are fixed, while the middle arm is attached to a "wiper" that selects a resistance proportional to its position. The two fixed ends are connected across a reference voltage, while the variable voltage appears at the moveable wiper arm. The output voltage is proportional to the applied voltage and the relative position of the wiper on the resistance element. A galvanometer connected between the wiper and an unknown voltage will register zero deflection when the known voltage from the potentiometer is equal to the unknown voltage. This condition is called the "null" state. But if these two voltages are unequal, then the PMMC "galvie" will deflect an amount proportional to the difference between the two potentials.

It is possible to build a self-nulling potentiometer. If a pen is connected to the potentiometer's wiper drive mechanism, then the applied voltage can be recorded on paper as the mechanism constantly seeks to null out the applied voltage from the signal source.

Figure 18-28 shows the basic DC potentiometer servorecorder system. The pen is attached to a string that is wound around two idler pulleys and a drive pulley on the shaft of a DC servomotor. The pen assembly is also linked with a potentiometer (R1) in such a way that the position of the wiper arm on the resistance element is proportional to the pen position.

The potentiometer element is connected across a stable, precision reference potential, E_{ref}, so potential E will represent the position of the pen. That is to say, E is the electrical analog of pen position. When the pen is at the left-hand side of the paper (in Figure 18-28), then E is zero, and when the pen is full-scale (i.e., at the right-hand side of the paper) E is equal to E_{ref}. The pen position is controlled by the DC servomotor, which is in turn driven by the output of the servoamplifier. This amplifier has differential inputs. The input signal E_{in} is applied to one amplifier input, and the position signal E is applied to the other amplifier input. The difference signal $(E_{in} - E)$ represents the error between the actual pen position and the position it should take in response to the command from E_{in}.

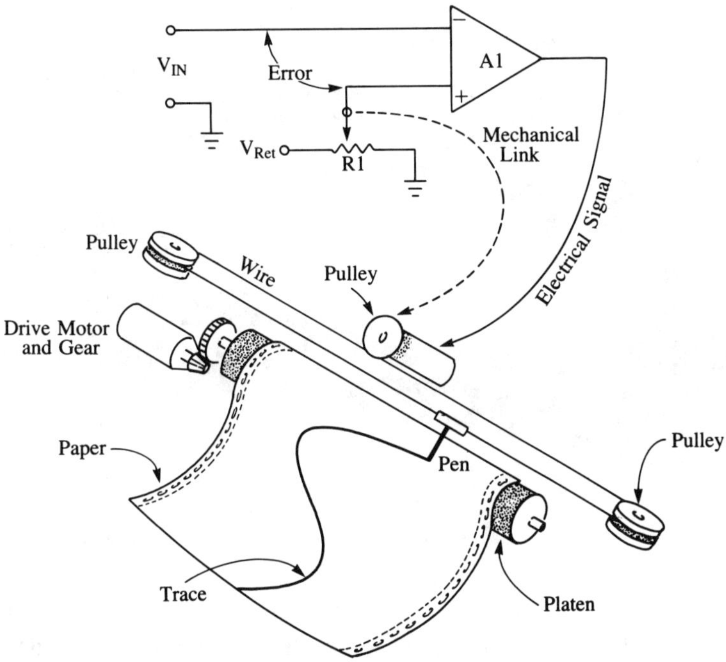

Figure 18-28 Servomechanism recorder.

If the error signal is zero, then the amplifier output is also zero, so the motor remains turned off. But if E is not equal to E@-{in}, then the amplifier sees an input signal and creates an output signal that turns on the motor.

The motor drives the pen potentiometer assembly in such a direction as to cancel the error signal. When the input signal and the pen position signal are equal, then the motor shuts off and the pen remains at rest.

A paper drive motor forms a time base because it pulls the paper underneath the pen at a fixed, constant rate. In most servorecorders a sprocket drive is used instead of the roller drive system popular in PMMC machines. The paper used in these machines (see Figure 18-29) has holes along each margin to accept the sprocket teeth, much like personal computer printer paper.

Most high-grade servorecorders use either a stepper motor that rotates only a few degrees for each pulse applied to its winding, or a continuously running motor. The stepper motor method is low in cost and can be very precise when a stable reference clock oscillator is used as the source of the electronic drive pulses.

The actual potentiometer resistance element used may be any of the following: slide wire, rectilinear potentiometer, or rotary potentiometer.

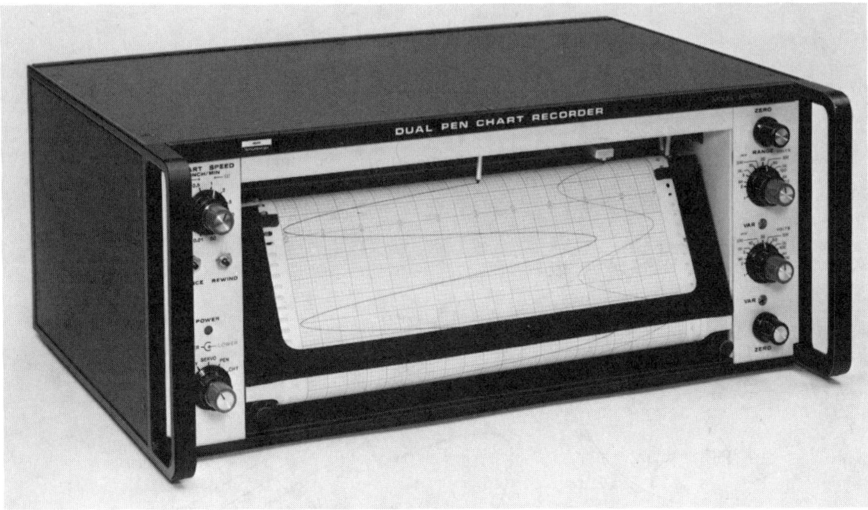

Figure 18-29 Typical servorecorder with sprocketed paper drive.

The slide wire system is very common because it can be built with less friction loss and no mechanical linkage. A shorting bar on the pen assembly serves as the wiper element on the slide wire resistance element; it also serves to connect the shorting bar to the wire. The letters A, B, and C in Figure 18-30A refer to the potentiometer terminals shown in Figure 18-30B. The shorting wire B serves as the terminal for the wiper.

X-Y RECORDERS AND PLOTTERS

An X-Y recorder uses two servomechanism pen assemblies mounted at right angles to one another. The vertical (or Y-plane) amplifier moves the pen vertically along the bar, while the X-plane amplifier moves the bar back and forth across the paper.

The paper itself does not move, and it is usually held in place by either clamps or, in high-grade machines, a vacuum drawn by a pump connected to a hollow chamber beneath the paper platform. Holes in the paper platform create the negative pressure needed to keep the paper in place.

One advantage to the X-Y recorder over either PMMC or servorecorders is that almost any type of paper may be used. Most X-Y recorders are designed to accept the standard-size graph papers used by scientists, engineers, and physicians.

The X-Y recorder can be made to record time-varying signals by applying a linear ramp voltage to the X-plane input. The ramp is adjusted so that it traverses the horizontal width of the paper in the desired length of time.

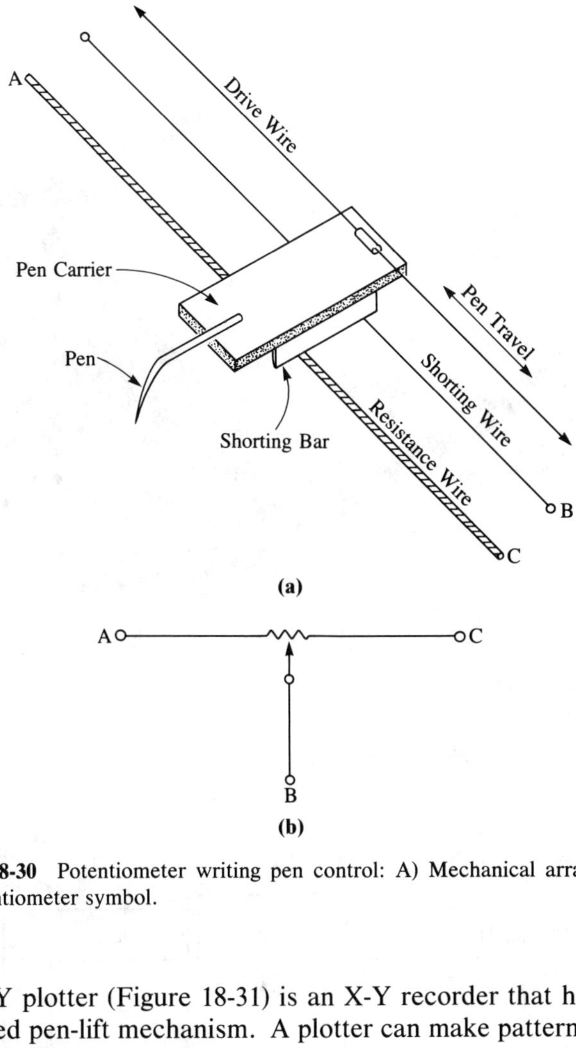

Figure 18-30 Potentiometer writing pen control: A) Mechanical arrangement, B) Potentiometer symbol.

An X-Y plotter (Figure 18-31) is an X-Y recorder that has an electrically operated pen-lift mechanism. A plotter can make patterns of complex shape, including alphanumerics. Many modern plotters are designed to do computer graphics, while others handle both analog and digital data.

DIGITAL RECORDERS

A relatively new class of medical analog recorder is the digital type (Figure 18-32). The input signal is applied to an A/D converter. The A/D device produces a binary output "data word" (of the sort used in computers) that is proportional to the signal amplitude. If a large number of successive "samples" of the analog waveform are taken and stored in a computerlike mem-

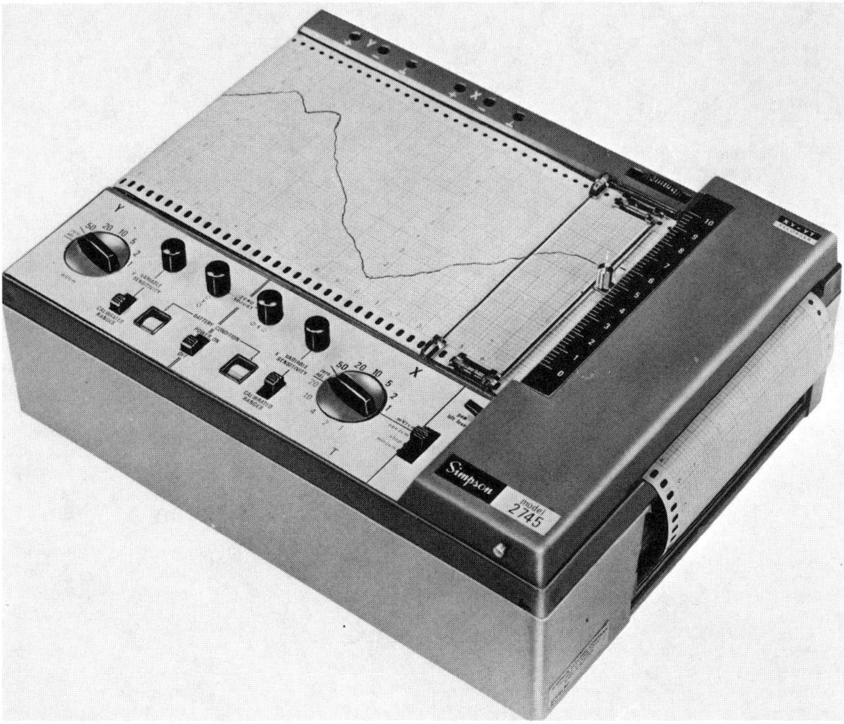

Figure 18-31 Plotter recorder.

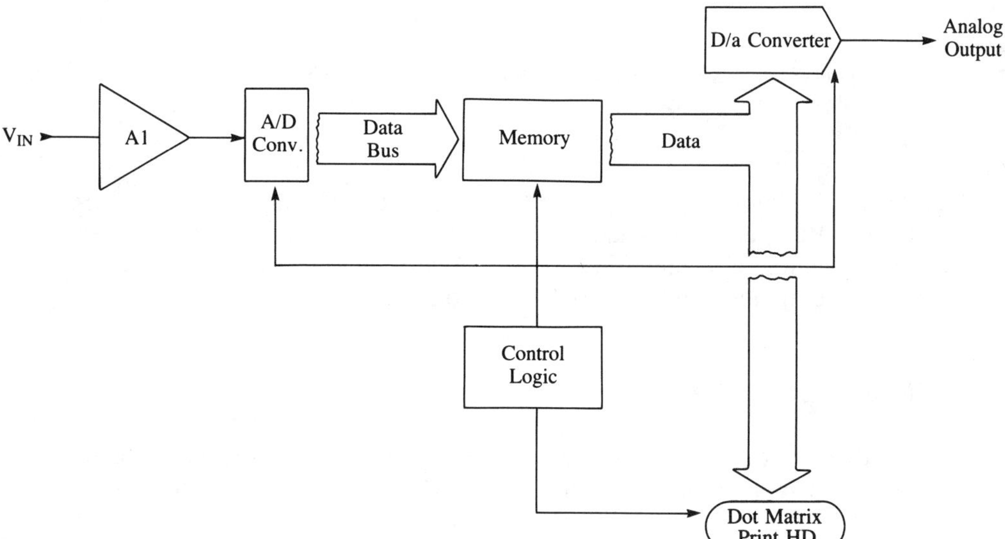

Figure 18-32 Memory recorder block diagram.

ory, then the pattern of digital data in the memory will represent the time-varying analog signal.

The data in memory can be scanned and played back to either a strip chart recorder or CRT screen for viewing. In that case, a D/A converter is used to recreate the analog voltage signal required by the display device.

An advantage of this system is that low-frequency (which also means low-cost) paper recorders can be used to record very high frequency analog signals simply by varying the time base. For example, suppose we want to record a 1000-Hz analog signal. According to a standard engineering convention the sampling rate of an analog signal should be at least twice the highest frequency, so we need to take $2 \times 1000 = 2000$ samples per second. These samples are stored in the computer memory. When we want to make a recording we need only reduce the sampling rate proportionally, say to 20/s, and play it back through a standard low-frequency machine.

This method is one means for standard hospital ECG machines to be used for higher-frequency studies. The same method is also used for producing the "nonfade" oscilloscopes in ECG monitors used in emergency medicine. Normally, the ECG waveform would fade almost immediately after it occurs. But with the digital recording oscilloscope it will remain on the screen until the memory is used up and is overwritten by newer data.

RECORDER PROBLEMS

The pen assembly always has a certain amount of inertia because of its mass, regardless of whether PMMC or servo systems are used. Because of its mass the pen assembly will not start moving until a certain minimum signal is applied. Figure 18-33 shows this so-called "deadband" phenomenon. The definition of the deadband is "the largest amplitude signal to which the instrument will NOT respond." In most quality instruments the deadband is not more than 0.05 to 0.1 percent of full-scale.

The deadband can cause severe distortion of the recorded trace, especially on low-amplitude signals whose level approximates the deadband itself. The solution to the problem of deadband is to slew the signal through the deadband as rapidly as possible.

Another problem is overshoot or undershoot of the pen in response to a step function (e.g., a square wave). This problem also affects both forms of mechanical writer. Figure 18-34 shows three different responses. In a properly damped recorder the pen will rise to the correct position with very little rounding of the trace. This situation is called "critical damping."

A subcritically damped signal will overshoot the correct position and then hunt back and forth across the correct position for a few cycles before it settles on the correct point. If a square wave is applied to such a system, then the recording will appear to be "ringing."

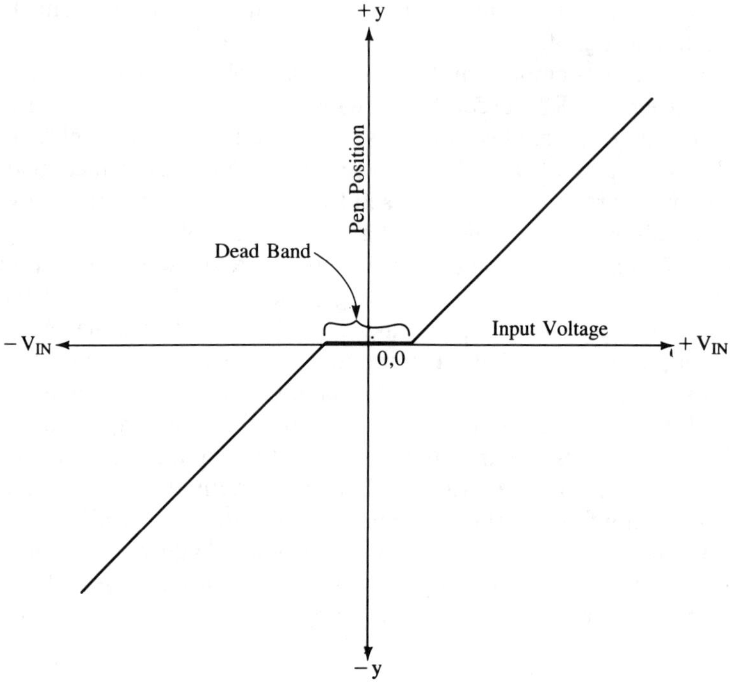

Figure 18-33 Dead-band problem.

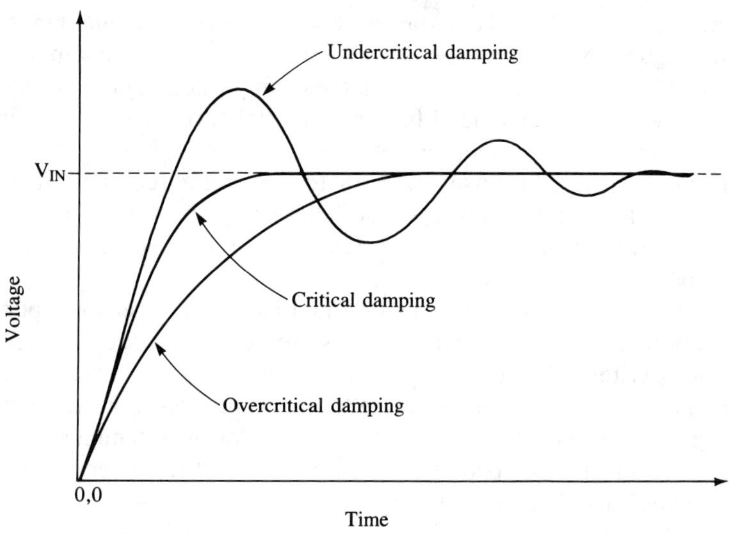

Figure 18-34 Effects of A) Undercritical, B) Critical, C) Overcritical damping.

An overcritically damped system approaches the final value very sluggishly. The pen in that type of system will appear to be very sluggish. A square wave applied to such a system will have rounded edges. Damping is an important parameter in medical recorders because it can distort the resultant traces and thereby mimmick pathological traces. Subcritical damping mimmicks the double or "notched" QRS complex of heart block, while overcritical damping mimmicks the trace often found in recent heart attack patients by altering the S-T segment of the waveform.

MAINTENANCE OF PMMC WRITING STYLI AND PENS

There are several common faults found on medical paper recorders. These are sometimes amenable to adjustment, while in other cases repair is necessary. Let's separately consider the ink pen and heated stylus types of recorder. Because they find only limited use in medical equipment, we will not consider some of the other types (even though you need to be aware of them conceptually). In this section we will take a look at some common maintenance actions that don't always need to be performed by trained engineering technicians only.

Figure 18-35 shows how to remove an ink blockage from an ink pen recorder that's been allowed to stand too long without being used. As a

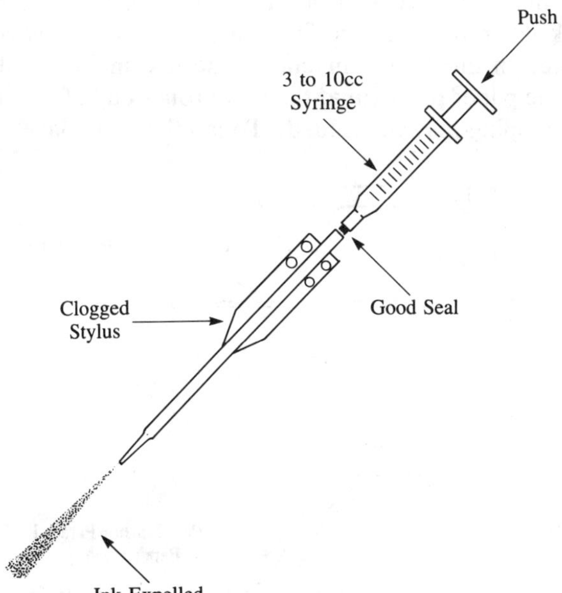

Figure 18-35 Cleaning an ink pen.

general rule, such recorders should be run for about five minutes or so once a week when not in regular service. Otherwise, the ink will dry up at the tip and prevent the recorder from writing. Fill a 3- to 10-cc syringe with water (or acetone if certain types of ink are used), and insert the needle end into the ink inlet on the rear end of the pen assembly. On most recorders the pen will have to be removed from the machine for this operation. The needle should be inserted up to the Luer-lock hub in order to make a good fluid seal. Quickly, and with a single sharp motion, drive the plunger "home" so that a high-pressure jet of water or acetone is forced into the pen. The ink clot should be forced under high pressure out the other end.

This procedure usually works quite well. Some precautions are in order, however. First, always wear protective goggles when doing this operation. Also, wear protective clothing, as ink and/or water may splatter everywhere. Second, make sure that the pen tip is aimed downward into a sink (there is still a laboratory at George Washington University Medical Center with a blue spot on the ceiling tiles over the sink where the author failed to observe this precaution). Finally, as always when dealing with needles, be careful not to stick yourself. Keep in mind that the thick, high-viscosity ink used in these machines stains everything it touches . . . and is the virtual dickens to clean.

Ink pen tips are designed to operate parallel to the paper surface (see inset in Figure 18-36). If the pen is worn, or when a new pen is installed, it is necessary to "lap" the tip in order to reestablish the parallelism. The symptom that lapping is needed will be either (or both) of the following: a) a "blob" of ink when the machine first starts recording a waveform, or b) a too-thick trace. In most such machines the ink should be dry before the paper leaves the paper platform at the drive roller end of the surface. If it is not dry, then lapping may be needed. To lap the pen, place a piece of very

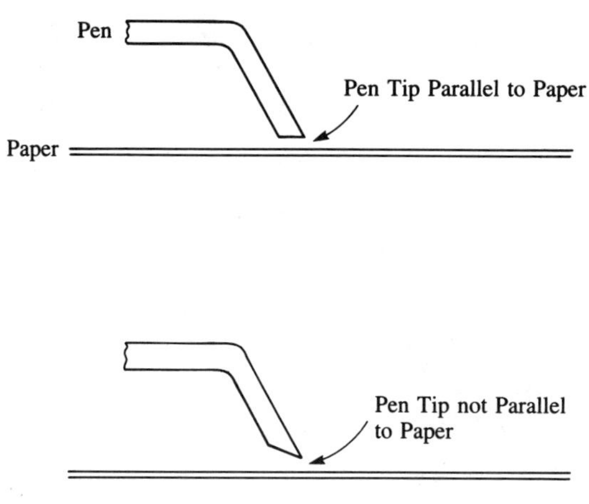

Figure 18-36 Proper placement of pen on paper.

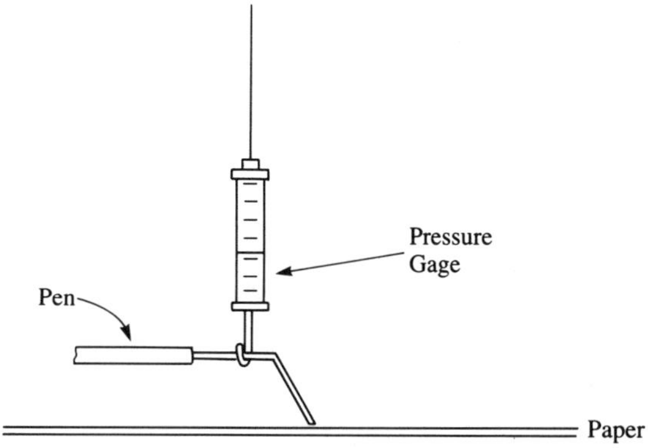

Figure 18-37 Using stylus pressure gage.

fine emery cloth (a sandpaperlike material available at any hardware store) under the tip. The pen tip is worked back and forth five to ten times to "sand" the tip parallel to the paper.

The pressure of the stylus or pen is also important. If the pressure is not correct, then the waveform may be distorted. On medical equipment it is possible to make a normally healthy Lead-I ECG signal look like it represents either a heart block or a recent myocardial infarction because of improperly adjusted stylus pressure. The manufacturer will specify a pressure in grams. These numbers vary from 1 to 10 grams depending upon the machine model (and its year of manufacture—older model machines tended to use heavier stylus pressures).

A stylus pressure gage (see Figure 18-37) is used to lift the pen or stylus from the paper as the machine is running until the trace just disappears. The pressure reading is then made from the barrel of the gage. Suitable stylus pressure gages can be purchased from ECG machine manufacturer's service or parts departments. Alternatively, the stylus pressure gage used for record player "tone arms" is also useful if the specified pressure is within their relatively limited range (usually 0 to 4 grams). The stylus pressure adjustment is made using a screw that is usually located on the rear of the stylus (or pen) or the assembly that holds it in place.

Figure 18-38 shows several different 1-mV calibration pulses from an ECG machine. These traces can be made by pressing the "1-MV" or "CAL" button on the front panel of the machine. The ideal shape is perfectly square, as shown in Figure 18-38A. But this ideal is almost never achieved in practical machines because of the inertia of the pen or stylus assembly. Usually we see the slightly rounded features of the pulse shown in Figures 18-38B. This waveform is usually acceptable. What is not acceptable, however, are the overdamped and underdamped waveforms of Figures 18-38C and 18-38D, respectively.

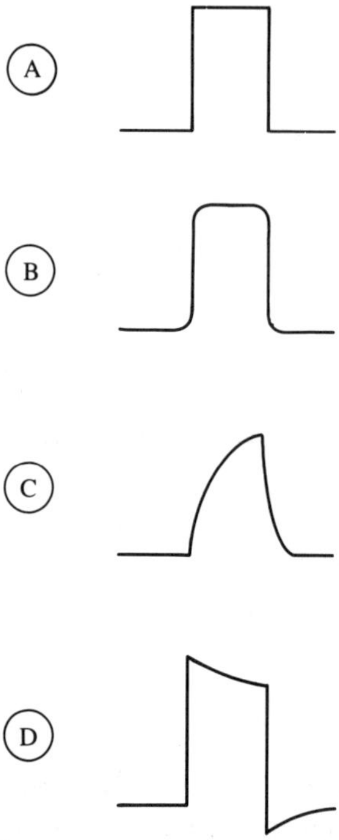

Figure 18-38 1-mV calibration signal: A) Ideal, B) Usual, C) Overdamped, D) Underdamped.

On all recorders there may be a damping control available for adjustment by a properly trained technician. This control is adjusted (usually internally to the machine) in order to compensate for problems. On ink pen machines the stylus pressure can affect the waveshape, especially if it is set to too high a value (it produces the overdamped waveform). On heated stylus machines both the stylus pressure and the heat can affect the waveform.

On some heated stylus machines the standard procedure is to set the pressure to a specified value, set the voltage applied to the stylus heating element to a specified value (usually either 5.00 or 7.00 V), and then adjust the internal damping control to pıoduce the waveform of Figure 18-38B in response to either a square-wave input or successive presses of the 1-MV or CAL button.

If the manufacturer of the machine did not provide a knob on the stylus heat control, then don't adjust heat control without the correct equipment (usually a stylus pressure gage and voltmeter) AND the manufacturer's service manual! The author remembers a time when the director of the emergency room called and snorted that he didn't really believe that the last 40 patients—only one of whom had a cardiac complaint—all had had recent myocardial infarctions. The problem was that an ambitious medical student (who also held an electrical engineering degree) thought he knew how to adjust the heat . . . but did not. He brought a small screwdriver to work and adjusted the heat to make the trace lighter "for reliability reasons." The trace was lighter, alright, but the machine would not recover from fast waveform transitions (such as an ECG QRS complex) fast enough. As a result, the extra inertia caused by insufficient melting of the paraffin on the paper made the ECG tracings all erroneously show S-T segment anomalies.

DOT MATRIX ANALOG RECORDERS

The dot matrix printer is long familiar to users of computer equipment. The dot matrix printer was developed in response to the high cost of traditional computer printers. At the time they became popular, the dot matrix printer cost about one-fifth to one-third the cost of daisy wheel, Selectric, or similar printer mechanisms. The original dot matrix printers (mid-1970s) used a 5 × 7 matrix of dots to form alphanumeric characters.

The dot matrix machine used a print head (Figure 18-39B) to cause the correct dot elements to be energized to make a mark on the paper. Two different methods were once popular, although one has since faded almost to obscurity. Some of the earliest machines were thermally based. The dots were thermally connected to heating coils and could be heated when needed. Special temperature-sensitive paper was used to receive the text. This method is no longer used widely. The second method used an array of seven print hammers (actually pins). The pins would either extend or retract depending upon whether or not that particular dot was active for the character being printed. An advantage of the pin method is that ordinary paper can be used. The pins impact an inked ribbon to leave the impression. Although the original low-resolution printers were 7-pin models, there are now available higher resolution models with 9-, 18-, and 24-pin print heads.

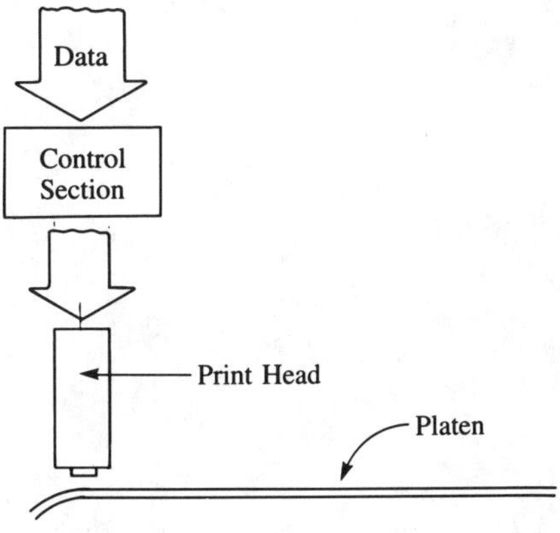

Figure 18-39 Dot matrix analog recorder.

It was rapidly discovered in the computer world that clever programmers could make a dot matrix print head do graphics as well as alphanumerics. The spate of newsletters, church and school bulletins, and other low-cost (but often creatively done) publications are a testimony to the graphics capability of modern dot matrix technology.

Dot matrix printing can be used to make analog recorders that outperform most older mechanical analog recorders. Figure 18-39 shows the concept schematically. A dot matrix print head with a very large number of pins is arrayed over a platten. The action depends upon whether thermal, electro-arc, or plain strip chart paper is used. Regardless of the particulars, however, the result is a strip chart recording that mixes analog and digital data on the same chart.

19

CONTROLLING EXTERNAL CIRCUITS

The ability to control external circuitry makes the microcomputer a much more useful device. Certain calculation or signal-processing chores can be performed in the machine and then used to control external circuits. The simplest forms of external control are on-off switches that are controlled by a single bit of a computer output port. More complex control applications use devices such as amplifiers, digital-to-analog converters, and so forth. Extremely complex feedback control systems have been implemented using computers. The availability of microcomputers has accelerated the process and has, in an interesting way, made the design of computerized control circuits less a problem for arcane areas of engineering and more a game for all.

Figure 19-1 shows two methods for interfacing electromechanical relays to the microcomputer. Why would one want to interface an electromechanical relay, which is a century-old device, to a space-age device like a microcomputer? The old relay may well be the best solution to many problems, especially where a certain degree of isolation (e.g., for safety) is needed between the computer and the controlled circuit. An example might be 115-V AC applications, especially those that may require heavy current loads. A typical application might be turning on and off 115-V AC lamps around a building. The computer can be used as a timer and will turn on and off the lights according to a programmed schedule. Another application might be to use the computer to monitor burglar alarm sensors and then turn on a lamp if one of them senses a break-in.

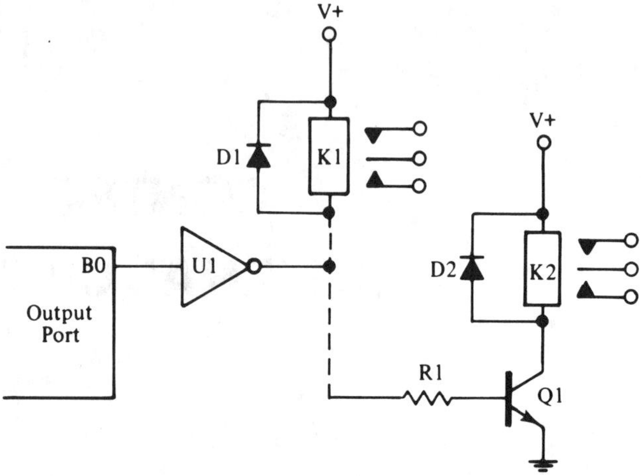

Figure 19-1 Interfacing computer to external relays.

Figure 19-1 shows two methods for connecting the relay to the computer. Control over the relay is maintained by using 1 bit of the computer output port, in this case B0. Since only one bit is used, the other 7 bits are available for other applications, which may be displays, other relays, or certain other devices.

Most microcomputer outputs are not capable of driving heavy loads. Some devices will have a fan-out of 10 (i.e., will drive 18 mA at +5 V), while others have a low fan-out, typically 2 (3.6 mA). To increase the drive capacity and to provide a mechanism for control, we use an open-collector TTL inverter stage, U1. One end of relay coil K1 is connected to the inverter output, and the other end of the coil is connected to the V+ supply. Some TTL devices (7406, 7407, 7416, and 7417) will operate with potentials greater than +5 V DC on the output, so we can use 6-, 12-, or 28-V relays. The package DC potential applied to the inverter is still the normal +5 V required by all TTL devices. These inverters are actually hex inverters and so will contain six individual inverter circuits in each package. All six inverters can be operated independently of each other.

The operation of the circuit revolves around the fact that the relay (K1) coil is grounded when the inverter output is LOW and ungrounded when the inverter output is HIGH. As a result, we can control the on-off states of the relay by applying a HIGH or LOW level to the input of the inverter. If the inverter input is LOW, for example, the output is HIGH, so the relay coil is not grounded. In that case, the relay coil is not energized because both ends are at the same electrical potential. When a HIGH is applied to the input of the inverter (i.e., when B0 of the output port is HIGH), the inverter output is LOW, which makes the "cold" end of the relay coil grounded. The relay will be energized, which closes the contacts. We may turn the relay on, then, by writing a HIGH (logical 1) to bit B0 of the output port, and turn if off by writing a LOW (logical 0) to the output port.

The inverter devices cited previously have greater output current capability than some TTL devices, but it is still low compared with the current requirements of some relays. High-current relays, for example, may have coil current requirements of 1 to 5 A. If we want to increase the drive capability of the circuit, we may connect a transistor driver such as Q1 in Figure 19-1.

In the case of relay K2, the "cold" end of the coil is grounded or kept high by the action of transistor Q1. This relay driver will ground the coil when the transistor is turned on (i.e., saturated) and will unground the coil when the transistor is turned off. As a result, we must design a method by which the transistor will be cut off when we want the relay off, and saturated when we want the relay on.

For circuits such as K2, the TTL interface with the computer output port (U1) may be an inverter or a noninverting TTL buffer. The on-off protocol will be different for the two. Also, we need not use an open-collector inverter for U1, as was the case in the preceding. If we want to use an open-collector device, however, we supply a 2.2-k@ pull-up resistor from the inverter output to the +5-V DC power supply. The idea in this circuit is to use the inverter or buffer output to provide a bias current to transistor Q1. The value of the base resistor (R1) is a function of the Q1 collector current and the beta of Q1. This resistor should be selected to safely turn on the transistor all the way to saturation when the output of U1 is HIGH.

The relay will be energized when the output of Q1 is HIGH. Therefore, the B0 control signal should also be HIGH if U1 is a noninverting buffer and LOW if U1 is an inverter.

Both relays K1 and K2 in Figure 19-1 use a diode in parallel with the relay coil. This diode is used to suppress the inductive kick spike created when the relay is deenergized. The magnetic field surrounding the coil contains energy. When the current flow is interrupted, the field collapses, which causes that energy to be dumped back into the circuit. The result is a high-voltage counter electromotive force (CEMF) spike that will possibly burn out the semiconductor devices or, in the case of digital circuits, create glitches (pulses that should not occur). The diode should be a rectifier type with a peak inverse voltage rating of 1000 V and a current of 500 mA or more. The 1N4007 diode has a 1000 PIV rating at 1 A. This diode will suffice for all but the heaviest relay currents.

Figure 19-2 shows a method for driving a relay from a low fan-out output port bit without the use of the inverter. The transistor driver is a pair of transistors connected in the Darlington amplifier configuration. Such a circuit connects the two collectors; the base of Q1 becomes the base for the pair; the emitter of Q2 becomes the emitter for the pair. The advantage of

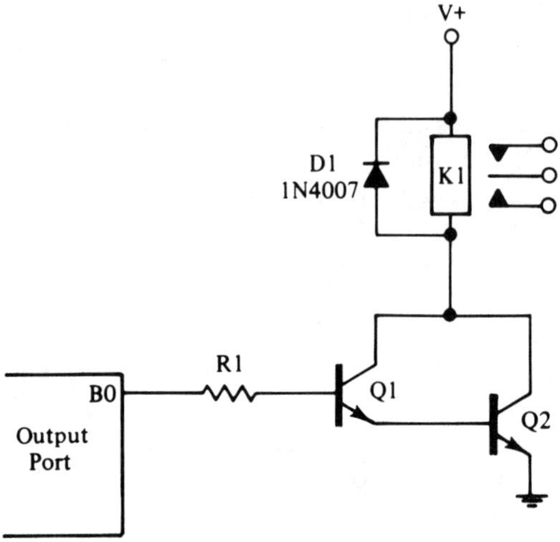

Figure 19-2 High-gain relay interface circuit.

the Darlington amplifier is that the current gain is greatly magnified. Current gain, *beta*, is defined as the ratio of the collector current to base current (I_c/I_b). For the Darlington amplifier, the beta of the pair is the product of the individual beta ratings:

$$\beta_{1\text{-}2} = \beta_{Q1} \times \beta_{Q2} \tag{19-1}$$

or,

$$\beta_{1\text{-}2} = \beta^2 \tag{19-2}$$

Equation 19-2 is used when the two transistors are identical. Since the total beta is the product of the individual beta ratings, when two identical transistors are used this figure is the beta squared.

You can use either a pair of discrete transistors to make the Darlington pair or one of the newer Darlington devices that houses both transistors inside one TO-5, TO-66, or TO-3 power transistor case.

Another method of isolating dangerous or heavy-duty loads from the microcomputer output port is shown in Figure 19-3. In this case we use an optoisolator as the interface media. The optoisolator uses light flux between an LED and a phototransistor to couple the on-off signal from input to output. The LED produces light when a current is caused to flow in it, while the phototransistor is turned on (saturated) when light falls on the base and is off when the base is dark (LED off). The transistor and LED are housed together, usually in a six-pin DIP package.

The LED in the optoisolator is connected to the output of an open-collector TTL inverter. The cathode end of the LED is grounded and the

LED thereby turned on whenever the output of the inverter is LOW. Thus the LED is turned on whenever bit B0 of the output port is HIGH. At the instant that the LED is turned on, transistor Q1 comes saturated, so collector-emitter current flows in resistor R4, thereby causing a voltage drop that can be used for control purposes.

The voltage drop across resistor R4 can be used to drive another NPN transistor that actually controls the load. Or we can create an RC differentiator (R2/C1) and use the leading edge of the voltage across R4 (as it turns on) to trigger some other device. In Figure 19-3, for example, we are using a triac to control the AC load. A *triac* is a full-wave silicon-controlled rectifier (SCR) and will gate on when a pulse is received at the gate (G) terminal. Most triacs or SCRs will not turn off with gate signals, so some means must be provided to reduce the cathode-anode current to near zero when we want to turn off the device. That is the purpose of switch S1. When we want to turn the circuit off, switch Sl is opened long enough to allow the triac/SCR to revert to its off condition. Some devices allow turn-off as well as turn-on by external pulses.

A method for interfacing the microcomputer with display devices such as an oscilloscope or a strip chart paper recorder is shown in Figure 19-4. There are instances when these devices are the most appropriate means of display, so we will want to provide some means to convert binary data to analog voltages for the oscilloscope or recorder. In Figure 19-4, the conversion is made by a digital-to-analog converter (DAC). The DAC produces an output potential V_0 that is proportional to the binary output. Various coding

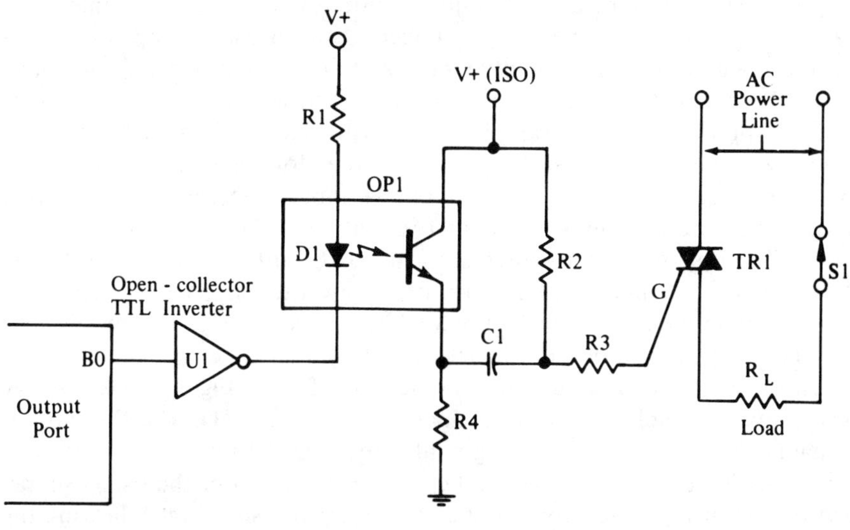

Figure 19-3 Optically isolated AC power controller.

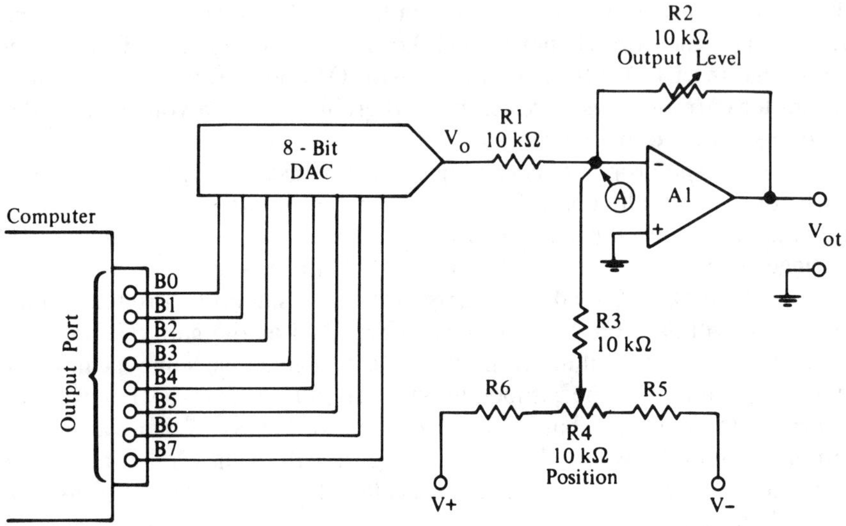

Figure 19-4 DAC interface to oscilloscope or recorder.

schemes are available (they will not be discussed here). We will assume for the purposes of our discussion that straight binary coding is used in which the 0-V state is represented by a binary word of 00000000, and full-scale output is represented by the binary word 11111111. States in between zero and full-scale are represented by proportional binary words; half-scale, for example, is represented by 10000000.

We will want to be able to scale the output potential V_{ot} to some value that is compatible with the display device. Not all oscilloscopes or paper recorders will accept any potential we apply, so some other must be introduced. Some oscilloscopes used in special medical, scientific, or industrial monitor applications, for example, come with fixed 1-V inputs. These instruments are often the most likely to be selected for applications involving a computer, yet they lack the multivoltage input selector of engineering models. For these we must select a DAC output voltage V_0 that will match the oscilloscope input requirements. If the DAC output is somewhat higher (0 to 2.56 V is common), some form of output attenuation is needed. The operational amplifier used in Figure 19-4 provides that attenuation.

The voltage gain of an ordinary operational amplifier connected in the inverting follower configuration, as in the case of A1 in Figure 19-4, is set by the ratio of feedback to input resistances (i.e., R2 and R1). For this circuit, the gain is $(-R2/R1)$; the minus sign indicates polarity inversion. The inversion means that we must design either the DAC output or the oscilloscope/recorder input to be negative. We can reinvert the signal by following the amplifier with another circuit that is identical except that R2 is a fixed resistor rather than a potentiometer. In that case, R1 = R2 = 10KΩ (or any other value that is convenient). The product of two inversions is the same as if none had taken place; V_{ot} will be inphase with V_0.

A position control is provided by potentiometer R4. In this circuit, we are producing an intentional output offset potential around which the waveform V_0 will vary. The effect of this potential is to position the waveform on the oscilloscope screen or chart paper where we want it. Sometimes the base-line (i.e., zero-signal) position will be in the center of the display screen or paper; in other cases it will be at one limit or the other.

An alternative system that would allow positioning of the base line under program control is to connect a second DAC (with its own R1) to point A, which is the operational amplifier summing junction. The program can output a binary word to this other DAC that represents the desired position on the display. That position can be controlled automatically by the program or manually in response to some keyboard action by the operator. This approach requires the investment of one additional DAC, but IC DACs are relatively cheap these days.

If the DC load driven by the DAC-computer combination is somewhat more significant than an oscilloscope input, the simple op-amp method of Figure 19-4 may not suffice. For such applications we may need a power amplifier to drive the load.

A power amplifier is shown in Figure 19-5. Here we have a *complementary symmetry* class B power amplifier. A *complementary pair* of power transistors is a pair, one NPN and the other PNP, that are electrically identical except for polarity. When these transistors are connected with their respective bases in parallel and their collector-emitter paths in series, the result is a simple push-pull class B amplifier. When the DAC output voltage V_0 goes positive, transistor Q1 will tend to turn on, and current flowing through Q1 under the influence of V+ will drive the load also positive. If, on the other hand, the output voltage of the DAC is negative, PNP transistor Q2

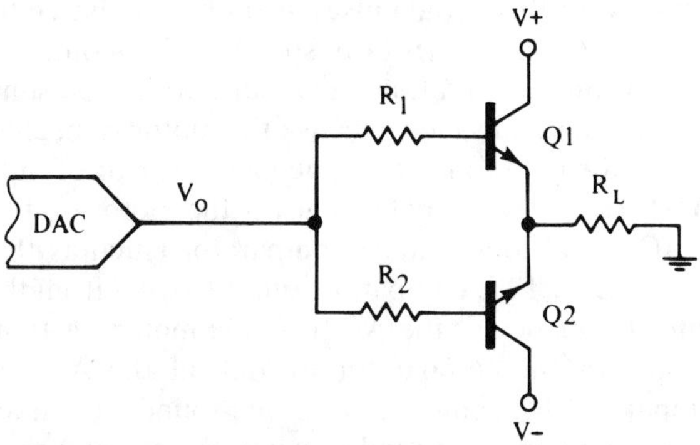

Figure 19-5 High-power DAC output.

will turn on and the load will be driven by current from the V− power supply. Since each transistor turns on only on one-half of the input signal, the result is full wave power amplification when the two signals are combined in the load.

The load in Figure 19-5 can be any of several different devices. If it is an electrical motor, for example, the DAC output voltage will vary the speed of the DC motor; hence the computer will control the speed because it controls V_0. If we provide some means for measuring the speed of the motor, the computer can be used in a negative feedback loop to keep the speed constant or change it to some specific value at will.

There is also a method by which the motor can be controlled without the DAC. If we use a transistor driver to turn the motor on and off, we can effectively control its speed by controlling the relative duty cycle of the motor current. By using a form of pulse-width modulation, we can set the motor speed as desired.

Pulse-width modulation of the motor current works by setting the total percentage of unit time that the motor is energized. The current will always be either all on or all off, never at some intermediate value. If we vary the length of time during each second that current is applied, we control the total energy applied to the motor and hence its speed. If we want the motor to turn very slowly, we arrange to output very narrow pulses through the output port to U1 to the motor control transistor. If, however, we want the motor speed to be very fast, long-duration pulses, or a constant level, is applied to the output port.

Can you spot the most common programming error that will be made when you actually try to implement this circuit? It occurs at turn-on. The DC motor has a certain amount of inertia that keeps it from wanting to start moving when it is off. As a result, if we want to start the motor at a slow speed, the pulse width may not be great enough to overcome inertia, and the motor will just sit there dormant. The solution is to apply a quick, one-time, long-duration pulse to get the motor in motion anytime we want it to turn on from a dead stop. After this initial pulse, the normal pulse coding will apply.

If we want to actively control the speed of the motor, we will need a sensor that converts angular rotation into a pulse train. On some motors this problem is made less of a nuisance because the motor is mechanically linked with an AC alternator housed in the same case. A pair of output terminals will exhibit an AC sine-wave signal whenever the motor shaft is rotating. If we apply this AC signal to a voltage comparator (such as the LM-311 device), we will produce a TTL-compatible output signal from the comparator that has the same frequency as the AC from the motor. A typical case uses the inverting input of the comparator to look at the AC signal, and the noninverting input of the comparator is at ground potential. Under this condition, there will be an output pulse generated every time the AC signal crosses the 0-V base line. Such a circuit is called a *zero-crossing detector,* appropriately enough.

If there is no alternator, some other means of providing the signal must be designed. One popular system is shown in Figure 19-6. A wheel with holes in the outer rim is connected to the motor output shaft. An LED and phototransistor are positioned such that light from the LED will fall on the phototransistor whenever a hole in the wheel is in the path; otherwise, the light path is interrupted. Flashes of light produced when the wheel rotates trigger the transistor to produce a signal that is in turn applied to the input port bit, as shown. A program can then be written to sample this input port bit and determine the motor speed from the frequency of the pulses or, as is more likely with some microprocessors, the time between successive pulses.

The sensor shown in Figure 19-6 may be constructed from discrete components, if desired, but several companies make such sensors already

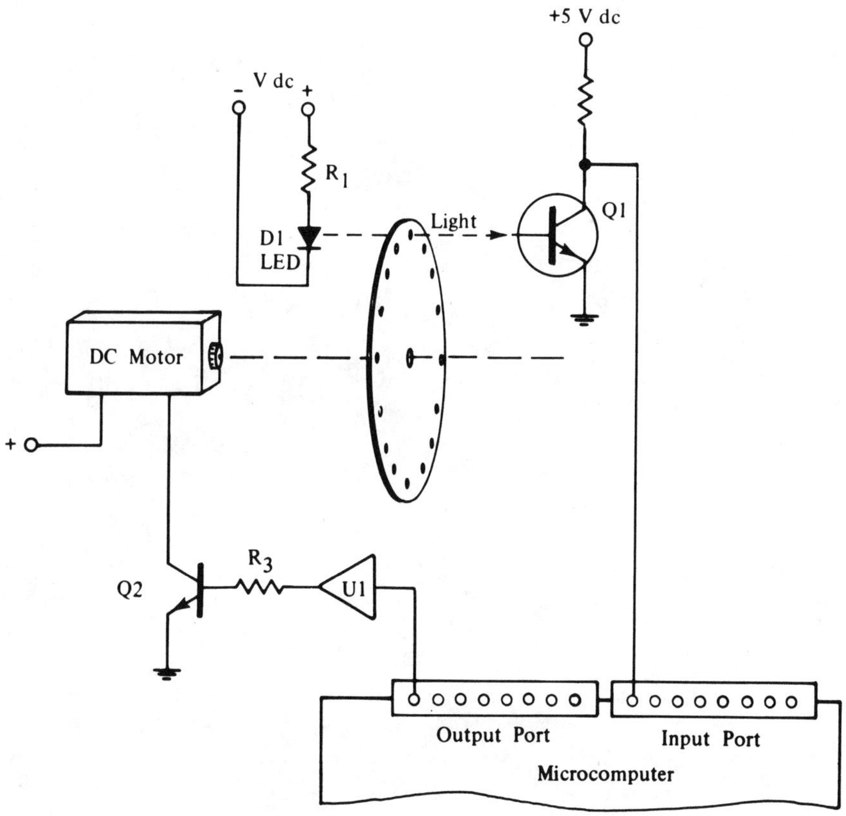

Figure 19-6 DC motor controller.

built into a plastic housing. A slot is provided to admit the rim of the wheel to interfere with the light path.

The methods shown in this chapter are intended to be used as guides, and you may well come up with others that are a lot more clever. The computer does not need much in the way of sophisticated interfacing in most cases, as can be seen from some of the foregoing examples.

20

DC POWER SUPPLIES FOR SMALL COMPUTER SYSTEMS

The DC power supply is a mundane component and is thus overlooked all too often by designers who should know-better. The reliability and proper functioning of the computer-based instrument is determined in no small measure by the performance and reliability of the DC power supply. Consider that the DC power supply accounts for the largest number of failures in systems, regardless of the size or complexity. In addition, operating components too close to (or above) their ratings will force reliability problems and possible secondary damage to the computer or instrument circuit.

The purpose of this chapter is not to make you into a power supply designer. Although most power supplies tend to be quite simple, the subject of professional power supply design—especially modern high-efficiency switching types—is too complex for glib treatment in a single, short chapter. Readers who wish to delve deeper into the subject are referred to any number of texts on the design of DC supplies. The purpose of this chapter is to offer a couple of circuits, with certain options that you must select from, that will cover the majority—perhaps vast majority—of cases.

Most small digital computers, microprocessor-based instruments, and assorted digitial electronics instruments require three power supplies: +5 V (regulated) at 1 to 5 A, +12 V at 1 A, and −12 V at 1 A. In some cases, the +/−12-V supplies are replaced with +/−15-V supplies, but the principles remain the same. In the sections below we will look at both forms of DC power suppy and offer you some guidelines for selecting components. Be sure to read the entire discussion; some of the principles are common among all of the supplies and so will not be presented redundantly. The criterion for selection of these circuits included availability of components. If you are an individual user, or represent a small laboratory or a school, then it is unlikely that industrial sources of components are open to you . . . at least not economically. Therefore, the components need to be easily available through mailorder sources or local electronics parts distributors.

+5-VOLT DC POWER SUPPLIES

The standard voltage for digital computers is +5 V DC. The +5-V DC potential must be tightly regulated between tolerance limits of +4.90 and +5.1 V. Although modern logic families such as CMOS (and other MOS forms) can operate over a number of different power supply potentials, the TTL family requires these tight limits, so they form the design driver in most digital instruments. As a result, the manufacturers of microprocessor chips routinely elect to use the +5-V DC potential for their chips.

The currents specified tend to be 1 A for very small systems to as much as 5 or 7 A for larger systems. These values are for instruments based on a single-board computer, not for full-sized microcomputers such as the IBM PC. For those size machines, power supplies in the 10- to 25-A range are used, and that requirement makes switching power supplies more efficient than fixed regulator supplies. The IBM PC/XT/AT level of supply can be purchased at relatively low cost from a large number of import and domestic sources. A typical 135-watt IBM XT clone power supply now costs less than $60. These supplies offer multiple +5-V DC, +12-V DC, and −12-V DC outputs, so they can also be used as the basis for other forms of instrument or computer than the IBM PC family of computers.

Figure 20-1 shows the circuit for a simple +5-V DC power supply capable of delivering maximum output currents of 1 to 5 A. The same circuit, with some alteration, would also serve to provide maximum currents of

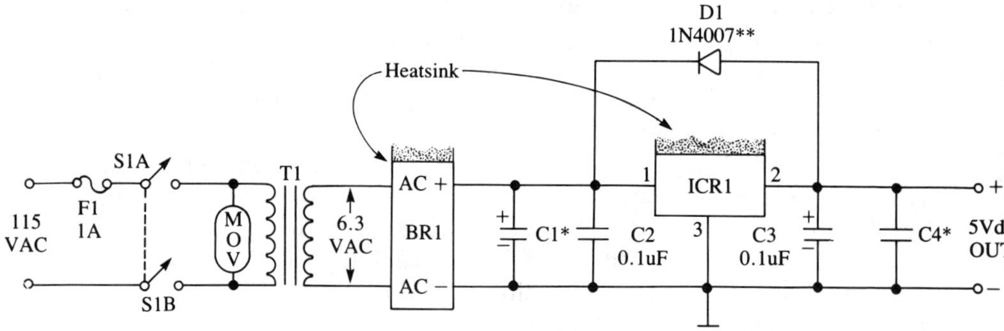

Figure 20-1 Typical +5-V DC power supply.

100 mA, 500 mA, or greater than 5 A. But more of that later. For now let's spend some time looking at the various components and how to make selection of some of the options.

Fuse. The job of a fuse (F1 in Figure 20-1) is to interrupt the circuit when excessive electrical current flows. It therefore protects the circuit from further damage when a fault or output short circuit occurs. The value of the fuse should be 1.5 to 2 times the normal maximum current flow, according to one popular rule of thumb. Place the fuse in the hot lead from the AC power lines, not the neutral. This may require the use of a polarized AC line cord to ensure that the fuse remains in the hot line.

Power switch. The on/off power switch (S1A/B in Figure 20-1) is used to turn the power supply, thus also the circuit that it serves, on and off. Although only a single-pole single-throw (SPST) switch is actually needed, I prefer to use a double-pole single-throw (DPST) switch in the power supply. The DPST switch allows us to open both the hot and neutral sides of the AC power mains and is thus an extra safety feature. Be sure to use a switch that is rated at the correct current and voltage for the application.

MOV. A *metal oxide varistor* (MOV) is a special device that provides a measure of protection against high-voltage transients that arrive on the AC power lines. According to some studies, a large number of these transients (>1,500-V peak) can occur on residential and light industrial power systems. On some heavier systems, including those of schools and universities, the number is increased because of heavy electrical machinery in the building. These transients are often the cause of seemingly random "bomb-outs" and other odd pathological behavior of some digital computers.

The MOV clips those transients to some maximum voltage that is well within the range tolerated by the computer and its power supply. Although the MOV is considered optional by some people, I consider it essential if proper operation of the computer is a must . . . and besides, it's cheap protection.

Transformer. The power transformer (T1 in Figure 20-1) steps the 115-V AC power line voltage down to the level where it is most useful for the electronic circuits. In general, the voltage regulators (more about these in a moment) require an input voltage that is 2.5 or 3 V higher than the rated output voltage. Thus a +5-V DC regulator needs to see +8 V DC as a minimum input voltage. This voltage sets the required secondary voltage of the transformer.

When transformer secondary voltages are specified, it is the *rms* voltage that is used. On a sine wave (which the AC power line supplies) the peak voltage is 1.414 times the rms voltage ($V_p = 1.414V_{rms}$). The output of the transformer should be sufficient to supply a peak voltage just a small amount above that which will produce the input voltage that the regulator needs. One of the standard transformer secondary voltages is 6.3 V AC, a holdover from vacuum days but still useful. The peak voltage of this transformer is 1.414 × 6.3 V AC = 8.9 V. Thus for a +5-V DC power supply, a 6.3-V AC "filament" (vacuum tube term) transformer is sufficient.

One does not want to use a higher voltage transformer secondary rating, even though the regulator can usually tolerate up to 30, 35, or even 40 V. The power dissipation of the regulator is the product of the load current and the *difference* between input and output voltages. Thus in order to reduce internal power dissipation we need to keep the input voltage to a point just above the minimum value. That's why it's preferable to use a 6.3-V AC transformer rather than a 12.6-V AC transformer for a +5-V DC power supply.

The current rating of the transformer secondary mst be sufficient to produce the required output current under constant conditions without burning up *plus* a small reserve for reliability purposes. The rule of thumb is to have a reserve capability of 20 percent, although I prefer at least a 40 percent reserve. Thus a 1-A power supply should have a transformer with a secondary current of 1.2 A or more.

There is a little "glitch" in the rating of some transformers that is overlooked by many people. If the transformer is center-tapped, then it is most likely rated for a standard full-wave bridge circuit (Figure 20-2) rather than the bridge rectifier (BR1) shown in Figure 20-1. The same transformer can be used for the bridge circuit if the center tap is not grounded. However, check the *primary volt-ampere* rating of the transformer to determine whether the output current rating is still good. Otherwise, as a rule of thumb, *halve* the secondary current rating of the transformer *unless* the product of the entire secondary voltage and the secondary current rating is less than or equal to the primary VA rating.

Rectifier. The job of the rectifier (BR1 in Figure 20-1 and D1/D2 in Figure 20-2) is to convert bidirectional alternating current into a form of unidirectional current called *pulsating DC*. There are two ratings of rectifiers that must be considered: *forward current* and *peak inverse voltage* (PIV). The forward current rating tells us the maximum current that the rectifier can pass without harm. Again, apply a 20 percent or more tolerance for reliability's sake. The PIV rating is the maximum reverse voltage that the rectifier can tolerate without being harmed. For a filtered DC power supply such as we are discussing here, the PIV voltage of the rectifier should be not less than 2.83 times the applied rms (i.e. twice the applied peak voltage). For +5-V DC power supplies this requirement is not inordinately difficult to meet because the lowest value commonly seen is 50 V PIV.

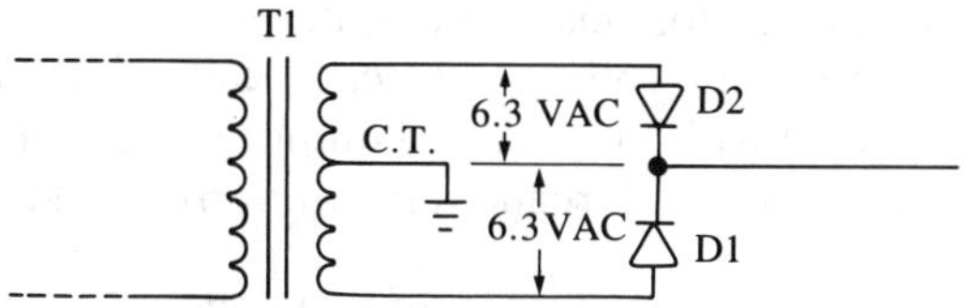

Figure 20-2 Rectifier/transformer circuit.

Filter capacitor. The job of the filter capacitor (C1 in Figure 20-1) is to smooth the pulsations of the pulsating DC and make them into nearly pure DC. For nonregulated power supplies one uses the maximum ripple factor that the circuits being served can tolerate to select the capacitance. In regulated power supplies such as Figure 20-1, however, the manufacturer of the regulator (ICR1) will set a recommended value of at least 1,000 μV per ampere of output current. Other authorities ask for 2,000 μF/A. Thus, the chart in Figure 20-1 shows values of at least 1,000 μF/A for C1.

The working voltage rating, usually called *working voltage DC,* or WVDC, is the maximum constant voltage that the capacitor will accommodate. Again, apply at least a 20 percent tolerance . . . and again I prefer a wider tolerance than the "rule of thumb." For a 5-V DC power supply that uses a 6.3-V AC transformer, the output of the rectifier will be close to 8.9 V. With a reasonable safety factor, the WVDC rating of Cl will be 15 WVDC or more. Therefore, for Cl select a 2,000 μF/15 WVDC or more capacitor.

Noise capacitors. Capacitors C2 and C3 in Figure 20-1 are used for suppression of transient noise that sometimes enters the DC power supply from external sources. These noise pulses can interrupt the operation of the regulator (ICR1) and also affect its reliability. To prevent this problem, C2 and C3 (0.1 to 1.0 μF) are mounted as close as possible to the pins of the regulator device . . . indeed they are often mounted directly to the device.

IC voltage regulator. The job of the voltage regulator (ICR1 in Figure 20-1) is to keep the output voltage constant despite a) changes in input voltage, and b) changes in output load current requirements. At one time, voltage regulators were somewhat more complex and difficult to design, so they were rarely used. Today, however, modern three-terminal IC regulators are used extensively. They are easy to use and are quite cheap. The complexity is still present, but it is internal to the IC and therefore transparent to the user.

There are only a few different general classes of three-terminal IC voltage regulator that we need to be aware of; four of the most popular are the 78xx and LM-340n-xx for positive output voltages, and 79xx and LM-320n-xx for negative output voltages. In each case, the "xx" term in the type number is replaced with the required output voltage. For example, the 7805 and LM-340n-05 are +5-V regulators, while 7812 and LM-340n-12 are +12-V regulators.

The "n" term in the LM-340n-xx and LM-320n-xx type numbers indicate the package style, hence the current rating as well. An "H" in place of "n" represents a small plastic package similar to the TO-92 transistor package, or a metal package similar to the TO-5 transistor package. These regulators pass up to 100 or 150 mA of current, depending on type. A "T" (e.g., LM-340T-12) denotes a plastic TO-220 package. Some spec sheets rate these regulators at 1A, but a rating of 750 mA without a heatsink is more reasonable. A "K" in the type number (e.g., LM-340K-12) denotes a diamondshaped metal TO-3 package, and the ability to pass up to 1 A (or more in certain cases).

In addition to the LM-340, LM-320, 78xx, and 79xx regulators, one also sees the LM-309 device (+5 V, 1 A, in the K package) and LM-323 device (+5 V, 3 A, in the K package). Lambda Electronics offers a series of high-current +5-V regulators under the nomenclature L-1905.

Heatsinks. The voltage regulator (ICR1) in Figure 20-1 is shown heatsinked, as is the rectifier bridge stack. These parts are often overlooked but are nonetheless quite important. The heatsink is designed to carry away the heat generated inside the part, and it thus affects reliability. In general, a 10° C decrease in a semiconductor's junction temperature *doubles* the mean time between failures. Thus the heatsink is, in my opinion, not optional. For the regulator select a heatsink that matches the package style. For the bridge rectifier, use a heatsink for rectifiers rated at 3 A or more.

Output capacitor. Capacitor C4 in Figure 20-1 is used to smooth out variations in the output voltage caused by sudden changes in the load current demand. It takes the regulator only a very short time to respond to the change, but that time is long enough to create a glitch in the power supply voltage. Capacitor C4 creates a small reservoir of charge that is dumped into the circuit when the glitch occurs, smoothing it out. Select a value for C4 according to the rule 100 μF/A of output current.

Protection diode. Some power supplies use a reverse polarity protection diode (D1 in Figure 20-1) to prevent damage to the regulator in case the output voltage becomes larger than the input voltage. This situation can occur at turn-off when a) capacitor C4 is used, or b) the load contains a large capacitance. For 1-A and 3-A DC supplies, the 1N4007 is sufficient (1000-V PIV @ 1-A). For larger supplies use a 1000-V PIV @ 3-A diode for D1.

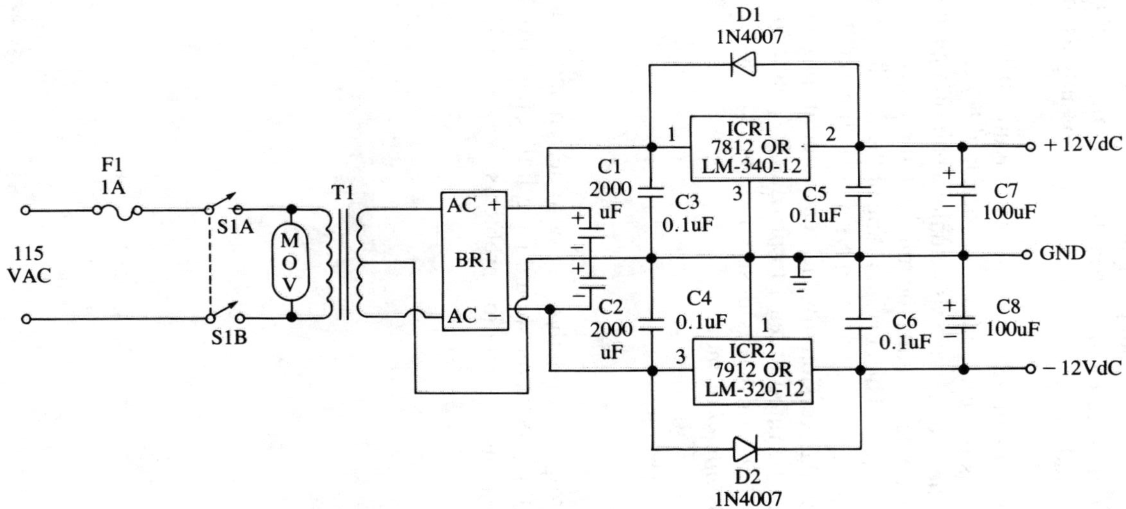

Figure 20-3 Dual-polarity DC power supply for ±12 V DC.

+/−12-VOLT DC POWER SUPPLY

The other DC power supply often needed in small computer projects is the +/−12-V DC @ 1-A bipolar supply. This supply (Figure 20-3) provides two output voltages (measured with respect to ground): +12 V and −12 V. The current rating should be at least 500 mA, and possibly as much as 3 A, per output.

The circuit of Figure 20-3 is similar to Figure 20-1 except that it uses a center-tapped transformer. Although a standard "bridge" rectifier stack is used for the rectifier, the rectifier is actually a pair of regular (nonbridge) full-wave rectifiers in tandem. Thus the full current rating of the transformer can be used for this supply. The current rating of the rectifier is set to at least the maximum output current of the supply, plus a 20 percent safety margin.

For the popular version in which each output provides up to 1 A, the transformer must be rated at least to provide 2 A and, with safety margins considered, up to 2.5 or 3 A is desirable. The secondary voltage rating is 25-V AC *center-tapped* or more.

CONCLUSION

The DC power supply is relatively easy to design and build, at least for the purposes stated herein. You should have little or no trouble applying this material to actual instrumentation problems.

APPENDIX A

FAST FOURIER TRANSFORM (FFT) PROGRAM FOR SMALL COMPUTERS

The *Fast Fourier Transform* (FFT) revolutionized certain instrumentation chores because it allowed more rapid determination of frequency components than was previously possible. In previous times, the scientist or engineer who needed to know the constituent frequency components of a waveform had to use either a mammoth analog filter system (which was both expensive and unreliable) or a mainframe computer to calculate the Fourier components. The FFT is a relatively recent development (made 20 years or so ago) that, when coupled with microcomputer technology, has made possible a large number of new and improved instruments, as well as allowing laboratory scientists to perform experiments that were previously impossible for all but the wealthy and well-equipped.

The FFT program in this appendix was developed by C. E. McCullough, a personal friend and professional colleague of the author's, from earlier published sources and his own ingenuity. The original FFT project was to permit analysis of medical electroencephalograph (EEG) real-time and evoked potentials signals in the Anesthesiology Research Laboratory of the George Washington University Medical Center.

The assembly language program published herein is configured to run on Apple II 6502-based microcomputers with Applesoft and Apple DOS 3.2 (or higher). It may easily be rewritten for other 6502 (and related) systems.

Figure A-1 shows the flowchart for the FFT program. This chart is included to facilitate understanding of the program and to make it easy to modify it for other systems.

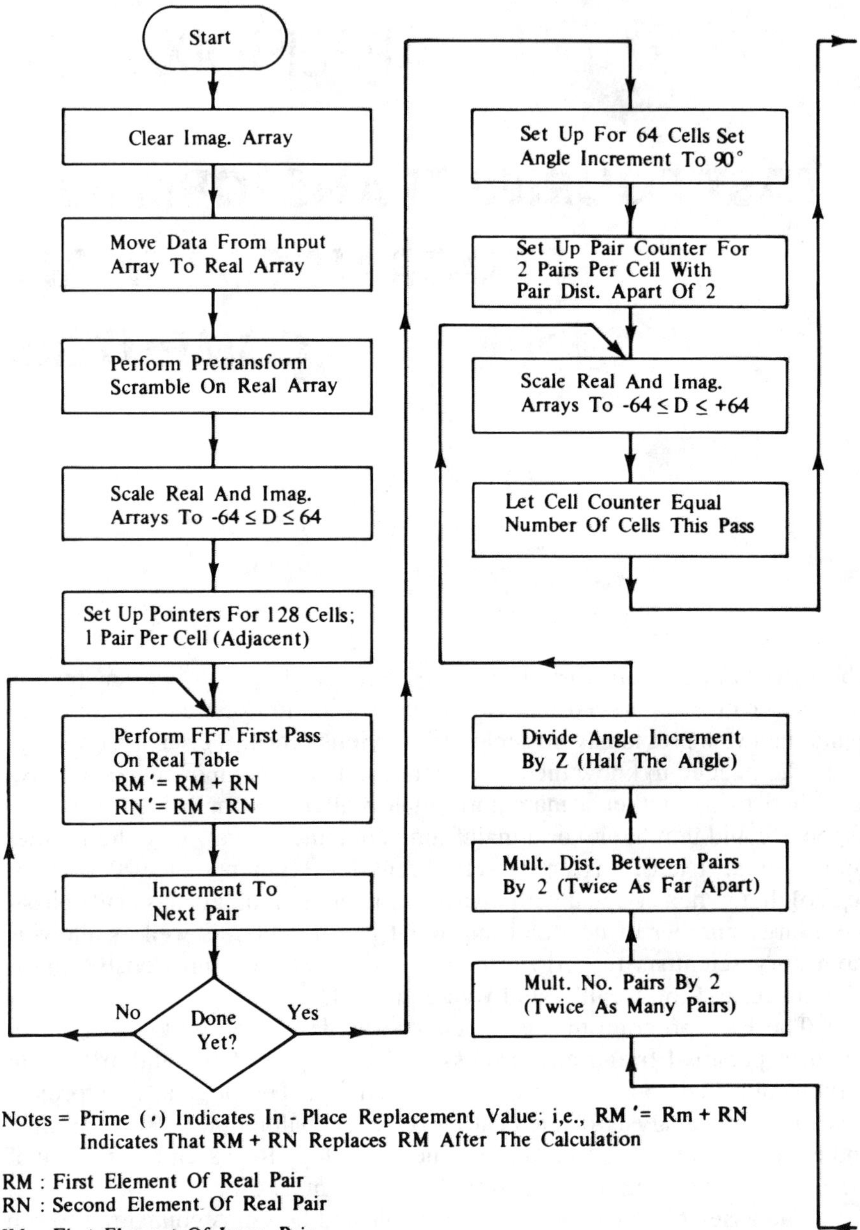

Notes = Prime (′) Indicates In - Place Replacement Value; i,e., RM ′= Rm + RN Indicates That RM + RN Replaces RM After The Calculation

RM : First Element Of Real Pair
RN : Second Element Of Real Pair

IM : First Element Of Imag. Pair
IN : Second Element Of Imag. Pair

Figure A-1 FFT flow chart.

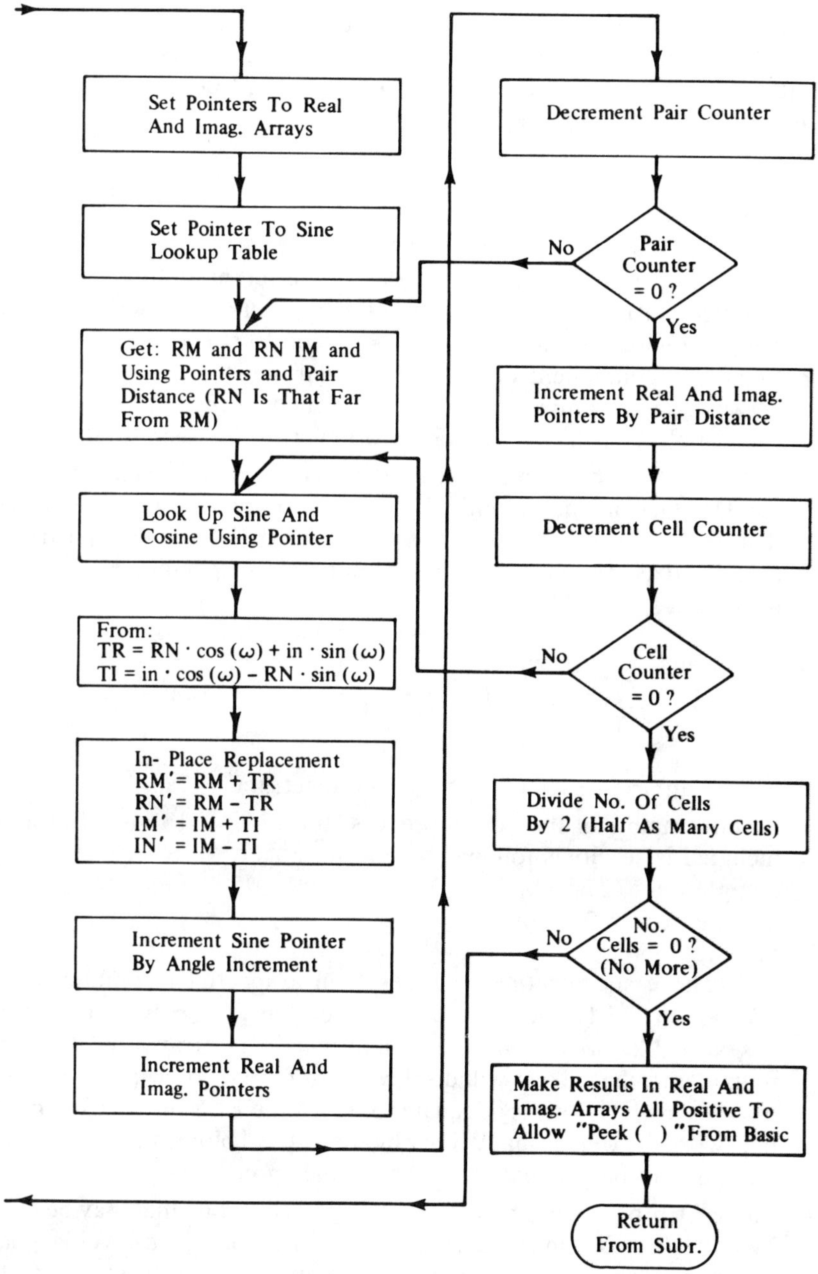
Set Pointers To Real And Imag. Arrays
Set Pointer To Sine Lookup Table
Get: RM and RN IM and Using Pointers and Pair Distance (RN Is That Far From RM)
Look Up Sine And Cosine Using Pointer
From:
TR = RN · cos (ω) + in · sin (ω)
TI = in · cos (ω) – RN · sin (ω)
In- Place Replacement
RM' = RM + TR
RN' = RM – TR
IM' = IM + TI
IN' = IM – TI
Increment Sine Pointer By Angle Increment
Increment Real And Imag. Pointers
Decrement Pair Counter
Pair Counter = 0 ?
No
Yes
Increment Real And Imag. Pointers By Pair Distance
Decrement Cell Counter
Cell Counter = 0 ?
No
Yes
Divide No. Of Cells By 2 (Half As Many Cells)
No. Cells = 0 ? (No More)
No
Yes
Make Results In Real And Imag. Arrays All Positive To Allow "Peek ( ) "From Basic
Return From Subr.

An eight-channel analog-to-digital (A/D) converter designed for use on Apple II computers is shown in Figure A-2. The circuit is extremely simple, and it (or its equivalent) may be built on a standard plug-in interfacing or prototyping card to fit Apple II.

FFT NOTES

1. For A/D conversion, an 8-bit A/D converter must input 256 two's complement numbers into the input sample area. To design the A/D sampling hardware and software, you need to consider the maximum frequency component of the input signal and the desired resolution between spectral lines. You are constrained by the fact that you need 256 samples, so this and other factors will set the record length.[1]
2. The output will be found in the two areas of memory designated REAL and IMAG. The theory of the FFT and the in-place algorithm are such that the first and last 128 points are redundant. The power spectrum is calculated as the sum of the squares of the corresponding 128 reals and imaginaries. In BASIC,

```
FOR I = 0 TO 127
FFT(I) = PEEK(REAL + I)^2 + PEEK(IMAG = 1)^2
NEXT I
```

   will fill an array FFT with the power spectrum.
3. As an example of the design process, the FFT A/D sampler program included is set up as follows:
   Sample rate: 127.5 Hz
   Sample period: 2 seconds
   Samples: 256
   These specifications will result in a spectral resolution of 1/2-seconds or 0.5 Hz between lines. Since the FFT power spectrum is based on 128 pairs from the real and imaginary arrays (the 128 being redundant), the system will display 128 × 0.5 Hz, or a spectral display of DC to 64 Hz. This is adequate for real-time EEG and evoked potentials work (as well as many other biological and biomedical waveforms) but may not be adequate for other signals processing tasks.
4. The FFT program is a machine lanquage subroutine that may be called by either another machine language program or a high-level language program. Control is returned to the main program by executing the RTS instruction.

[1] See William Stanley, *Digital Signal Processing,* Reston Publishing Company (Prentice-Hall), Reston, Va.

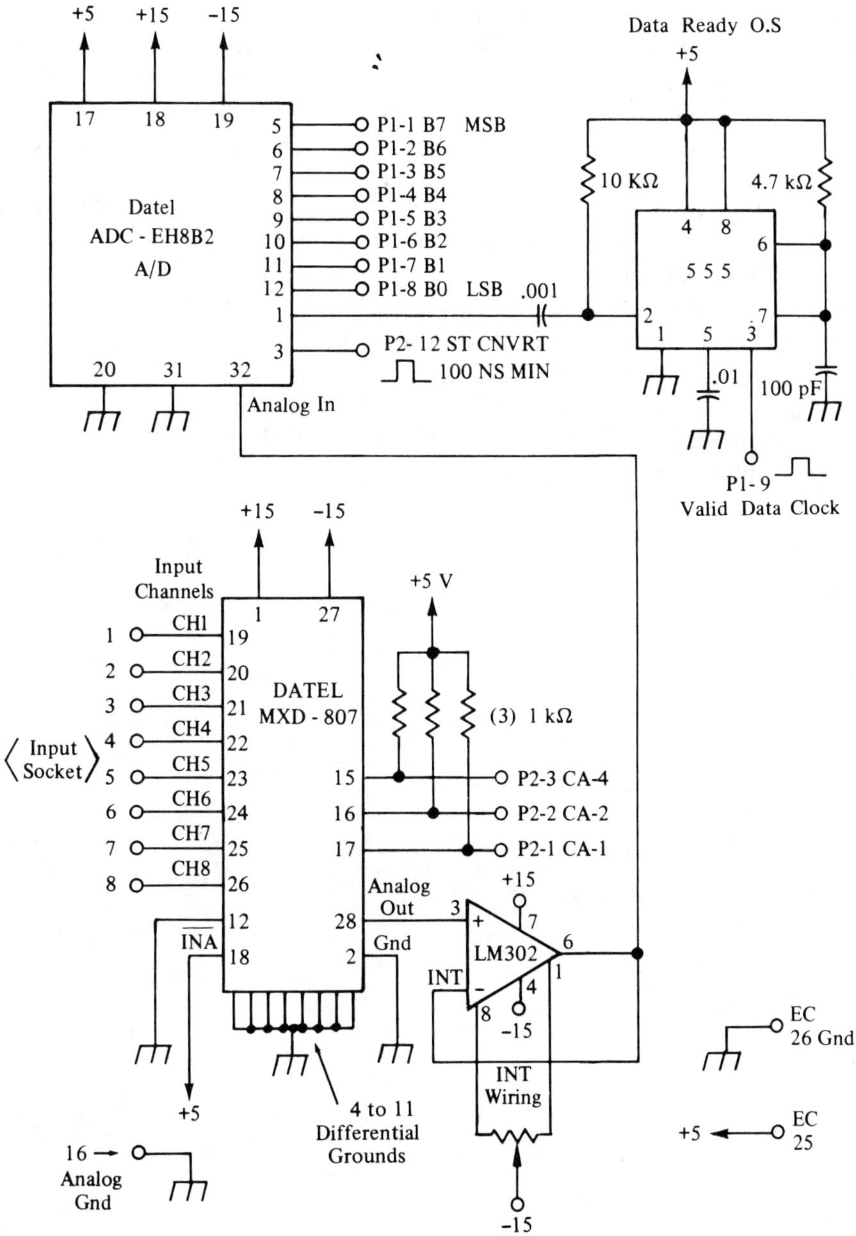

Figure A-2 Interface circuit.

5. The FFT program and driver software for the A/D converter was written to run on Apple II (DOS 3.2) and fits into the memory space between Applesoft and DOS. The A/D routine is completely hardware-dependent, so the A/D converter circuit is provided in Fig. A-2.
6. The whole program can be reassembled to run anywhere in memory on any 6502 computer system, but keep in mind the following:
   a. The input data, sine look-up table, and REAL and IMAG areas must be located on page boundaries and contiguously in memory.
   b. Note that the look-up table (see program listing) must be resident in memory at all times; it is not generated by the FFT program listing.
7. Note that a routine was added after the FFT was calculated that makes all the resulting data positive, because Applesoft's PEEK statement can only interpret straight binary data (decimal 0 to 255), not two's complement. You square these numbers and add them in order to calculate the power spectrum.

```
             1000 *     6502 ASSEMBLY LANGUAGE
             1010 *     FAST FOURIER TRANSFORM SUBROUTINE
             1020 *     VERSION 1.41  10/22/79
             1030 *
             1040 *
             1050 *
             1060 * * *   H I M E M :  3 6 3 5 2   T O   M A K E   R O O M
                                                          F O R   F F T * * *
             1070 *
             1080 *
             1090 *
             1100 *
             1110 *   M E M O R Y   M A P
             1120 *
             1130 *   HEX ADDRESS    USE
             1140 *   -----------    ---
             1150 *   8E01 - 8EFF    POINTER AND SCRATCHPAD AREA
             1160 *   8F00 - 8FFF    INPUT DATA TABLE
             1170 *   9000 - 90FF    SINE LOOKUP TABLE
             1180 *   9100 - 91FF    REAL DATA TABLE
             1190 *   9200 - 92FF    IMAG DATA TABLE
             1200 *   9300 - 95FF    SUBROUTINE CODING
             1210 *
             1220 *
             1230 *   JSR $9300 OR CALL 37632 TO RUN FFT OF INPUT DATA
             1240 *
             1250 *   JSR $8E1A OR CALL 36378 TO SAMPLE INPUT
             1260 *
             1270 *  LDA $C0B1 WITH 0 THRU 7 TO SELECT CHANNEL
             1280 *  OR POKE 49329,CHAN    WHERE CHAN= 0 THRU 7
             1290 *
             1300 *
             1310 *   DATA AREAS
             1320 *
             1330 INPT .EQ $8F00      INPUT DATA TABLE
             1340 RELT .EQ $9100      REAL DATA TABLE
             1350 IMGT .EQ $9200      IMAG DATA TABLE
             1360 SINT .EQ $9000      SINE LOOKUP TABLE
             1370 *
             1380 *
             1390 *
             1400 *   POINTER AREA
             1410 *
             1420      .OR $8E01
             1430      .TA $0801
8E01- 00 00  1440 RPT1 .DA *-*        "REAL" DATA POINTERS
8E03- 00 00  1450 RPT2 .DA *-*
8E05- 00 00  1460 IPT1 .DA *-*        "IMAG" DATA POINTERS
8E07- 00 00  1470 IPT2 .DA *-*
8E09- 00 00  1480 SNPT .DA *-*        SINE TABLE POINTER
```

```
8E0B- 00        1490 CLNM .DA #*-*       CELLS FOR THIS PASS
8E0C- 00        1500 CLCT .DA #*-*       CELL COUNTER FOR PASS
8E0D- 00        1510 PRNM .DA #*-*       PAIRS/CELL
8E0E- 00        1520 CLDS .DA #*-*       CELL OFFSET (DISTANCE)
8E0F- 00        1530 DLTA .DA #*-*       ANGLE INCREMENT
8E10- 00        1540 SCFT .DA #*-*       SCALE FACTOR COUNTER
8E11- 00        1550 COSA .DA #*-*       TEMPORARY COSINE
8E12- 00        1560 SINA .DA #*-*       TEMPORARY SINE
8E13- 00        1570 TREL .DA #*-*       TEMP. REAL DATA
8E14- 00        1580 TMAG .DA #*-*       TEMP. IMAG DATA
8E15- 00        1590 LSBY .DA #*-*       PRODUCT LSBY
8E16- 00        1600 MSBY .DA #*-*       PRODUCT MSBY
8E17- 00        1610 CAND .DA #*-*       MULTIPLICAND
8E18- 00        1620 PLYR .DA #*-*       MULTIPLIER
8E19- 00        1630 DEC  .DA #*-*       DECIMAL MODE FLAG
                1640 *
                1650 *
                1660 *
                1670 *   A-D SAMPLING SUBROUTINE
                1680 *   2 SECONDS AT 127.5 HERTZ
                1690 *   OR 256 SAMPLES TAKEN.
                1700 *
                1710 CVRT .EQ $C0B4      START CONVERT STROBE
                1720 IPRT .EQ $C0B0      INPUT PORT ADDRESS
                1730 CHAN .EQ $C0B1      CHAN. SELECT PORT
                1740 *
8E1A- A2 00     1750 SMPL LDX #$00
8E1C- 8D B4 C0  1760 SPL  STA CVRT       FIRE START CONVERT ONE SHOT
8E1F- A9 04     1770      LDA #$04       DELAY 7.84 MSEC
8E21- A0 66     1780 POT  LDY #$66
8E23- EA        1790 INS  NOP
8E24- EA        1800      NOP
8E25- EA        1810      NOP
8E26- EA        1820      NOP
8E27- EA        1830      NOP
8E28- EA        1840      NOP
8E29- EA        1850      NOP
8E2A- 88        1860      DEY
8E2B- D0 F6     1870      BNE INS
8E2D- 38        1880      SEC
8E2E- E9 01     1890      SBC #$01
8E30- EA        1900      NOP
8E31- EA        1910      NOP
8E32- EA        1920      NOP
8E33- EA        1930      NOP
8E34- EA        1940      NOP
8E35- D0 EA     1950      BNE POT        END OF DELAY
8E37- AD B0 C0  1960      LDA IPRT       READ A/D
8E3A- EA        1970      NOP
8E3B- 9D 00 8F  1980      STA INPT,X     PUT IN INPUT TABLE
```

```
8E3E- E8        1990      INX             NEXT SAMPLE
8E3F- E0 00     2000      CPX #$00        DONE?
8E41- D0 D9     2010      BNE SPL         NO, GET ANOTHER
8E43- 60        2020      RTS             RETURN TO PROG.
                2030 *
                2040 *
                2050 *
                2060 *
                2070 *
                2080 *   START OF TRANSFORM
                2090 *
                2100      .OR $9300
                2110      .TA $0D00
9300- 08        2120      PHP             PUSH PSW
9301- 68        2130      PLA             PULL PSW
9302- 29 08     2140      AND #$08        MASK OFF DECI. FLAG
9304- 8D 19 8E  2150      STA DEC
9307- D8        2160      CLD             CLEAR DEC. MODE
9308- 4C 13 93  2170      JMP STRT        JUMP AROUND PARAMETERS
                2180 *
                2190 *
                2200 *
                2210 *   ADDRESS LOOKUP TABLE
                2220 *
930B- 00 8F     2230 INPD .DA INPT        SET UP DATA AREAS
930D- 00 91     2240 REAL .DA RELT
930F- 00 92     2250 IMAG .DA IMGT
9311- 00 90     2260 SINE .DA SINT
                2270 *
                2280 *
                2290 *
                2300 *   INPUT DATA SET UP
                2310 *
9313- A9 00     2320 STRT LDA #00
9315- 8D 10 8E  2330      STA SCFT
9318- A2 00     2340 CLER LDX #00
931A- 9D 00 92  2350 CLR1 STA IMGT,X      CLEAR IMAG AREA
931D- CA        2360      DEX
931E- D0 FA     2370      BNE CLR1
9320- A2 00     2380      LDX #00         MOVE INPUT DATA
9322- BD 00 8F  2390 MOV1 LDA INPT,X      TO REAL ARRAY
9325- 9D 00 91  2400      STA RELT,X
9328- CA        2410      DEX
9329- D0 F7     2420      BNE MOV1
                2430 *
                2440 *
                2450 *
                2460 *   PRE-TRANSFORM SCRAMBLER ROUTINE
                2470 *
932B- AD 0D 93  2480      LDA REAL        SET UP DATA POINTERS
932E- 8D 01 8E  2490      STA RPT1
```

```
9331- 8D 03 8E 2500      STA RPT2
9334- AD 0E 93 2510      LDA REAL+1
9337- 8D 02 8E 2520      STA RPT1+1
933A- 8D 04 8E 2530      STA RPT2+1
933D- A0 08    2540 BREV LDY #8        SET BIT COUNTER
933F- AD 01 8E 2550      LDA RPT1      GET POINTER 1
9342- 6A       2560 BRV1 ROR           REVERSE BIT ORDER
9343- 2E 03 8E 2570      ROL RPT2      FOR SECOND POINTER
9346- 88       2580      DEY           COUNT BITS
9347- D0 F9    2590      BNE BRV1
9349- AD 03 8E 2600      LDA RPT2      GET REVERSED BYTE
934C- CD 01 8E 2610      CMP RPT1      COMPARE WITH #1
934F- 90 16    2620      BCC SWP1      BRANCH IF SWAPPED
9351- AE 01 8E 2630 SWAP LDX RPT1      GET POINTER 1
9354- BD 00 91 2640      LDA RELT,X    GET VAL 1
9357- AE 03 8E 2650      LDX RPT2      GET POINTER 2
935A- BC 00 91 2660      LDY RELT,X    GET VAL 2
935D- 9D 00 91 2670      STA RELT,X    REPLACE WITH VAL 1
9360- AE 01 8E 2680      LDX RPT1      GET FIRST POINTER
9363- 98       2690      TYA           COMPLETE THE
9364- 9D 00 91 2700      STA RELT,X    SWAP.
9367- EE 01 8E 2710 SWP1 INC RPT1      DO NEXT POINT PAIR
936A- D0 D1    2720      BNE BREV      UNLESS ALL ARE DONE
               2730 *
               2740 *
               2750 *
               2760 *   FFT FIRST PASS
               2770 *
936C- 20 E4 94 2780 PAS1 JSR SCAL      SCALE IF ANY OVER-RANGE DATA
936F- AE 0D 93 2790      LDX REAL      GET REAL POINTER
9372- BD 00 91 2800 PA1  LDA RELT,X    GET RM
9375- A8       2810      TAY           HOLD RM
9376- 18       2820      CLC
9377- 7D 01 91 2830      ADC RELT+1,X  LET RM'=RM+RN
937A- 9D 00 91 2840      STA RELT,X    STORE RM'
937D- 98       2850      TYA           GET OLD RM
937E- 38       2860      SEC
937F- FD 01 91 2870      SBC RELT+1,X  LET RN'=RM-RN
9382- 9D 01 91 2880      STA RELT+1,X  STORE RN'
9385- E8       2890      INX           MOVE TO
9386- E8       2900      INX           NEXT PAIR
9387- D0 E9    2910      BNE PA1       KEEP GOING TILL DONE
               2920 *
               2930 *
               2940 *
               2950 *   FFT COMPUTATION:  PASS 2 THRU N
               2960 *
9389- A9 40    2970 FPAS LDA #64       SET UP PARAMETERS
938B- 8D 0B 8E 2980      STA CLNM      FOR CELL COUNT
938E- 8D 0F 8E 2990      STA DLTA      AND ANGLE
9391- A9 02    3000      LDA #2        AND FOR
```

```
9393- 8D 0D 8E 3010      STA PRNM      PAIRS/CELL
9396- 8D 0E 8E 3020      STA CLDS      DIST. BETWEEN PAIRS
9399- 20 E4 94 3030 NPAS JSR SCAL      KEEP DATA IN RANGE
939C- AD 0B 8E 3040      LDA CLNM      GET NUMBER OF CELLS
939F- 8D 0C 8E 3050      STA CLCT      PUT IN COUNTER
93A2- AD 0D 93 3060      LDA REAL      SET UP POINTERS
93A5- 8D 01 8E 3070      STA RPT1
93A8- 8D 03 8E 3080      STA RPT2
93AB- AD 0E 93 3090      LDA REAL+1
93AE- 8D 02 8E 3100      STA RPT1+1
93B1- 8D 04 8E 3110      STA RPT2+1
93B4- AD 0F 93 3120      LDA IMAG
93B7- 8D 05 8E 3130      STA IPT1
93BA- 8D 07 8E 3140      STA IPT2
93BD- AD 10 93 3150      LDA IMAG+1
93C0- 8D 06 8E 3160      STA IPT1+1
93C3- 8D 08 8E 3170      STA IPT2+1
93C6- AD 11 93 3180 NCEL LDA SINE
93C9- 8D 09 8E 3190      STA SNPT
93CC- AD 12 93 3200      LDA SINE+1
93CF- 8D 0A 8E 3210      STA SNPT+1
93D2- AC 0D 8E 3220      LDY PRNM      GET PAIRS/CELL COUNTER
93D5- AD 01 8E 3230 NC1  LDA RPT1      GET POINTER 1 LSBY
93D8- 18       3240      CLC
93D9- 6D 0E 8E 3250      ADC CLDS      ADD PAIR OFFSET
93DC- 8D 03 8E 3260      STA RPT2      SET BOTH POINTER 2'S
93DF- 8D 07 8E 3270      STA IPT2
93E2- 98       3280      TYA           SAVE PAIR COUNTER
93E3- 48       3290      PHA
93E4- AE 09 8E 3300      LDX SNPT      SET UP SINE LOOKUP
93E7- BD 00 90 3310      LDA SINT,X    GET COS OF ANGLE
93EA- 8D 11 8E 3320      STA COSA      SAVE COS
93ED- BD 40 90 3330      LDA SINT+64,X GET SINE
93F0- 8D 12 8E 3340      STA SINA      SAVE SINE
93F3- AE 03 8E 3350      LDX RPT2      GET REAL POINTER 2
93F6- BD 00 91 3360      LDA RELT,X    GET RN
93F9- 48       3370      PHA           SAVE IT
93FA- AC 11 8E 3380      LDY COSA      GET COS
93FD- 20 17 95 3390      JSR MPY       MAKE RN*COS(A)
9400- 8D 13 8E 3400      STA TREL      SAVE IT
9403- 68       3410      PLA           RESTORE RN
9404- AC 12 8E 3420      LDY SINA      GET SINE
9407- 20 17 95 3430      JSR MPY       RN*SIN(A)
940A- 8D 14 8E 3440      STA TMAG
940D- AE 07 8E 3450      LDX IPT2      GET IMAG POINTER 2
9410- BD 00 92 3460      LDA IMGT,X    GET IN
9413- 48       3470      PHA           SAVE IT
9414- AC 12 8E 3480      LDY SINA      GET SINE
9417- 20 17 95 3490      JSR MPY       IN*SIN(A)
941A- 18       3500      CLC
941B- 6D 13 8E 3510      ADC TREL      TR=RN*COS+IN*SINE
```

```
941E- 8D 13 8E          STA TREL
9421- 68                PLA             RESTORE IN
9422- AC 11 8E          LDY COSA        GET COS
9425- 20 17 95          JSR MPY         IN*COS(A)
9428- 38                SEC
9429- ED 14 8E          SBC TMAG        TI=IN*COS-RN*SINE
942C- 8D 14 8E          STA TMAG
942F- AE 01 8E          LDX RPT1
9432- BD 00 91          LDA RELT,X      GET RM
9435- A8                TAY             SAVE IT
9436- 18                CLC
9437- 6D 13 8E          ADC TREL        RM'=RM+TR
943A- 9D 00 91          STA RELT,X
943D- AE 03 8E          LDX RPT2
9440- 98                TYA
9441- 38                SEC
9442- ED 13 8E          SBC TREL        RN'=RM-TREL
9445- 9D 00 91          STA RELT,X
9448- AE 05 8E          LDX IPT1
944B- BD 00 92          LDA IMGT,X      GET IM
944E- A8                TAY             SAVE IT
944F- 18                CLC
9450- 6D 14 8E          ADC TMAG        IM'=IM+TI
9453- 9D 00 92          STA IMGT,X
9456- AE 07 8E          LDX IPT2
9459- 98                TYA
945A- 38                SEC
945B- ED 14 8E          SBC TMAG        IN'=IM-TI
945E- 9D 00 92          STA IMGT,X
9461- AD 09 8E          LDA SNPT        INCREMENT ANGLE
9464- 18                CLC
9465- 6D 0F 8E          ADC DLTA
9468- 8D 09 8E          STA SNPT
946B- EE 01 8E          INC RPT1        INCREMENT POINTERS
946E- EE 05 8E          INC IPT1
9471- 68                PLA             GET PAIR COUNTER
9472- A8                TAY             AND
9473- 88                DEY             DECREMENT IT
9474- F0 03             BEQ NOPE
9476- 4C D5 93          JMP NC1         DO NEXT PAIR
9479- AD 01 8E     NOPE LDA RPT1        GET POINTERS
947C- 18                CLC
947D- 6D 0E 8E          ADC CLDS        ADD CELL OFFSET
9480- 8D 01 8E          STA RPT1
9483- 8D 05 8E          STA IPT1
9486- CE 0C 8E          DEC CLCT        DECREMENT CELL COUNTER
9489- F0 03             BEQ NP1         NEXT PASS?
948B- 4C C6 93          JMP NCEL        NO, DO NEXT CELL
                   *
                   *
                   *
```

```
               4030 *   CHANGE PARAMETERS FOR NEXT PASS
               4040 *
948E- 4E 0B 8E 4050 NP1  LSR CLNM      HALF AS MANY CELLS
9491- F0 0C    4060      BEQ ABSO      NO MORE CELLS
9493- 0E 0D 8E 4070      ASL PRNM      TWICE AS MANY PAIRS
9496- 0E 0E 8E 4080      ASL CLDS      TWICE AS FAR APART
9499- 4E 0F 8E 4090      LSR DLTA      HALF THE ANGLE
949C- 4C 99 93 4100      JMP NPAS      DO NEXT PASS
               4110 *
               4120 *
               4130 *
               4140 *   END OF FFT ROUTINE
               4150 *
               4160 *
               4170 *
               4180 *   MAKE OUTPUT DATA ALL POSITIVE
               4190 *
               4200 *
949F- A2 00    4210 ABSO LDX #00
94A1- BD 00 92 4220 ONE  LDA IMGT,X    CHECK IMAG AREA
94A4- C9 00    4230      CMP #00       IS IT NEG?
94A6- 30 06    4240      BMI PLUS      IF SO, MAKE POS.
94A8- CA       4250      DEX
94A9- D0 F6    4260      BNE ONE       NEXT POINT
94AB- 4C BD 94 4270      JMP DOWN      DO REALS NEXT
94AE- 38       4280 PLUS SEC
94AF- A9 FF    4290      LDA #255      COMPLEMENT THE NUMBER
94B1- FD 00 92 4300      SBC IMGT,X
94B4- A8       4310      TAY
94B5- C8       4320      INY           ADD ONE
94B6- 98       4330      TYA
94B7- 9D 00 92 4340      STA IMGT,X    STORE THE POS. #
94BA- CA       4350      DEX
94BB- D0 E4    4360      BNE ONE       NEXT POINT
94BD- A2 00    4370 DOWN LDX #00       CHECK REAL AREA
94BF- BD 00 91 4380 TWO  LDA RELT,X
94C2- C9 00    4390      CMP #00       IS IT NEG?
94C4- 30 06    4400      BMI POS       IF SO, MAKE POS.
94C6- CA       4410      DEX
94C7- D0 F6    4420      BNE TWO       NEXT POINT
94C9- 4C DB 94 4430      JMP MDST      DONE, GO RESTORE DEC. MODE
94CC- 38       4440 POS  SEC
94CD- A9 FF    4450      LDA #255      COMPLEMENT THE NUMBER
94CF- FD 00 91 4460      SBC RELT,X
94D2- A8       4470      TAY
94D3- C8       4480      INY           ADD ONE
94D4- 98       4490      TYA
94D5- 9D 00 91 4500      STA RELT,X    STORE THE POS. NUMBER
94D8- CA       4510      DEX
94D9- D0 E4    4520      BNE TWO       NEXT POINT
94DB- AD 19 8E 4530 MDST LDA DEC       GET DEC. FLAG
```

```
94DE- C9 00              CMP #00       IF ZERO
94E0- F0 01              BEQ DONE      LEAVE ALONE
94E2- F8                 SED           OTHERWISE, SET IT
94E3- 60            DONE RTS           RETURN TO PROGRAM
                    *
                    *
                    *
                    *   OVER RANGE DATA SCALING ROUTINE
                    *   BY D. MACINTOSH
                    *
94E4- A2 00         SCAL LDX #00       INIT INDEX
94E6- BD 00 91      TEST LDA RELT,X    GET AN ARRAY VALUE
94E9- 18                 CLC           TEST FOR OUT OF RANGE
94EA- 69 40              ADC #$40      MOVE C0-3F TO 00-7F
94EC- 30 0C              BMI DIV       OUT OF RANGE VALUES NOW 80-FF
94EE- BD 00 92           LDA IMGT,X    DO THE
94F1- 18                 CLC           SAME TEST ON AN
94F2- 69 40              ADC #$40      IMAG ARRAY
94F4- 30 04              BMI DIV       VALUE
94F6- E8                 INX           POINT TO NEXT VALUE
94F7- D0 ED              BNE TEST      MORE? YES:  LOOP
94F9- 60                 RTS           NO:  RETURN
94FA- A2 00         DIV  LDX #00       BOTTOM OF TABLE
94FC- BD 00 91      DIV1 LDA RELT,X    GET AN ARRAY VALUE
94FF- 18                 CLC           ASSUME POSITIVE:  SIGN EXTEND=0
9500- 10 01              BPL DIV2      TEST IT: POS: YES,GO
9502- 38                 SEC           NEG: SIGN EXTEND=1
9503- 7E 00 91      DIV2 ROR RELT,X    SCALE IT
9506- BD 00 92           LDA IMGT,X    NOW DO
9509- 18                 CLC           THE SAME
950A- 10 01              BPL DIV3      FOR
950C- 38                 SEC           IMAG
950D- 7E 00 92      DIV3 ROR IMGT,X    ARRAY
9510- E8                 INX           POINT TO NEXT VALUE
9511- D0 E9              BNE DIV1      MORE:  YES, LOOP
9513- EE 10 8E           INC SCFT      NO....
9516- 60                 RTS           DONE SCALING.
                    *
                    *
                    *
                    *   TWO'S COMPLEMENT MULTIPLY SUBROUTINE
                    *   ROUTINE FROM EDN MAGAZINE 9/5/79
                    *   BY A. D. ROBISON
                    *
9517- 8D 18 8E      MPY  STA PLYR      STORE A AT MULTIPLIER
951A- 8C 17 8E           STY CAND      STORE Y AT MULTIPLICAND
951D- 4A                 LSR           SHIFT OUT FIRST BIT OF MULTIPLIER
951E- 8D 15 8E           STA LSBY      STORE IN LOW BYTE IF PROD.
9521- A9 00              LDA #$00      CLEAR HIGH BYTE OF PROD.
9523- A2 07              LDX #$07      INITIALIZE LOOP
9525- 90 04         CHEK BCC SHFT      TEST MULTIPLIER
```

```
9527- 18         5050      CLC          IF 1, ADD MULTIPLICAND
9528- 6D 17 8E  5060      ADC CAND       TO PARTIAL PROD.
952B- 6A         5070 SHFT ROR          SHIFT PROD. AND SHIFT IN POSSIBLE
952C- 6E 15 8E  5080      ROR LSBY       CARRY FROM MULTIPLICAND ADDITION
952F- CA         5090      DEX          REPEAT LOOP UNITIL ALL 7 LOWER BITS
9530- D0 F3      5100      BNE CHEK       OF MULTIPLIER
9532- 90 07      5110      BCC ADJ      TEST LAST BIT OF MULTIPLIER.
9534- ED 17 8E  5120      SBC CAND     IF 1, SUBTRACT MULTIPLICAND
9537- 6A         5130      ROR          COMPLEMENT CARRY, SINCE THE 6502
9538- 49 80      5140      EOR #$80       DEFINES BORROW AS THE COMPLEMENT
953A- 2A         5150      ROL            OF THE CARRY FLAG.
953B- 6A         5160 ADJ  ROR          SHIFT PRODUCT ONCE MORE AND SHIFT IN
953C- 6E 15 8E  5170      ROR LSBY       POSSIBLE BORRPW FROM SUBTRACTION
953F- AE 17 8E  5180      LDX CAND     TEST SIGN OF MULTIPLICAND.
9542- 10 04      5190      BPL EXIT     IF MULTIPLICAND IS NEGATIVE,
9544- 38         5200      SEC            THEN SUBTRACT MULTIPLIER FROM
9545- ED 18 8E  5210      SBC PLYR       HIGH BYTE OF PRODUCT
9548- 8D 16 8E  5220 EXIT STA MSBY     STORE PRODUCT MSB
954B- 2E 15 8E  5230      ROL LSBY     SCALE IT UP
954E- 2A         5240      ROL
954F- 60         5250      RTS          RETURN WITH PRODUCT IN A.
                 5260 ***********  E N D  **************
                 5270      .EN
```

SYMBOL TABLE

```
INPT   8F00   RELT   9100   IMGT   9200
SINT   9000   RPT1   8E01   RPT2   8E03
IPT1   8E05   IPT2   8E07   SNPT   8E09
CLNM   8E0B   CLCT   8E0C   PRNM   8E0D
CLDS   8E0E   DLTA   8E0F   SCFT   8E10
COSA   8E11   SINA   8E12   TREL   8E13
TMAG   8E14   LSBY   8E15   MSBY   8E16
CAND   8E17   PLYR   8E18   DEC    8E19
CVRT   C0B4   IPRT   C0B0   CHAN   C0B1
SMPL   8E1A   SPL    8E1C   POT    8E21
INS    8E23   INPD   930B   REAL   930D
IMAG   930F   SINE   9311   STRT   9313
CLER   9318   CLR1   931A   MOV1   9322
BREV   933D   BRV1   9342   SWAP   9351
SWP1   9367   PAS1   936C   PA1    9372
FPAS   9389   NPAS   9399   NCEL   93C6
NC1    93D5   NOPE   9479   NP1    948E
ABSO   949F   ONE    94A1   PLUS   94AE
DOWN   94BD   TWO    94BF   POS    94CC
MDST   94DB   DONE   94E3   SCAL   94E4
TEST   94E6   DIV    94FA   DIV1   94FC
DIV2   9503   DIV3   950D   MPY    9517
CHEK   9525   SHFT   952B   ADJ    953B
EXIT   9548
```

```
'a hex dump of the sine lookup follows:

9000.90FF

9000- 7F 7F 7F 7F 7F 7F 7E 7E
9008- 7D 7D 7C 7B 7A 79 78 77
9010- 76 75 73 72 71 6F 6D 6C
9018- 6A 68 66 65 63 61 5E 5C
9020- 5A 58 56 53 51 4E 4C 49
9028- 47 44 41 3F 3C 39 36 33
9030- 31 2E 2B 28 25 22 1F 1C
9038- 19 16 12 0F 0C 09 06 03
9040- 00 FD FA F7 F4 F1 EE EA
9048- E7 E4 E1 DE DB D8 D5 D2
9050- CF CD CA C7 C4 C1 BF BC
9058- B9 B7 B4 B2 AF AD AA A8
9060- A6 A4 A2 9F 9D 9B 9A 98
9068- 96 94 93 91 8F 8E 8D 8B
9070- 8A 89 88 87 86 85 84 83
9078- 83 82 82 81 81 81 81 81
9080- 81 81 81 81 81 81 82 82
9088- 83 83 84 85 86 87 88 89
9090- 8A 8B 8D 8E 8F 91 93 94
9098- 96 98 9A 9B 9D 9F A2 A4
90A0- A6 A8 AA AD AF B2 B4 B7
90A8- B9 BC BF C1 C4 C7 CA CD
90B0- CF D2 D5 D8 DB DE E1 E4
90B8- E7 EA EE F1 F4 F7 FA FD
90C0- 00 03 06 09 0C 0F 12 16
90C8- 19 1C 1F 22 25 28 2B 2E
90D0- 31 33 36 39 3C 3F 41 44
90D8- 47 49 4C 4E 51 53 56 58
90E0- 5A 5C 5E 61 63 65 66 68
90E8- 6A 6C 6D 6F 71 72 73 75
90F0- 76 77 78 79 7A 7B 7C 7D
90F8- 7D 7E 7E 7F 7F 7F 7F 4C
*
```

INDEX

C

D

E

F

G

H

I

J

K

Q

R

S

T

U

V

W

X

Z